AF248973

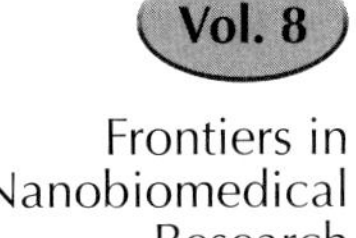

Frontiers in
Nanobiomedical
Research

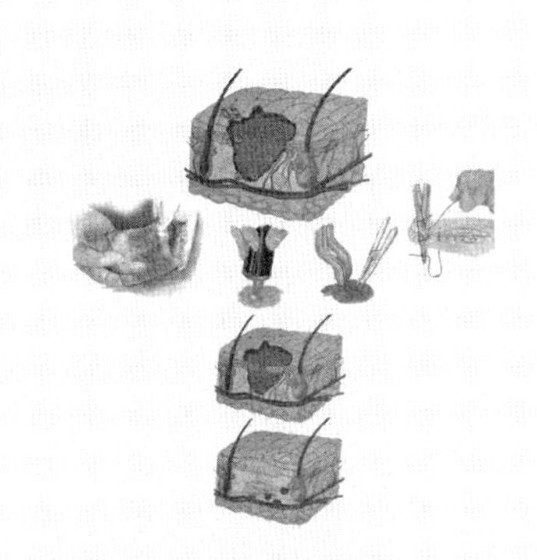

BIOENGINEERING IN WOUND HEALING

A Systems Approach

Frontiers in Nanobiomedical Research

ISSN: 2251-3965

Series Editors: Martin L. Yarmush *(Harvard Medical School, USA)*
Donglu Shi *(University of Cincinnati, USA)*

Published

Vol. 5   The World Scientific Encyclopedia of Nanomedicine and Bioengineering I:
Nanotechnology for Translational Medicine: Tissue Engineering,
Biological Sensing, Medical Imaging, and Therapeutics (A 4-Volume Set)
*edited by Yu Cheng (University of Chicago, USA), Yarong Liu (University of
Southern California, USA), Jia Huang and Bingbo Zhang (Tongji University
School of Medicine, China)*

Vol. 6   Handbook of Immunological Properties of Engineered Nanomaterials
(In 3 Volumes)
*edited by Marina A. Dobrovolskaia and Scott E. McNeil
(Leidos Biomedical Research Inc., USA)*

Vol. 7   Multiscale Technologies for Cryomedicine: Implementation from Nano
to Macroscale
*edited by Xiaoming He (The Ohio State University, USA) and
John C. Bischof (University of Minnesota, USA)*

Vol. 8   Bioengineering in Wound Healing: A Systems Approach
*edited by Martin L. Yarmush (Rutgers University, USA & Harvard Medical
School, USA) and Alexander Goldberg (Tel Aviv University, Israel)*

Vol. 9   The World Scientific Encyclopedia of Nanomedicine and Bioengineering II:
Bioimplants, Regenerative Medicine, and Nano-Cancer Diagnosis and
Phototherapy (A 3-Volume Set)
*edited by Donglu Shi (University of Cincinnati, USA),
Maoquan Chu (Tongji University, China) and
Jiang Chang (Chinese Academy of Sciences, China)*

Forthcoming title

Vol. 10   Tissue Engineering and Nano Theranostics
*edited by Donglu Shi (University of Cincinnati, USA) and
Qing Liu (Tongji University, China)*

The complete list of titles in the series can be found at
http://www.worldscientific.com/series/fnbmr

Vol. 8

Frontiers in
Nanobiomedical
Research

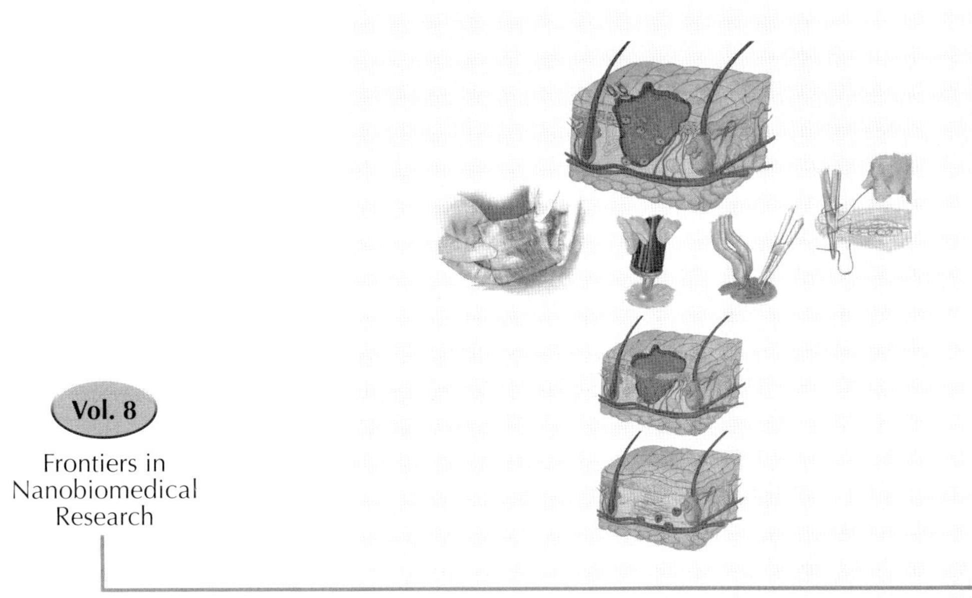

BIOENGINEERING IN WOUND HEALING

A Systems Approach

editors

Martin L Yarmush

Rutgers University and Harvard Medical School, USA

Alexander Golberg

Tel Aviv University, Israel

World Scientific

NEW JERSEY · LONDON · SINGAPORE · BEIJING · SHANGHAI · HONG KONG · TAIPEI · CHENNAI · TOKYO

Published by

World Scientific Publishing Co. Pte. Ltd.

5 Toh Tuck Link, Singapore 596224

USA office: 27 Warren Street, Suite 401-402, Hackensack, NJ 07601

UK office: 57 Shelton Street, Covent Garden, London WC2H 9HE

Library of Congress Cataloging-in-Publication Data
Names: Yarmush, Martin L., editor. | Golberg, Alexander, editor.
Title: Bioengineering in wound healing : a systems approach / [edited by]
    Martin L. Yarmush, Harvard Medical School, USA, Alexander Golberg, Tel Aviv University, Israel.
Description: New Jersey : World Scientific, 2016. | Series: Frontiers in nanobiomedical research ; volume 8 |
    Includes bibliographical references and index.
Identifiers: LCCN 2016027968 | ISBN 9789813144576 (hardcover : alk. paper)
Subjects: LCSH: Wound healing. | Bioengineering.
Classification: LCC RD94 .B54 2016 | DDC 610.28--dc23
LC record available at https://lccn.loc.gov/2016027968

British Library Cataloguing-in-Publication Data
A catalogue record for this book is available from the British Library.

Desk Editor: Catherine Yeo

Typeset by Stallion Press
Email: enquiries@stallionpress.com

Printed in Singapore

Dedicated to the memory of Dr. Alan Jay Fischman (May 2, 1950–February 12, 2016).

Dedicated Doctor, Mentor, and Friend

Preface

Wounds are an essential part of the life cycle of organisms due to their interaction with the environment, which can sometimes be very harsh. Development of new tissues in place of a wound is a complex set of processes controlled by myriad internal and external factors. Understanding and exercising control over this overall set of processes could transform a very large segment of patient care.

Although, wound healing has puzzled humanity from early days and has gradually emerged from art and witchcraft to modern medical procedures, the detailed mechanisms underlying what results in normal and abnormal wound healing are not well established. This gap in knowledge exists because of the tremendous complexity of the overall wound healing process with results from interactions among the fields of biology, chemistry, and physics (i.e. biochemical, mechanical, and electrical phenomena) all occurring in space and time. Despite this gap in our understanding of detailed mechanisms, new methods to treat wounds continue to appear, bringing promising new solutions to patients.

For many years, the field of wound healing was mostly dominated by investigators in the medical and life science communities with the goal of elucidating the fundamental mechanisms of healing, tissue regrowth, and tissue development. In more recent years, problems in wound healing have also attracted the attention of bioengineers. Thus, the modern study of wound healing and the development of new technologies to treat wounds requires broad and interdisciplinary knowledge, rarely available through text books. In this book, we try to close this gap by providing bioengineers and applied scientists with a broad, extensive, and multidisciplinary introduction to the field of wound healing.

To cover this broad spectrum of topics, we invited experts from dermatopathology, physiology, metabolism, developmental biology, material science, medical optics, bioelectrical engineering, and tissue engineering. We hope that the combination of these fields will enable newcomers to rapidly gain basic insights in the fundamentals of wound healing and the bioengineering fundamentals that can be brought to bear on this field.

This book is divided into four major parts. The first part of the book (Chapters 1–6) provides a biological, anatomical, metabolic, and physiological background on wound healing. The second part of the book (Chapters 7 and 8) shows the use of medical optics for structural and functional imaging of wound healing. The third part (Chapters 9 and 10) describes the use of material science in wound healing. The final, fourth part (Chapters 11–13) describes recently emerged engineering platforms such as laser welding, bioprinting, and electroporation to treat wounds.

Alex Golberg, Tel Aviv
Martin Yarmush, New Jersey
2016

Contents

1. Scarless Tissue Regeneration

Alexander Golberg

Porter School of Environmental Studies,
Tel Aviv University, Tel Aviv
6997801, Israel

Abstract

Wounds and wound healing are parts of every organism's life. What is a wound, how does it heal and what can we do to speed the healing, but decrease scarring? All these fundamental questions occupy the minds of people through all known history. Wound healing is an extremely complex phenomenon, which involves both the organism and its environment. Moreover, chemical, biological, mechanical and electrical forces drive this phenomenon. This chapter attempts to combine various aspects of wound healing phenomena with an emphasis on scarless regeneration. The system approach to wound healing is needed to develop new therapies. This chapter starts with the general overview on what is a wound from the system perspective. Then it describes the high-level events that take place during normal and abnormal wound healing. Next, it gives a historic perceptive on the history of wound healing in humans. Then it describes the fundamentals of scar biology, going to the details of known biochemical and cellular events involved in the process of scar formation. It then deals with the mechanical and electrical events that affect wound healing process. Next, complex events such as inflammation and wound-bacteria interaction are discussed, followed by descriptions of the *in vitro* and *in vivo* models of scarless wound healing. This chapter concludes with open questions and future perspectives of scarless wound healing.

1. The Problem of the Abnormal Wound Healing and Scarring

Wounds are a part of life. A living system at any level: cell, tissue, organ, or organism is defined by the boundary with the external environment. The disruption of this boundary by environmental hazards is a wound. Indeed, a wound can occur on the cellular, organ, and whole organism level (Figure 1). Once the injury takes place, all the organism's are mobilized to close the wound and restore the boundary that separates it from the external environment; otherwise, the leakage of nutrients and imbalance of energetic expenditures will lead to the organism's death. The process of this boundary restoration between the basic unit of life and its environment is known as a wound healing process (Figure 1).

In mammals, wound healing is a dynamic, chronic process that is divided to four overlapping phases: hemostasis, inflammation, proliferation and remodeling.[1–3] During hemostasis, constriction of the damaged vessels and clot formation physically limit blood loss. During the inflammatory phase, leukocytes and then monocytes accumulate to combat infection in the wounded tissue. In this phase, multiple cytokines and growth factors are released to the wound area and contribute to the fibroblast migration, differentiation, and activity. During the proliferative phase, fibroblasts deposit new extracellular matrix (ECM) and collagen and differentiate into myofibroblasts. In the final remodeling phase, re-organization of the closed wound environment occurs until repair is completed.

To describe this complex dynamic process, Robson *et al.*[4] introduced the concept of wound-healing trajectory (Figure 2), which demonstrates the time-dependent cumulative effects of these multiple processes that occur from injury though healing.[4] According to the healing trajectory curve, normally healed tissues are characterized by complete restoration of function and structure (Figure 3).[5] In contrast, chronic wounds are characterized by incomplete restoration of structure and function.[5] In proliferative scarring, however, the healing process does not stop as it should and the tissue fails

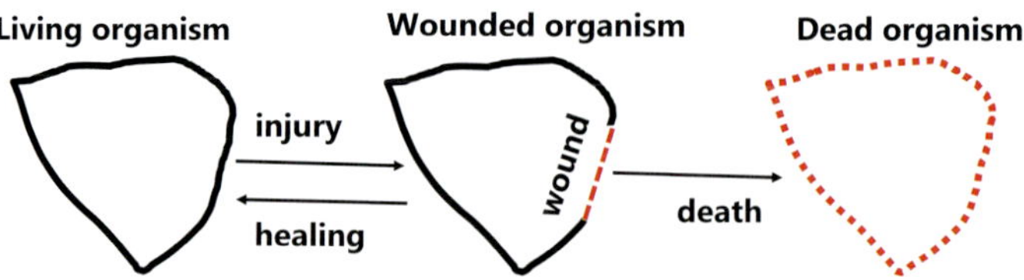

Figure 1.   Wound healing processes.

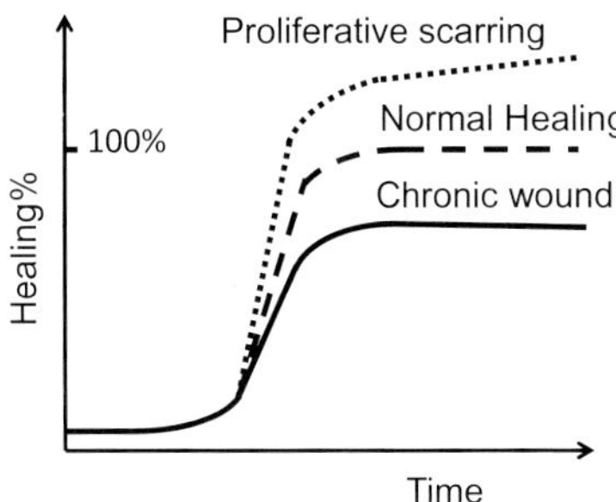

Figure 2.   Wound healing trajectories. Figure adapted with permission from Ref. 8.

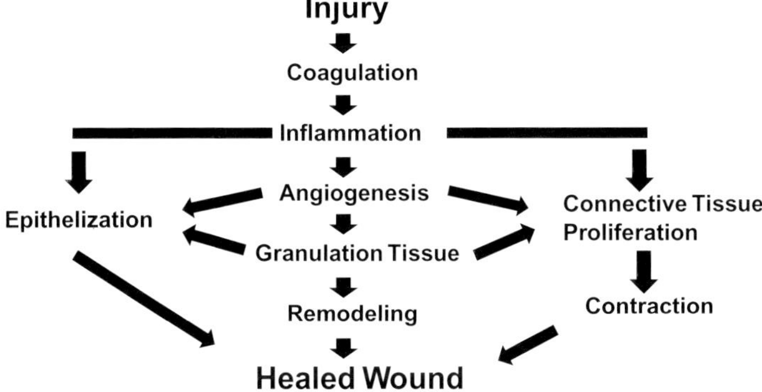

Figure 3.   Scheme of the wound healing. Major dynamic and intercalated processes that happens after in the injury till the wound is healed are shown. Figure adapted with permission from Ref. 2.

to reach a normal cell density and a balance between collagen deposition and degradation.[4] Proliferative scarring and chronic wounds in humans have adverse physical, aesthetic, functional, psychological, and social consequences.[6,7] The major aim of all wound healing technologies and treatments is to completely restore the function and structure of the wounded site to its initial condition.

2. Historical Perspectives on the Wound Healing in Humans

Treatment of wounds has been the center of human attention from the beginning of humanity. One of oldest evidences for documents protocols for attending wounds is a clay tablet from 2200 BC, which described the three phases of wound attending: washing the injured area, plastering, and putting a bandage on a wound.[9] Recipes for treating wounds were developed by all ancient healers from Middle East, India, China, Korea, and Africa.[10] Whole chapters in the Ayurvedic and traditional Chinese medicine are devoted treatment of wound.[10] Over the ages, wound healing transformed from magic, to a systematic text of wound care and surgery from Hippocrates and Celsus.[11]

Early recipes to treat wounds include the use of available raw or processed resources. Sumerians prescribed the following: "Pound together fur-turpentine, pine-turpentine, tamarisk, daisy, flour of inninnu strain; mix in milk and beer in a small copper pan; spread on skin; bind on him, and he shall recover."[9,11] In India, Sushruta Samhita describes more than 100 plants for treatment of wounds, including medical composition to reduce abnormal healing such as keloids.[10] In China, Traditional Chinese Medicine took an artistic and holistic approach to wound healing. In the classical concept of Yin and Yang, which describes two opposing and, at the same time, complementary aspects of any one phenomenon, the Yin tends is associated with poor circulation, stagnation, and poor healing; while the Yang is associated with being overheated or having a scarring.[10] Medical Qi Gong, an acupuncture technique without needles, herbal medicines, breathing exercises, massage, and Feng Shui, an art of living in harmony with nature, are shown to improve wound healing.[10] In Africa, some of the cultural practices and beliefs such as smoke, hot metal burns, dung, animal fur, saliva, soil, and local herbs are practiced from the early times of humanity till.[10]

The Egyptians developed a complex system for wound classification. 48 different types of wounds have been described in the 1650 BC Edwin Smith Surgical Papyrus, a copy of a much older document.[10] The Egyptians were the first to apply the bandages and treated the symptoms of infection and inflammation with such innovative materials as honey.[10]

The importance of cleanliness was particularly stressed by the Greeks.[9] They washed the wound with boiled water, vinegar, and wine. Importantly, the Greeks also classified acute and chronic wounds.[8] Hippocrates washed ulcers with wine, softened them by oil, and dressed them with fig leaves.[12] In Rome, Galen was first to recognize that pus from wounds inflicted by the gladiators preceded wound healing.[12]

Following the collapse of the Roman Empire, the wound healing practices in the West regressed to "magic", which was the major basis for healers during the Middle Ages. It was not until the 19th century that the use of antiseptics in surgery was revived by Dr. Lister, who attended the wounds with dressings soaked with carbolic acid.[11] Further advances were made by Dr. Carl Reyher who recommended adding a more extensive mechanical wound cleansing which he termed debridement.[11] However, full credit for the modern practice of débridement belongs to Belgian army surgeon Antoine Depage, who showed the importance removing of contaminated tissue and necrotic flesh.[11]

The advances in the biotechnology, tissue engineering, nanotechnology, and medical physics in the 20th century led to the progress in all frontiers of the wound healing. There are more than 5,000 wound care products in the market.[9] This high number, however, also shows that there is a no single magic solution to the complex process of wound healing.

3. Scar Biology Research Overview

The process of wound healing is extremely complex and combines several processes that occur after the injury (Figure 3). These processes include coagulation, inflammation, angiogenesis, fibropalacia, epithelization, contraction, and remodeling. The final result, if the organisms survive, is the healed wound. Importantly, these processes are highly nonlinear and depend on multiple local, system, and environmental characteristics of the wounded organism. However, does the healed wound restore completely the function and structure of the original tissue? This is a fundamental question of regenerative medicine and scar biology research.

When the healed wound completely restores the anatomy and function of the initial tissue, the regeneration is complete and scarless. However, fundamental conditions required for tissue regeneration are not favorable in mammals — the wound must close fast to enable survival and thus, in most cases in mammals, a tick fibrous tissue — a scar — is formed at the injuried site.

In most types of injuries, blood vessels are damaged. The immediate organism response is a clot formation by circulating platelets, which release various cytokines for further wound repair mechanism activation. Scarring, or fibrosis, is an abnormal tissue healing, first described in the Smith papyrus of 1700 BC. Although several systemic and genomic studies have identified potential cellular and extracellular factors that mediate the formation of proliferative scar (Figure 4), the exact mechanism that induces proliferative scar tissue formation instead of a healthy tissue is not known. Current data show that alterations in coagulation, inflammation, angiogenesis, fibroplasia, contraction, remodeling, mechanical tension correlate with the formation of scars.[2,13–16]

Wound healing in mammals is often accompanied by the presence and activity of highly specialized cell types, most notably fibroblastic phenotypes, we will call from now on fibroblasts, which are responsible for the replacement of original tissue components with scar tissue. Fibroblast, a

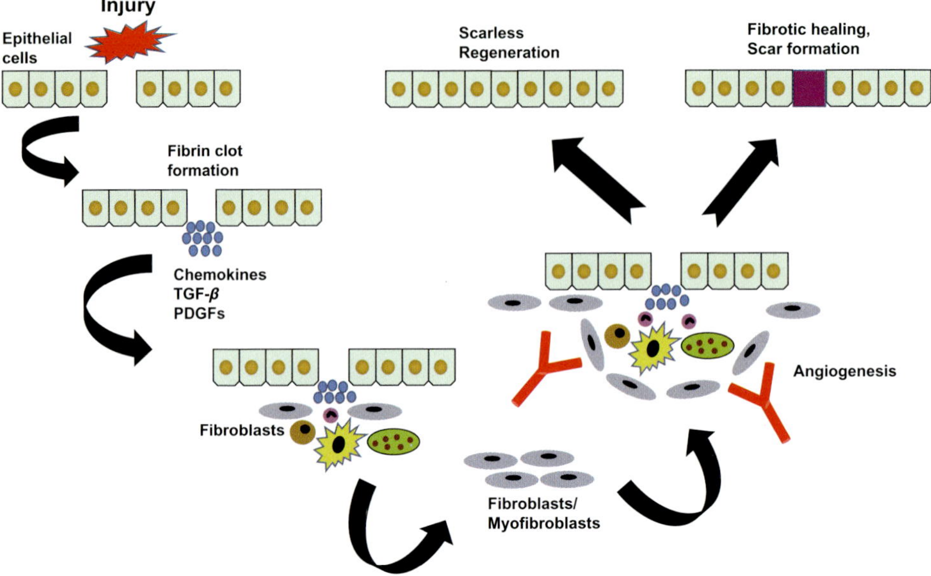

Figure 4.   Wound healing trajectories: true tissue regeneration or fibrosis and scarring. Following tissue injury, epithelial and/or endothelial cells release inflammatory mediators that initiate an antifibrinolytic-coagulation cascade, which triggers blood clot formation.[17] This is followed by an inflammatory and proliferative phase, when leukocytes are recruited and then activated and induced to proliferate by chemokines and growth factors.[17] The activated leukocytes secrete profibrotic cytokines such as transforming growth factor-β (TGF-β). Stimulated epithelial cells, endothelial cells, and myofibroblasts also produce matrix metalloproteinases (MMPs), which disrupt the basement membrane, and additional cytokines and chemokines that recruit and activate immune system cells that are critical for injury response, organism survival, and wound healing. The activated immune system cells such as macrophages and neutrophils clean up tissue debris, dead cells, and kill and clean invading organisms. Shortly after the initial inflammatory phase, myofibroblasts produce ECM components, and endothelial cells start angiogenesis. As the wound heals by regeneration pathway, collagen fibers become more organized, blood vessels are restored to normal, and scar tissue is eliminated. Persistent inflammation, tissue necrosis, and infection lead to chronic myofibroblast activation and excessive accumulation of ECM components, which promotes the formation of a permanent fibrotic scar.

remarkably complex cell, is the most important organ in the wound healing and scar formation. Fibroblast metabolism, which is responsible for all activities of these cells during wound healing and scarring, is predicated by genetics, mechanical, chemical, and electrical factors (Figure 5).

Signaling which affects fibroblast metabolism is different in individuals who suffer from proliferative scarring from those who do not.[4,18,19] The major role of fibroblasts in wound healing is to replace the fibrin-based provisional

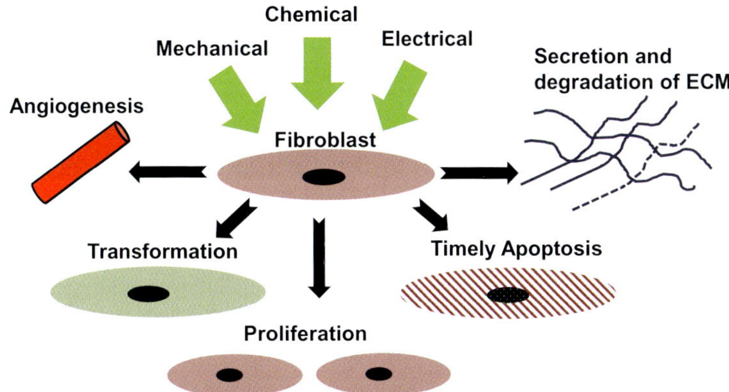

Figure 5.   Fibroblast function in wound healing and scar formation. Fibroblasts are sensitive to multiple mechanical, chemical, and electrical stimuli that are activated during different stages of the wound healing. Multiple events accompany fibroblast metabolisms that could lead to abnormal transformation to contractive myofibrolasts, abnormal angiogenesis, uncontrolled proliferation, delayed apoptosis, and abnormal regulation of secretion and degradation of ECM.

matrix established during the inflammatory phase of wound healing with collagen-rich granulation tissue. The behavior of fibroblasts in the wound is highly dynamic and varies at each healing phase.[4] Fibroblasts reach the wound during the second or third day after the injury.[4] Four days after the injury, fibroblasts are usually the major cell type in the developing granulation tissue.[4] The wound fibroblast number increases initially through migration from nearby non-injured tissue and then through cell proliferation. Fibroblast density in the wound reaches its maximum between 7 and 14 days after injury. When the anatomic function of the tissue is mostly restored, the maturing granulation tissue undergoes remodeling leading to reduction of fibroblast density by apoptosis.[4] Interestingly, clinical observations showed that in patients with proliferative scarring the apoptosis inhibitor - bcl-2 proto-oncogene is elevated; however, the apoptosis effector-interleukin-converting enzyme is decreased.[20] Theses findings suggest that the apoptosis mechanism is altered in patients with proliferative scarring.[20]

MMPs are zinc-depended proteinases that play a critical role in tissue remodeling. MMPs are involved in the proteolytic cleavage of collagen and degradation of other elements of the ECM.[19] MMP activity is controlled by tissue inhibitors of metalloproteinases (TIMPs) which bind to them in the 1:1 ratio.[21] Alterations of MMPs and TIMPs expression and ratio result in an imbalance between ECM production and degradation, which could lead to either chronic, non-healing wounds, or scarring.[19] Several studies showed the reduction in MMP-1 expression in scars. Smad interacting protein 1 (SIP1) is currently thought to regulate the skin fibrosis by controlling the expression levels of MMP-1 and collagen type I α2 (COL1A2).[22] Transfection of SIP1 to the hypertrophic scar derived fibroblasts upregulated the expression levels of MMP-1 expression and decreased the COL1A2.[22] However, the knock-down of SIP1 in normal fibroblasts up-regulated COL1A2 levels induced by TGF-β1.[22] Decreased levels of MMP-2, MMP-9, and increased levels of TIMP-1 were found in patients with hypertrophic scars.[14,23] Another study showed that MMP-1, MMP-2, and MMP-9 are up-regulated by the interaction between fibroblasts and keratinocytes, indicating the role of keratinocytes as additional regulators of ECM formation and degradation.[24,25] Keratinocyte derived stratifin was shown to stimulate MMP-1 expression in dermal fibroblasts through c-fos and p38 mitogen-activated protein kinase (MAPK) pathway[26]; however, this simulation was reversed by insulin.[27]

An important factor in the wound-healing and remodeling phase is the TGF-β.[28,29] TGF-β is a secreted protein that exists in three distinct isoforms in mammals, TGF-β1, TGF-β2, and TGF-β3. TGF-β1 and TGF-β2 are secreted from degranulated platelets, as well as monocytes and macrophages, TGF-β3 is produced by keratinocytes.[19] In normal tissue, the TGF-β isoforms exist as latent precursors where they are bound to latent TGF-β1 binding proteins. Neutralization of both TGF-β1 and TGF-β2 at the same time with neutralizing antibodies led to scar reduction in adult rat incisional wounds.[30] However, TGF-β3 inhibited scarring, indicating the TGF-β3 might be the antagonist of the other two TGF-β isoforms. Smad, intracellular regulatory signal transduction proteins, is the best studied pathway that is regulated by TGF-β factors.[31,32] MAPK, extracellular signal-regulated kinases (ERK) and the c-Jun N-terminal kinases (JNK) signaling pathway have also been involved in the TGF-β activated pathways, but, different from Smads, the exact mechanisms are not fully understood.[19,31,33–35]

4. Mechanical Tension in the Scar Formation

Mechanical tension is known to be critical for the formation of hypertrophic scars and indeed, the well-established surgical treatments — Z-plasty or W-plasty[36] — work by relieving the tension along the scar. Decreased cellular apoptosis was observed in murine models in which mechanical stress was applied for healing wounds.[37] Recent studies reported that focal adhesion kinase (FAK), activated after skin injury and regulated by mechanical

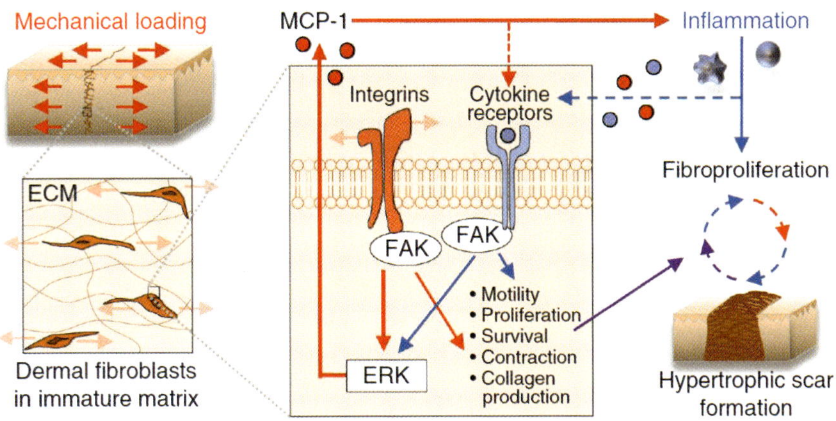

Figure 6. Schematic of the proposed vicious cycle of hypertrophic scarring driven by mechanical activation of local and systemic fibroproliferative pathways through fibroblast FAK. Figure adapted with permission from Ref. 16.

loading, acts through ERK to mechanically trigger the secretion of monocyte chemoattractant protein-1 (MCP-1, or CCL2), a chemokine that is linked to human fibrosis and scarring (Figure 6).[16] A follow-up study suggested a mechanomodulatory polymer device that effectively off-loaded high tension wounds and blocked pro-fibrotic pathways and scar formation in the pig model.[15,38]

5. Endogenous Electric Fields in the Wound Healing

The role of both endogenous and externally applied electric fields on the wound healing and scar formation has a long history of both real applications and quackery.[39] In the 17th century, gold leaf was used to prevent scarring from small pox.[39] Disruption of an epithelial layer of every multicellular species instantaneously generates endogenous electric fields, which are important in wound healing.[40–42] Mechanistic studies show that electric stimulation with magnitudes close to these endogenous electric fields activated Src and inositol–phospholipid signaling, which polarized in the direction of cell migration.[43] Genetic disruption of phosphatidylinositol-3-OH kinase-γ (PI(3)Kγ) decreased electric-field-induced signaling and abolished directed movements of healing epithelium in response to electric signals.[43]

Several examples on the impact of electric field simulation are described follows. Specific parameters of pulsed electric field increased the syntheses of DNA and protein in human fibrobalsts (16–20 Vcm^{-1}, 20 min stimulation with 100 μs pulses, 60–120 Hz). However, higher fields (33 Vcm^{-1}) had an inhibitory effect. No pH or significant temperature changes were observed in these studies.[44] Activation of (Na+K+) ATPase by electric field and activated syntheses of ATP were analyzed in Ref. 45.

In an additional study, fibroblasts entered the cell cycle after continuous stimulation by an electric field. In this study, the authors reported on incorporation of [3H]thymidine to fibroblast DNA at 41 mVm^{-1} amplitude, 10 Hz in the dermal wound model. The authors concluded that electric field stimulation increased the kinetics of cell entrance to the cell cycle by facilitating G0/G1 fast transition.[46]

Another study on electric field control of cell cycle reported that the physiological electric field of 200 mVmm^{-1} inhibited proliferation of cultured vascular endothelial cells. This field did not induce apoptosis, arrested the cells at G1 stage, reduced the expression of cycline E, increased the expression of p27^{kip1}, and did not influence the expression of Cdk2. Lower fields (50 and 100 mV mm^{-1}) did not show the proliferation's inhibitory effect.[47]

The details of molecular mechanisms of interaction between cells and electric fields in the wound models have been reported in Ref. 43. Electric fields activated Akt (Ser 473), Src (Tyr 416), ERK, and p38 in primary cultures of mouse keratinocyte and mouse peritoneal neutrophils in serum-free medium (200 mV mm^{-1}) and induced cell migration through the PI(3)K/Akt pathway.[43] Table 1 summarizes the proteins that regulate cell metabolism due to exposure to the electric fields.

6. Inflammation in the Scar Formation

Inflammation plays an important role in the organism's response to the injury and in all stages of the wound healing.[52] The inflammatory response includes the response of local inflammatory cells and recruitment of inflammatory cells from the circulation.[52] Local immune cells include resident macrophages and mast cells.[52] Macrophages are activated by proinflammatory mediators released in response to injury, as well as Damage Associated Molecular Pattern molecules.[53] Mast cells and circulation derived platelets degranulate rapidly after the injury.[54] Wound hypoxia also stimulates multiple inflammatory cells to secrete inflammatory mediators and fibroblasts ECM.[55] These events lead to increased recruitment of leukocyte recruitment.

Neutrophils are the first leukocytes that infiltrate the wound area. Circulating monocytes migrate to the wound in parallel with neutrophils

Table 1. Proteins involved in cell migration and proliferation during wound healing due to the applied electric field.

Protein	Manipulation	Phenotype	Reference
PTEN	p110γknockout	keratinocytes increased cathode-directed electrotaxis	43
Phosphatidylinositol-3-OH kinase	deletion	keratinocytes decreased cathode-directed electrotaxis	43
Phosphatidylinositol-3-OH kinase	p110γknockout	dermal fibroblast decreased anode-directed electrotaxis	48
EGFR/ErbB1	Deletion and overexpression	MTLn3 cells Electric-field-enhanced directional anodic migration of MTLn3 cells correlates with the expression level of EGF receptor	49
Protein kinase C	Chemical inhibition	Golgi apparatus reduced polarization. CHO cells reduced electric fields induced directional migration at 1–2 h	50
Glycogen synthase kinase-3b	Chemical inhibition	Golgi apparatus reduced polarization. CHO cells reduced electric fields induced directional migration at 2–3 h	50
Rho kinase (ROCK)	Inhibition by Y-27632	hiPS cells significantly decreased the directedness of migration, but increased the speed of migration in	51

and then differentiate into macrophages. Mast cells are mostly recruited from the adjacent tissue.[54] T lymphocytes that are found in the wound bed at later phase of healing impact the remodeling of the wound. With the resolving of inflammation, the number of leukocytes diminishes; yet, multiple studies show that inflammation continues to play a critical role in wound remodeling, probably leading to the scar formation (Figure 7).[56]

7. Bacteria in the Wound Healing

Wounding creates damage at the system's protective layer and provides a new, nutrient rich ground for microorganisms. Bacteria and fungi endotoxins lead to the modulation of the wound-healing process by increasing the levels

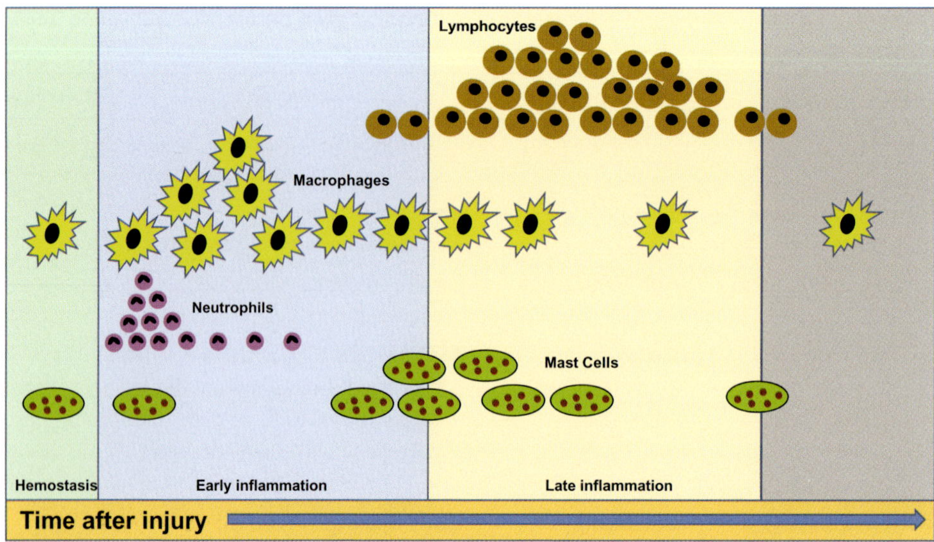

Figure 7. The pattern of leukocyte infiltration into wound. In hemostasis, the tissue is populated with rare mast cells and macrophages. After the injury, at the early inflammation period, neutrophils are rapidly recruited from circulation. They are followed by macrophages and then lymphocytes and mast cells at the late inflammation period. At the resolution and remodeling phase, the wound is mostly populated by mast cells and rare macrophages again.[52]

of pro-inflammatory cytokines such as interleukin-1 (IL-1) and TNF-α.[57] *Pseudomonas*, *Staphylococcus*, *Peptostreptococcus*, and *Candida* are the dominant species.[58,59] Microbiota play a role in both wound-healing abnormalities: chronic, non-healing wounds and hypertrophic scars, and also healing improvement.

The bacteria in infected wounds form biofilms, which are complex structures composed of bacteria embedded in a self-secreted extracellular polysaccharide matrix, which creates a unique protective microenvironment.[58] This matrix probably protects the bacteria from the inflammatory neutrophils and antibiotics and prevents the wounds from healing.[60] Besides attracting the inflammatory cells directly, multiple bacteria secrete exotoxins, which attack multiple types of cells leading to tissue necrosis.[61]

Although the etiology of hypertrophic scarring is unknown, the association of scarring with infection has been established.[62] A retrospective study of the association between hypertrophic burn scarring and bacterial colonization showed the association between *Staphylococcus aureus* and *Escherichia coli* colonization and scarring.

However, the bacteria not only slows the healing process or leads to scarring, they also could have beneficial properties. For instance, a recent exciting study has shown that supplementing the gut microbiome with lactic acid microbes in drinking water accelerated the wound healing process in an animal model.[63] The supplied bacterium, *Lactobacillus reuteri*, improved wound healing by up-regulation of the neuropeptide hormone oxytocin, a factor integral in social bonding and reproduction, by a vagus nerve-mediated pathway.[63]

8. Scarless Regeneration in Animal Models

Wound healing is an extremely complex process and therefore, physiological and anatomical animal models are required for fundamental and applied research.[64] Salamanders can regenerate a range of body parts throughout all stages of life.[65] Probably since the formal description of Spallanzani in 1768, salamanders have served as the most widely used model of scarless healing and complete regeneration of wounds in skin, heart, brain, spinal cord, jaws, retina, and lens.[65] Different from salamenders, regeneration is restricted to early stages of development in frogs. As shown in multiple studies, regenerative capacity is inversely correlated with the maturation of the immune system[66] (Figures 8A and 8B) The success of scarless regeneration in salamanders could be explained by the complex interactions between stem cell progenitors and immune cell subsets, where specific to salamanders inflammatory response drives cell differentiation towards original phenotypes[65] (Figure 8C).

Although salamanders and frogs are the center of regenerative biology research for decades, these studies are limited by the lack of genome sequence and arsenal of molecular biology tools available for model organisms. Recently, it was shown that adult zebra fish, an emerging model organism in molecular biology, heals rapidly and without scars.[67]

Murine models are widely used to study fundamental mechanisms in wound healing and also to test therapeutic compounds in pathophysiological environment. Although mice are a remarkable asset to study wound healing because of the availability of knockout strains, and transgenic tools, there are major anatomical and physiological differences between mice and human skin that must be considered (Table 2). An interesting recent study showed that some types of mice, found in Africa and not usually used in laboratory research, heal without scars.[68]

Because of their bigger size than mice, availability of various strains and costs, rats are often used as wound-healing models. As with mice, rat skin is

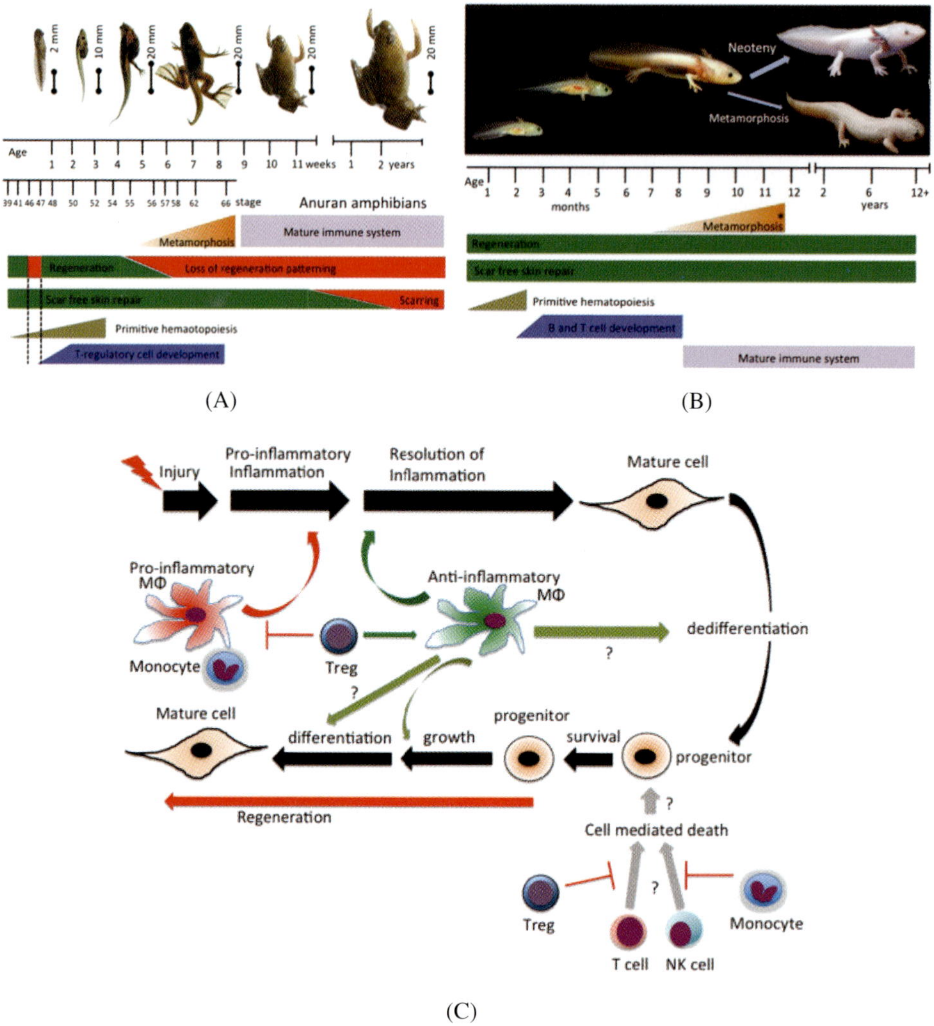

(A)

(B)

(C)

Figure 8. Scar-free healing and regeneration is inversely correlated with development in anuran amphibians but is maintained in adult life in urodeles (A). The development of sophisticated adult adaptive immunity begins at the onset of frog metamorphosis and is associated with a progressive loss of patterned regeneration where the number of digits that can be regenerated gradually declines between stages 55 and 60. While amputation at stages 53 results in a perfectly patterned limb, at stage 60 regeneration results in a "non-patterned spike". Tail regeneration is transiently lost during a "refractory period" (stages 45–47, marked with dotted line), which coincides with early immune cell development in the absence of regulatory T cells (T-regs). Restoration of functional regenerative potential in the tail is correlated with T-reg cell development. Scar-free wound healing is lost in adult anuran amphibians (B). Scar-free wound healing and regeneration is maintained throughout all life stages in salamanders. Metamorphosis is linked to relatively minor changes in adaptive immunity and regenerative capacity is maintained. Most salamander species transition

Figure 8.   (*Continued*) through metamorphosis; however, some species like the axolotl may have neotenic life cycles that do not normally transition through metamorphosis but can be chemically induced to do so. Unlike frogs, scar-free healing is maintained in adult life in salamanders. Various axolotl natural pigment mutants are available (C). Hypothetical model for monocyte/macrophage (*M*Φ) regulation of progenitor cell activation, survival, growth, and differentiation in the salamander. Within the influx of various white blood cells types that invade the damaged tissue, monocyte/macrophage cell types are critical for regenerative success. Monocyte/macrophage populations regulate the resolution of inflammation, provide growth factors, and signaling molecules important for angiogenesis and may influence the differentiation of progenitor cells. Monocyte/macrophage populations may also play a role in facilitating dedifferentiation and the regulation of cell mediated lysis. Figure adapted with permission from Ref. 65 under a Creative Commons License.

Table 2.   Mice and human wound-healing differences. Table is adapted with permission from Ref [69] under Creative Commons Attribution License.

	Mouse	**Human**
Hair cycle	Approximately 3 weeks	Highly variable, region-dependent
Epithelial architecture	No rete ridges	Rete ridges present
Apocrine sweat glands	Not present in skin, extensive in mammary glands	Present in axilla, inguinal, and perianal skin regions
Biomechanical properties	Thin, compliant, loose	Thick, relatively stiff, adherent to underlying tissues
Hypodermal thickness	Hair cycle-dependent	Less variable
Subcutaneous muscle layer	Present throughout as panniculosus carnosus	Present only in neck region as platysma
Major method of wound healing	Contraction	Granulation tissue formation and re-epithelialization

anatomically and physiologically different from humans, and heals mostly by contraction.[70] To generate a better model for human hypertrophic scars in mice and rats mechanical tension devices have been recently developed.[37,71] Application of tension on the healing incisional wound led to the formation of hypertrophic scars in rats and mice.[37,71]

As rodents usually do not form hypertrophic scars in standard wound models, alternative approaches have been investigated. In the rabbit ear wounds, the wounds heal without contraction; thus, epithelialization is delayed and a hypertrophic scar is formed.[72] This approach has been used to test various pharmaceuticals for scar treatments.

Pigs are probably the best, yet not perfect, currently available model for adult human skin wounds.[64] Yorkshire pigs and red Duroc pigs are used.[73] Wounds on the backs of red Duroc pigs show expression of cytokines, transcription factors, growth factors, and receptors similar to human scars.[73] Yet, the scars tend to diminish and the model can be used up to 5 months.[74] Although pig scars are the closest animal model to human scars, the skin structure in pigs and humans is not identical: (1) pig epidermis has only three layers as opposed to five in humans, (2) pigs have a thick and compact stratum corneum, (3) the distribution of apocrine and eccrine sweat glands are different, (4) the architecture of hair follicles is different. Moreover, working with these models requires special infrastructures and is of a high cost.[64,74]

9. Fetal Scarless Regeneration

Fetuses of mice, rats, pigs, monkeys, and humans regenerate without scars.[75] This ability, however, is age dependent in mammals.[76] In humans, wounds start to heal with scars at 24 weeks of gestation.[77] In mice, scarring of wounds begins on embryonic day 18.5.[77] Fetuses at this critical period completely regenerate the collagen matrix architecture and skin appendages such as sebaceous glands and hair follicles.[77] Observed differences between the ECM structure, inflammatory response, cellular mediators, differential gene expression, and stem cell function in fetal and post-natal wounds are linked to the mechanisms of scarless regeneration.[77] Interesting, human fetal skin heals without scars when it is transplanted into the subcutaneous tissue of adult athymic mice,[78] indicating that scarless healing is an intrinsic property of the fetal skin, which is not affected by environment.

There is a significant difference in the kinetics of gene expression changes between fetal and post-natal skins. In the fetus skin which regenerates without scars there is rapid up-regulation of groups of genes involved in cell growth and proliferation.[79] By 24 h, fetal skin has more upregulated genes involved in DNA transcription, DNA repair, cell cycle regulation, protein homeostasis, and intracellular signaling than post-natal skin.[79] Twenty four hours after the wound, the expression profile changes and more genes are expressed in the post-natal skin than in fetus.[79]

Different from post-natal wounds, where tissue injury leads to platelet activation, cytokines secretion, and massive infiltration of neutrophils, macrophages, and mast cell, fetal wound that heal without scars usually lack inflammation.[77] Moreover, reduction of inflammation in the post-natal wounds also is associated with reduced scarring.[77] At the same time, induction of inflammation in fetal wounds leads to scar formation.[77]

Signaling patterns are different between fetal and post-natal wounds. TGF-β3 expression is increased while, TGF-β1 expression is unchanged in the fetal wounds; however, in the post-natal wounds TGF-β1 expression increases.[77] In addition, interleukin-6 and interleukin-8 expression are significantly lower in fetal wounds compared with post-natal wounds.[77] Decreasing the levels of interleukin-6 and interleukin-8 is possible by introduction of interleukin-10.[77]

Fetal and post-natal skin have different composition of ECM and different functionality of fibroblasts. Fetal fibroblasts synthesize more type III and IV collagens.[77] Adult skin features more collagen type I. Different from adult wounds, where collagen synthesis is delayed after fibroblasts proliferation, in the fetal wounds fibroblasts divide and secrete collagen simultaneously.[77] Fetal skin contains more hyaluronic acid, fibromodulin, less decorin and lysyl oxidase than adult skin.[77] In addition, fetal skin which heals without scars has a higher ratio of MMP to TIMP[80]; thus, probably leading to faster remodeling of collagen, probably preventing its accumulation.

10. Wound Healing and Scar Formation in *Ex Vivo* Models

Wound-healing processes happen at all scales of the organism: single cells, tissue, organs, and organism. Single cell wound healing is an active process that involves multiple components of cellular machinery, and not passive reorganization of the cell membrane.[81] Studies on the single cell wound-healing processes have led to progress in the understanding of several cortical cytoskeleton controls.[3]

In vitro models of the wound healing of tissue and organs can be divided into two major types of models. The first model uses cultivated cells, mostly fibroblasts, keratinocytes, or endothelial cells. In this model, cells are cultured in the dense monolayers.[82] Wounding is modeled by creating an artificial gap, 'scratch' in the cell layer. This model is useful to study collective cell migration.[82] It is also used to study gene expression and metabolism of cells that transfer from stationary to motile phenotype.[82] Although this model is labor-, and cost-effective, it does not provide an accurate model for the wound as the chemotaxis gradient from the complex wound environment is missing.[82]

In vivo, most of the cells in the wound live in a 3D environment. Therefore, simulation of the condition using 3D cultivation media reproduces the environmental conditions of the wound better than the standard 2D cultivation. 3D *in vitro* models for wound healing include collagen gel matrixes, various addition hydrogels, and microfluidics devices.[83–85] An interesting recent study has been reported on the *in vitro* tissue engineered model of human hypertrophic scar model.[86]

11. Open Questions and Future Research Direction Towards Scarless Skin Regeneration

Wound healing and tissue regeneration are extremely complex phenomena, understanding of which requires an understanding of the interplay of genetic, metabolic, and environmental mechanisms. To achieve scarless regeneration, meaning anatomical and functional restoration of cells, tissues and organs, there is a need to maintain a balance between environmental and genetic parameters of the wound. This requires personalized wound healing and scar treatment therapies. To manipulate the trajectories of wound healing, there are chemical (C), electric (E), and mechanical (M) signals that affect cell machinery.

Since C, E, and M gradients can be viewed as independent phenomena, we propose to use CEM Coordinate System to define the whole field of possible combination of these three independent parameters. This is similar to the 3D Cartesian coordinate where the space is divided into x, y, and z planes. Definition of a new coordinate system allows us to use the exiting mathematical rules to describe the conditions under which cells enter the division cycle. For instance, we can define a regeneration vector, $\mathbf{R}$ (Figure 9) and wound-healing space (Figure 9), which describe all combinations of chemical, electrical, and mechanical gradients which drive a regeneration of fibrotic trajectory of wound-healing (Figure 3). Regeneration and fibrotic trajectories can be presented as topological surfaces in this wound-healing space (Figure 9). All interventions should be directed toward the shifting the wound healing trajectory from fibrotic to scarless regeneration surfaces.

The mathematical formulation of the regeneration vector appears in Eq. (1).

$$\mathbf{R} = \begin{bmatrix} C \\ E \\ M \end{bmatrix} = \begin{bmatrix} C_1 c_1 + C_2 c_2 + C_3 c_3 \\ E_1 e_1 + E_2 e_2 + E_3 e_3 \\ M_1 m_1 + M_2 m_2 + M_3 m_3 \end{bmatrix} \tag{1}$$

where C is the chemically controlled regeneration unit vector. c_1, c_2, c_3 are the coordinates that include concentrations of growth factors, cytokines,

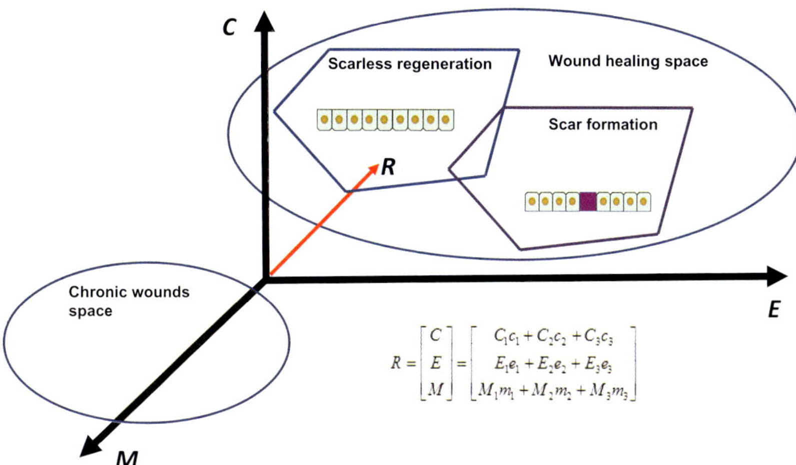

$$R = \begin{bmatrix} C \\ E \\ M \end{bmatrix} = \begin{bmatrix} C_1c_1 + C_2c_2 + C_3c_3 \\ E_1e_1 + E_2e_2 + E_3e_3 \\ M_1m_1 + M_2m_2 + M_3m_3 \end{bmatrix}$$

Figure 9. Chemo–electro–mechanical (CEM) coordinate system for wound-healing trajectories.

neurotransmitters, hormones, or chemical cell stressors such as reactive oxygen, sugars, and toxins, etc. E is the electrical field controlled regeneration that includes internal and external electric field in both DC and AC forms. e_1, e_2, e_3 are the coordinates that include the type and strength of the electric fields. M is the mechanics controlled unit vector. e_1, e_2, e_3 represent coordinates that include mechanical factors such as contact inhibition, cell geometry, whole organisms size, and cell energy balance. C_1, C_2, C_3, E_1, E_2, E_3, M_1, M_2, M_3 are the relevant coefficients.

Even though the control over all **R** vector coordinates could potentially lead to complete scarless healing, because of the large number of potential parameters such a control is almost impossible to achieve. Current challenges in the wound healing and scarless regeneration are in the identification of signaling pathways beyond TGF-β/Smad. Moreover, new molecular, electrical, and mechanical tools are needed to manipulate these signal pathways to redirect them from chronic wounds and fibrotic wound healing towards scarless regeneration.

Better understanding of the wound-healing models in animals and their relation to human wounds will significantly advance the translation of therapies from the bench to the bad side. Advances in *in vitro* models that capture at the same time multiple types of cells, their genetic background, patient age, and environmental conditions will further strengthen our understanding of wound-healing mechanisms.

Personalized medicine is the emerging approach to treat diseases in a systematic and not reductionist way. Currently, there are significant advances in personalized medicine.[87] Although broadly investigated for cancer, heart, and other organ and system diseases,[88] personalized medicine approaches for scarless regenerations are rare. However, multiple studies showed, that disease like hypertrophic scarring and keloids have a strong genetic component.[89] Development of personalized molecular diagnostics and adapting the treatment to the specific condition of the specific patient hold a great promise to solve the stubborn chronic wounds and hypertrophic scar abnormalities.

References

1. Diegelmann, R. F. and Evans, M. C. Wound healing: An overview of acute, fibrotic and delayed healing. *Front Biosci* **9**, 283–289 (2004).
2. Robson, M. C. Proliferative scarring. *Surg Clin N Am* **83**, 557–569 (2003).
3. Sonnemann, K. J. and Bement, W. M. Wound repair: Toward understanding and integration of single-cell and multicellular wound responses. *Annu Rev Cell Dev Bi* **27**, 237–263 (2011).
4. Robson, M. Steed, D. and Franz, M. Wound healing: Biologic features and approaches to maximize healing trajectories. *Curr Probl Surg* **38**, 72–140 (2001).
5. Lazarus, G. S. *et al.* Definitions and guidelines for assessment of wounds and evaluation of healing. *Arch Dermatol* **130**, 489–93 (1994).
6. Aarabi, S. Longaker, M. T. and Gurtner, G. C. Hypertrophic scar formation following burns and trauma: New approaches to treatment. *PLoS Med* **4**, 1464–1470 (2007).
7. Sheridan, R. L. and Tompkins, R. G. What's new in burns and metabolism. *J Am Coll Surg* **198**, 243–63 (2004).
8. Golberg, A. Bei, M. Sheridan, R. L. and Yarmush, M. L. Regeneration and control of human fibroblast cell density by intermittently delivered pulsed electric fields. *Biotechnol Bioeng* **110**, 1759–1768 (2013).
9. Shah, J. B. The history of wound care. *J Am Col Certif Wound Spec* **3**, 65–66 (2011).
10. Dorai, A. A. Wound care with traditional, complementary and alternative medicine. *Indian J Plast Surg* **45**, 418–24 (2012).
11. Broughton, G. Janis, J. E. and Attinger, C. E. A brief history of wound care. *Plast Reconstr Surg* **117**, 6S–11S (2006).
12. Bhattacharya, S. Wound healing through the ages. *Indian J Plast Surg* **45**, 177–179 (2012).
13. Leventhal, D. Furr, M. and Reiter, D. Treatment of keloids and hypertrophic scars: A meta-analysis and review of the literature. *Arch facial Plast Surg* **8**, 362–368 (2006).

14. Ulrich, D. Noah, E.-M. von Heimburg, D. and Pallua, N. TIMP-1, MMP-2, MMP-9, and PIIINP as serum markers for skin fibrosis in patients following severe burn trauma. *Plast Reconst Surg* **111**, 1423–1431 (2003).
15. Wong, V. W. *et al.* A mechanomodulatory device to minimize incisional scar formation. *Adv Wound Care* **2**, 185–194 (2013).
16. Wong, V. W. *et al.* Focal adhesion kinase links mechanical force to skin fibrosis via inflammatory signaling. *Nat Med* **18**, 148–152 (2011).
17. Wynn, T. A. Common and unique mechanisms regulate fibrosis in various fibroproliferative diseases. *J Clin Invest* **117**, 524–9 (2007).
18. Border, W. A. and Noble, N. A. Transforming growth factor beta in tissue fibrosis. *N Engl J Med* **331**, 1286–1292 (1994).
19. Zhu, Z. Ding, J. Shankowsky, H. A. and Tredget, E. E. The molecular mechanism of hypertrophic scar. *J Cell Commun Signal* **7**, 239–52 (2013).
20. Wassermann, R. J. *et al.* Differential production of apoptosis-modulating proteins in patients with hypertrophic burn scar. *J Surg Res* **75**, 74–80 (1998).
21. Stamenkovic, I. Extracellular matrix remodelling: The role of matrix metalloproteinases. *J Pathol* **200**, 448–464 (2003).
22. Zhang, Z.-F. *et al.* Smad interacting protein 1 as a regulator of skin fibrosis in pathological scars. *Burns* **37**, 665–72 (2011).
23. Simon, F. *et al.* Enhanced secretion of TIMP-1 by human hypertrophic scar keratinocytes could contribute to fibrosis. *Burns* **38**, 421–7 (2012).
24. Sawicki, G. Marcoux, Y. Sarkhosh, K. Tredget, E. E. and Ghahary, A. Interaction of keratinocytes and fibroblasts modulates the expression of matrix metalloproteinases-2 and -9 and their inhibitors. *Mol Cell Biochem* **269**, 209–16 (2005).
25. Ghahary, A. *et al.* Differentiated keratinocyte-releasable stratifin (14-3-3 sigma) stimulates MMP-1 expression in dermal fibroblasts. *J Invest Dermatol* **124**, 170–7 (2005).
26. Lam, E. Kilani, R. T. Li, Y. Tredget, E. E. and Ghahary, A. Stratifin-induced matrix metalloproteinase-1 in fibroblast is mediated by c-fos and p38 mitogen-activated protein kinase activation. *J Invest Dermatol* **125**, 230–238 (2005).
27. Lam, E. Tredget, E. E. Marcoux, Y. Li, Y. and Ghahary, A. Insulin suppresses collagenase stimulatory effect of stratifin in dermal fibroblasts. *Mol Cell Biochem* **266**, 167–74 (2004).
28. Penn, J. W. Grobbelaar, A. O. and Rolfe, K. J. The role of the TGF-β family in wound healing, burns and scarring: A review. *Int J Burns Trauma* **2**, 18–28 (2012).
29. Wang, R. *et al.* Hypertrophic scar tissues and fibroblasts produce more transforming growth factor-beta1 mRNA and protein than normal skin and cells. *Wound Repair Regen* **8**, 128–37.
30. Shah, M. Foreman, D. M. and Ferguson, M. W. Neutralising antibody to TGF-beta 1,2 reduces cutaneous scarring in adult rodents. *J Cell Sci* **107**, Pt 5, 1137–1157 (1994).

31. Pannu, J. Nakerakanti, S. Smith, E. ten Dijke, P. and Trojanowska, M. Transforming growth factor-beta receptor type I-dependent fibrogenic gene program is mediated via activation of Smad1 and ERK1/2 pathways. *J Biol Chem.* **282**, 10405–13 (2007).

32. Kopp, J. *et al.* Abrogation of transforming growth factor-beta signaling by SMAD7 inhibits collagen gel contraction of human dermal fibroblasts. *J Biol Chem* **280**, 21570–21576 (2005).

33. Moustakas, A. and Heldin, C.-H. Non-Smad TGF-beta signals. *J Cell Sci* **118**, 3573–84 (2005).

34. Zhang, Y. E. Non-Smad pathways in TGF-beta signaling. *Cell Res* **19**, 128–139 (2009).

35. Derynck, R. and Zhang, Y. E. Smad-dependent and Smad-independent pathways in TGF-beta family signalling. *Nature* **425**, 577–584 (2003).

36. English, R. S. and Shenefelt, P. D. Keloids and Hypertrophic Scars. *Dermatol Surg* **25**, 631–638 (1999).

37. Aarabi, S. *et al.* Mechanical load initiates hypertrophic scar formation through decreased cellular apoptosis. *Feb Am Soc Exp Biol J* **21**, 3250–3261 (2007).

38. Januszyk, M. *et al.* Mechanical offloading of incisional wounds is associated with transcriptional downregulation of inflammatory pathways in a large animal model. *Organogenesis* **10**, 186–93 (2014).

39. Thakral, G. *et al.* Electrical stimulation to accelerate wound healing. *Diabet Foot Ankle* **4**, (2013).

40. Barker, A. T. Jaffe, L. F. and Vanable, J. W. The glabrous epidermis of cavies contains a powerful battery. *Am J Physiol* **242**, R358–66 (1982).

41. Foulds, I. S. and Barker, A. T. Human skin battery potentials and their possible role in wound healing. *Br J Dermatol* **109**, 515–22 (1983).

42. McCaig, C. D. Rajnicek, A. M. Song, B. and Zhao, M. Controlling cell behavior electrically: Current views and future potential. *Physiol Rev* **85**, 943–78 (2005).

43. Zhao, M. *et al.* Electrical signals control wound healing through phosphatidylinositol-3-OH kinase-gamma and PTEN. *Nature* **442**, 457–60 (2006).

44. Bourguignon, G. J. and Bourguignon, L. Y. Electric stimulation of protein and DNA synthesis in human fibroblasts. *Feb Am Soc Exp Biol J* **1**, 398–402 (1987).

45. Tsong, T. Y. and Astumian, R. D. 863 — Absorption and conversion of electric field energy by membrane bound atpases. *Bioelectrochem Bioenerg* **15**, 457–476 (1986).

46. Cheng, K. and Goldman, R. J. Electric fields and proliferation in a dermal wound model: Cell cycle kinetics. *Bioelectromagnetics* **19**, 68–74 (1998).

47. Wang, E. Yin, Y. Zhao, M. Forrester, J. V and McCaig, C. D. Physiological electric fields control the G1/S phase cell cycle checkpoint to inhibit endothelial cell proliferation. *Feb Am Soc Exp Biol J* **17**, 458–60 (2003).

48. Guo, A. *et al.* Effects of physiological electric fields on migration of human dermal fibroblasts. *J. Invest. Dermatol* **130**, 2320–2327 (2010).

49. Pu, J. *et al.* EGF receptor signalling is essential for electric-field-directed migration of breast cancer cells. *J Cell Sci* **120**, 3395–3403 (2007).

50. Cao, L. Pu, J. and Zhao, M. GSK-3β is essential for physiological electric field-directed Golgi polarization and optimal electrotaxis. *Cell Mol Life Sci* **68**, 3081–3093 (2011).

51. Zhang, J. *et al.* Electrically guiding migration of human induced pluripotent stem cells. *Stem Cell Rev* **7**, 987–96 (2011).

52. Koh, T. J. and DiPietro, L. A. Inflammation and wound healing: The role of the macrophage. *Expert Rev. Mol. Med.* **13**, e23 (2011).

53. Zhang, X. and Mosser, D. M. Macrophage activation by endogenous danger signals. *J Pathol* **214**, 161–78 (2008).

54. Artuc, M. Hermes, B. Steckelings, U. M. Grützkau, and Henz, B. M. Mast cells and their mediators in cutaneous wound healing--active participants or innocent bystanders? *Exp Dermatol* **8**, 1–16 (1999).

55. Sen, C. K. Wound healing essentials: Let there be oxygen. *Wound Repair Regen* **17**, 1–18 (2009).

56. Martin, P. and Leibovich, S. J. Inflammatory cells during wound repair: The good, the bad and the ugly. *Trends Cell Biol* **15**, 599–607 (2005).

57. Guo, S. and Dipietro, L. a. Factors affecting wound healing. *J Dent Res* **89**, 219–229 (2010).

58. Edwards, R. and Harding, K. G. Bacteria and wound healing. *Curr Opin Infect Dis* **17**, 91–6 (2004).

59. Dowd, S. E. *et al.* Survey of fungi and yeast in polymicrobial infections in chronic wounds. *J Wound Care* **20**, 40–47 (2011).

60. Bjarnsholt, T. *et al.* Why chronic wounds will not heal: A novel hypothesis. *Wound Repair Regen* **16**, 2–10 (2008).

61. Ovington, L. Bacterial toxins and wound healing. *Ostomy Wound Manage* **49**, 8–12 (2003).

62. Baker, R. H. J. Townley, W. a, McKeon, S. Linge, C. and Vijh, V. Retrospective study of the association between hypertrophic burn scarring and bacterial colonization. *J Burn Care Res* **28**, 152–6 (2007).

63. Poutahidis, T. *et al.* Microbial symbionts accelerate wound healing via the neuropeptide hormone oxytocin. *PLoS One* **8**, e78898 (2013).

64. Ansell, D. M. Holden, K. a. and Hardman, M. J. Animal models of wound repair: Are they cutting it? *Exp Dermatol* **21**, 581–585 (2012).

65. Godwin, J. W. and Rosenthal, N. Scar-free wound healing and regeneration in amphibians: Immunological influences on regenerative success. *Differentiation* **87**, 66–75 (2014).

66. Godwin, J. W. and Brockes, J. P. Regeneration, tissue injury and the immune response. *J Anat* **209**, 423–432 (2006).

67. Richardson, R. *et al.* Adult zebrafish as a model system for cutaneous wound-healing research. *J Invest Dermatol* **133**, 1655–65 (2013).
68. Seifert, A. W. *et al.* Skin shedding and tissue regeneration in African spiny mice (Acomys). *Nature* **489**, 561–565 (2012).
69. Wong, V. W. Sorkin, M. Glotzbach, J. P. Longaker, M. T. and Gurtner, G. C. Surgical approaches to create murine models of human wound healing. *J Biomed Biotechnol* **2011**, 969618 (2011).
70. Dorsett-Martin, W. A. Rat models of skin wound healing: A review. *Wound Repair Regen.* **12**, 591–9
71. Lo, W. C. Y. *et al.* Longitudinal, 3D *in vivo* imaging of collagen remodeling in murine hypertrophic scars using polarization-sensitive optical frequency domain imaging. *J. Invest Dermatol* (2015).
72. Kloeters, O. Tandara, A. and Mustoe, T. a. Hypertrophic scar model in the rabbit ear: A reproducible model for studying scar tissue behavior with new observations on silicone gel sheeting for scar reduction. *Wound Repair Regen* **15** (1), S40–5 (2007).
73. Seo, B. F. Lee, J. Y. and Jung, S.-N. Models of abnormal scarring. *Biomed Res Int* **2013**, 423147 (2013).
74. Zhu, K. Q. Carrougher, G. J. Gibran, N. S. Isik, F. F. and Engrav, L. H. Review of the female Duroc/Yorkshire pig model of human fibroproliferative scarring. *Wound Repair Regen* **15** (1), S32–9
75. Colwell, A. S., Longaker, M. T. and Lorenz, H. P. Mammalian fetal organ regeneration. *Adv Biochem Eng/Biotechnol* **93**, 83–100 (2005).
76. Cass, D. L. *et al.* Wound size and gestational age modulate scar formation in fetal wound repair. *J. Pediatr. Surg* **32**, 411–5 (1997).
77. Larson, B. J. Longaker, M. T. and Lorenz, H. P. Scarless fetal wound healing: A basic science review. *Plast Reconstr Surg* **126**, 1172–1180 (2010).
78. Lorenz, H. P. *et al.* Scarless wound repair: A human fetal skin model. *Development* **114**, 253–259 (1992).
79. Colwell, A. S. Longaker, M. T. and Lorenz, H. P. Identification of differentially regulated genes in fetal wounds during regenerative repair. *Wound Repair Regen* **16**, 450–459 (2008).
80. Dang, C. M. *et al.* Scarless fetal wounds are associated with an increased matrix metalloproteinase-to-tissue-derived inhibitor of metalloproteinase ratio. *Plast Reconstr Surg* **111**, 2273–2285 (2003).
81. McNeil, P. L. Miyake, K. and Vogel, S. S. The endomembrane requirement for cell surface repair. *Proc Natl Acad Sci USA* **100**, 4592–7 (2003).
82. Liang, C.-C. Park, A. Y. and Guan, J.-L. *In vitro* scratch assay: A convenient and inexpensive method for analysis of cell migration in vitro. *Nat Protoc* **2**, 329–333 (2007).
83. Van Kilsdonk, J. W. J. *et al.* An *in vitro* wound healing model for evaluation of dermal substitutes. *Wound Repair Regen* **21**, 890–896 (2013).

84. Bell, E. Ivarsson, B. and Merrill, C. Production of a tissue-like structure by contraction of collagen lattices by human fibroblasts of different proliferative potential in vitro. *Proc Natl Acad Sci USA* **76**, 1274–8 (1979).
85. Li, X. J. Valadez, A. V. Zuo, P. and Nie, Z. Microfluidic 3D cell culture: Potential application for tissue-based bioassays. *Bioanalysis* **4**, 1509–25 (2012).
86. Van Den Broek, L. J. Niessen, F. B. Scheper, R. J. and Gibbs, S. Development, validation, and testing of a human tissue engineered hypertrophic scar model. *Altex Altern Anim Ex* **29**, 389–402 (2012).
87. Dowd, S. E. Wolcott, R. D. Kennedy, J. Jones, C. and Cox, S. B. Molecular diagnostics and personalised medicine in wound care: Assessment of outcomes. *J Wound Care* **20**, 232, 234–239 (2011).
88. Kirchhof, P. *et al.* The continuum of personalized cardiovascular medicine: A position paper of the European Society of Cardiology. *Eur Heart J* **35**, 3250–3257 (2014).
89. Bayat, A. Bock, O. Mrowietz, U. Ollier, W. E. R. and Ferguson, M. W. J. Genetic susceptibility to keloid disease and hypertrophic scarring: transforming growth factor beta1 common polymorphisms and plasma levels. *Plast Reconstr Surg* **111**, 535–543; discussion 544–546 (2003).

2. Anatomy of the Human Skin and Wound Healing

Amit Sharma*, Labib R. Zakka[†] and Martin C. Mihm Jr.[†]

*Department of Dermatology
Mayo Clinic, Scottsdale, Arizona
[†]Department of Dermatology
Brigham and Women's Hospital,
41 Avenue Louis Pasteur, Room 3117,
Boston, MA 02115

Abstract

A description of the histologic anatomy of the skin of the scalp begins with an overview of the entire skin specimen with specific structures indicated by arrows. The layers are identified. Then each layer is described according to its components and their significance. First, the epidermis is detailed along with its basement membrane zone. The dermal architecture as well as its connective tissue components are indicated. Their significance especially as far as involvement by either cutaneous disorders versus systemic diseases is reviewed. Lastly, the importance functionally and anatomically is elucidated. The vasculature of the skin is also described along with its importance in the role of inflammation of the skin. The different infiltrating cells in the skin are each enumerated, and their functions reviewed.

With this background, the morphologic features of wound healing are described and illustrated with both clinical images and appropriate microscopic features. The basic types of wound healing, both primary and secondary, are explained. The four phases of wound healing include: hemostasis, inflammation, proliferation, and remodeling. Each of these phases is described and illustrated. Examples of the different clinical conditions that lead to different type of ulcers are enumerated. Also, abnormal healing responses are described as well as their hypothetical causes.

Emphasis is placed on the pathophysiology of wound healing with a diagram that illustrates the structures and the cytokines that affect them and produce healing.

1. Anatomy of the Human Skin and Wound Healing

The skin is divided into three layers: the epidermis, the dermis, and the subcutaneous fat (Figure 1). The epidermis and dermis represent the two most important features of the skin with numerous functions. The subcutaneous fat on the other hand has limited function and makes up the bottom layer of the skin. The epidermis is composed of four cell types. The most common cell is the keratinocyte (Figure 2) responsible for making keratin, the skin barrier, photoprotection, home to immune related cells, the Langerhans' cells, and other functions. The epidermis also contains melanocytes (Figure 2), Merkel cells, and lymphocytes. The Langerhans' cells (Figure 2) are of great importance in presenting antigens for sensitization to lymphocytes in the skin and in the lymph nodes. Unmyelinated axons penetrate the skin. Hair follicles extend from the dermis into the subcutaneous fat (Figure 2). The dermis has two basic layers. The thin papillary dermis (Figure 3) lies beneath the epidermis

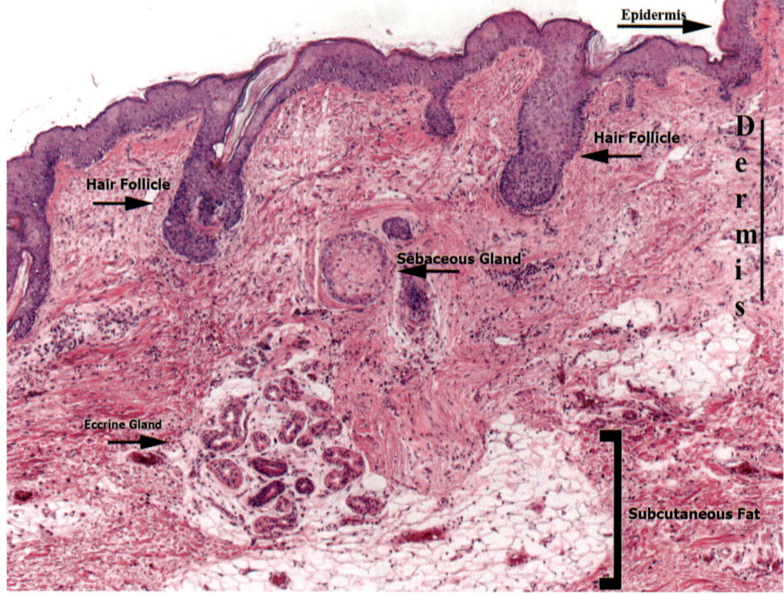

Figure 1.   Low power displays of the entire skin including epidermis, dermis, and subcutaneous fat. There are two hair follicles with hair present in them (green arrows). Note the sebaceous gland and the eccrine gland. H&E 4×.

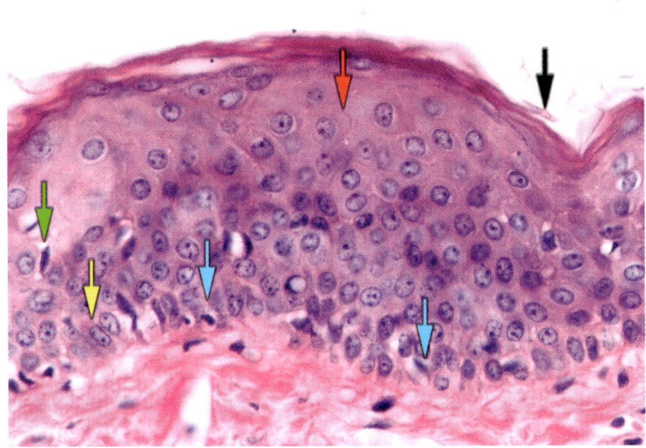

Figure 2.   Stratum corneum shows irregular flaking (black arrow). The squamous cells have a polygonal shape (red arrow) and show bridges. The basal cells have an oval shape (yellow arrow) and there are melanocytes (blue arrows) interspersed among the basal cells. Langerhans' cells (green arrow) are scattered in the epidermis. H&E 40×.

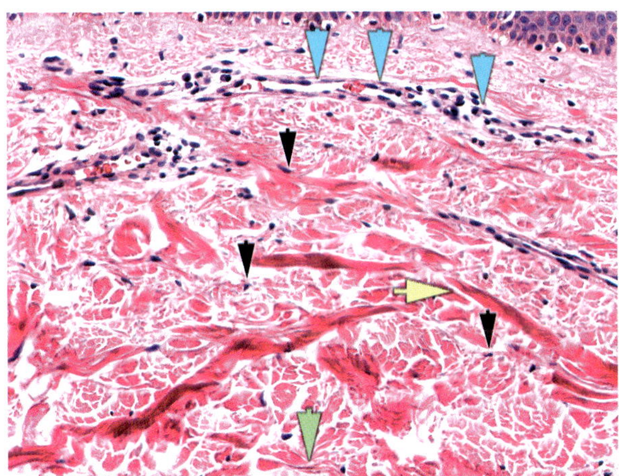

Figure 3.   Papillary dermis is composed of fine fibers. The reticular dermis exhibits coarse fibers (yellow arrow). Note the scattered fibroblasts in the reticular dermis (black arrows). Elastic fibers (green arrow) are smaller, short, thinner than reticular dermal fibers, and have often irregular shapes. The superficial venular plexus exhibits a mild perivascular infiltrate that includes lymphocytes and a rare plasma cell (blue arrows). The plexus lies at the junction of the papillary and reticular dermis. H&E 40×.

and is similar to the connective tissue that invests the appendages. Beneath the papillary dermis is the main portion of the dermis known as the reticular dermis (Figure 3). The superficial vascular plexus lies at the junction of the papillary and reticular dermis and is the site of most cutaneous inflammatory

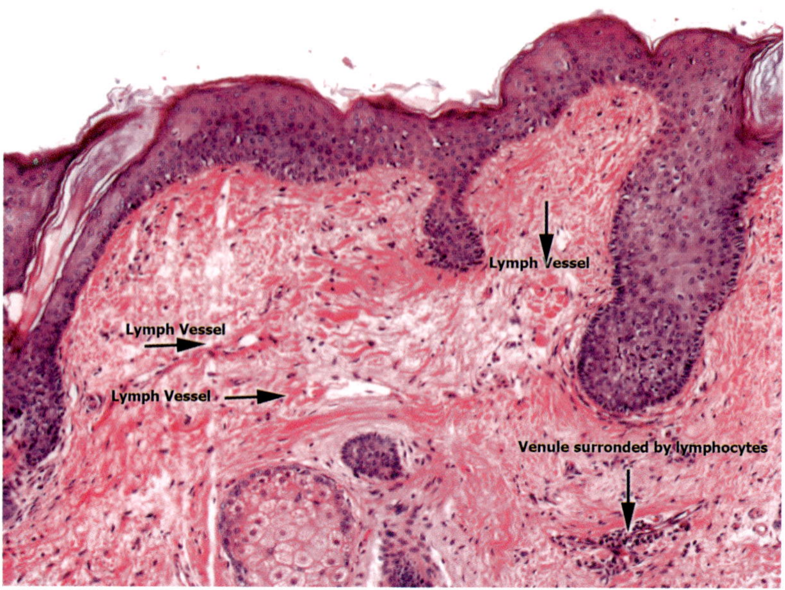

Figure 4.   There are small lymphatics in the papillary dermis. A superficial venule is surrounded by lymphocytes. H&E 10×.

disorders (Figure 3). Vessels, nerves, lymphatics, and the appendages course through the reticular dermis (Figure 4). From the vessels, there is the egress of the lymphocytes and other inflammatory cells that are involved in cutaneous inflammation, infection, immune disorders, and in response to trauma among other conditions. For a detailed discussion of skin anatomy, including immunohistochemical features and electron microscopy, the reader is referred to the first chapter of the book, *Lever's Histopathology of the Skin.*[1]

2. The Epidermal Anatomy

2.1 *Keratinoctyes*

The epidermis has four principal layers of keratinocytes. The surface is composed of the horny layer of keratinocytes; beneath is the granular cell layer, overlying the squamous layer, and finally at the bottom the basal or germinative layer.

2.2 *Basal cells*

These cells are disposed in single cell array. They appear to be cuboidal to ovoid and lie along the basement membrane zone (BMZ). They have a round nucleus with tiny nucleolus. With hematoxylin and eosin (H&E) staining they have a bluish tint. The nuclei appear dark blue. Above the nucleus there are numerous melanin granules, transferred from adjacent melanocytes, useful in photoprotection of the nuclear DNA. The basal cells are the site of mitoses in the normal skin. The keratinocytes all contain tono-filaments that are intermediate keratin filaments and form the cytoskeleton of the cell. The keratinocytes are all connected to each other by attachment plaques known as desmosomes. The basal layer attaches to the basement membrane by altered, so-called hemidesomsomes.[2,3]

2.3 *Squamous cells*

The squamous cell layer constitutes the largest portion of the epidermis and is composed of polyhedral keratinocytes. These cells stain more amphopilic to pink with H&E. They possess a prominent nucleus with nucleolus. Because of the polyhedral nature of the cells, there is a spinous appearance, resulting the name spinous layer. The cells contain keratofilaments that attach to the desmosomes in one portion and in the opposite end form the cytoskeleton. The desmosomes are easily visible on high power microscopy lying in the space between the cells. This space is the so-called intercellular zone that is normally composed of acid and neutral mucopolysaccharides that form a gel-like intercellular cement. This substance, along with desmosomes, result in cell adherence.[4,5] However, water and soluble substances such as nutrients can easily permeate the intercellular environment. The desmosomes are complex structures that are held together by several important protein moities called cadherins. The two very important ones are desmoglein I and desmoglein II. Blocking the function of these proteins by specific antibodies results in separation of the cells and a blistering disease, known as pemphigus.[6] Toward the top of the spinous layer the cells become gradually spindle shaped or flattered and give rise to the next layer.

2.4 *The granular layer*

The granular cell layer is composed of markedly flattened cells that contain very deeply staining blue granules, so-called keratohyaline granules. The size of this layer depends on the site. In areas of marked keratin formation as palms and soles, the layer is several cells thick. In the skin elsewhere it is only two or three layers thick. The keratohyaline ganules are important in the process of keratinization that results in the horny layer.[7] In the granules there is an abundance of sulfur that results in bonding by disulfide groups.

This cementing of the filaments is different and results in a so-called soft keratin. The tonofilaments contain very little sulfur. In hair and nails there are no keratohyaline granules and bonding of tonofilaments results in hard keratin. Soft keratin can be leached out of the tissue and can easily desquamate. In contrast, hard keratin of nails is formed by compacting of tonofilaments that are non-flexible and difficult to extract; hence, the necessity of cutting the nails. Also present in the superficial squamous cells and granular cell layer are the Odland bodies. They are extruded into the extracellular space. Their content is involved in the formation of the skin barrier.[8,9] The granular cell layer is the site in which there is autolysis of the nucleus of the cell and other intracellular components accomplished by lysosomal enzymes not present in the squamous or basal cell layers.[10]

2.5 *Stratum corneum or horny layer*

This layer is the top of the skin. It is composed of anucleate cells that are stacked one upon the other. The cells are anulceate and stain eosinophilic with H&E. In the process of fixation of tissue, much of the soft keratin cement is washed away and the residual compacted keratinocytes partially separate from one another and result in a so-called characteristic basket-woven appearance under the microscope.[11]

2.6 *Keratinocyte maturation and special functions*

The principal function of the epidermis, the making of keratin and the terminal maturation necessary for this process, involves a process of transformation of the cells. To pass from the basal layer to the stratum corneum takes an average of 30 days. Remarkably, in some cutaneous disorders this time can be shortened. For example, in psoriasis the turnover is approximately five days. There is constant formation of scales and desquamation.

One of the main functions of the epidermis is the formation of a barrier that is impermeable to water and prevents water loss. Dr. Thomas Fitzpatrick described the importance of this function without which we "would wash away in the bath water". This barrier lies deep in the area of the horny layer.[8,12,13]

Another very important function relates to the release of many cytokines by the keratinocytes that affect immune response in the patient.[1,14,15]

2.7 *Melanocytes*

Melanocytes are neural crest-derived cells that are present along the basal layer of the epidermis. They characteristically exhibit dendrites that extend between the keratinocytes in the extracellular spaces. The principal effect on melanocytes is sun protection through the production of melanin granules. Melanin granules are protein structures that contain tyrosine-derived melanin.[16] This melanin pigmentation is transferred from the melanocyte to the keratinocyte. The pigment lies above the nucleus of the keratinocyte where it can protect the DNA from ultraviolet light. The melanocyte has no desmosomes similar to the characteristic of the keratinocyte.[17] Therefore, it is a cell that can migrate in the epidermis and into the dermis. The dendrites of the melanocyte extend to keratinocytes and it is estimated that 1 melanocyte provides melanin to 54 keratinocytes.[1,18,19]

2.8 *Merkel cell*

Another cell in the epidermis is the Merkel cell. These cells are found at the tips of the rete and are thought to be either of neural crest origin or derived from stem cells.[20–22] It is considered that these cells may represent rudimentary touch receptors as they have a neural terminal in the basal portion of each cell.

2.9 *Langerhans cell*

The Langerhans cells are denditric cells that present antigen and are present in the epidermis where they form a network with their prominent dendrites. These cells are bone marrow derived.[23–25] They are positive for HLA-DR antigen and S100 protein, and cola. The Langerhans cell contains a specific electron microscopic granule, the so-called Birbeck granule that has a shape resembling a tennis racket. These cells are very important for antigen presentation in the skin. All three of these cells in the epidermis can form different tumors: the melanocyte, melanoma; the Merkel cell, the Merkel cell tumor; Langerhans cell, histiocytosis X; or Langerhans cell histiocytosis.

2.10 *Hair follicles*

Hair follicles begin in early embryonic life,[26] in the third month as small bulges in the epithelium that are called hair germs identified as aggregated basophilic cells in the basal layer. As these "buds" extend into the dermis where there is a striking aggregation of mesenchymal spindle cells that subsequently go on to become the hair papilla.[27,28] The germ continues to grow and the hair matures and is present at birth. During embryonic life, there are three other peg-like areas that occur in the developing hair. One gives origin to the sebaceous gland, another is the attachment site for the arrector pili muscle that each hair is associated with it and the third is for the development of apocrine gland.[1,29] The latter development is most commonly found in the axilla and groin areas as well as occasionally in the scalp. The hair follicle, when fully mature, gives rise to hair. Formation of the hair begins at the depth of the hair follicle along with the hair matrical cells that then gradually keratinize as the cells pursue an upward course of growth. In the lower portion of the follicle, the hairs are ensheathed in the inner root sheath that keratinizes forming hair. These cells extend up to the area of the attachment of the sebaceous gland. This area is called the isthmus that extends from the attachment of the arrector pili muscle to the entrance of the sebaceous duct. Above the isthmus is the infundibulum which is the superficial portion of the hair follicle lined by stratified squamous epithelium identical to the epidermis itself. This epithelium is part of the external root sheath that actually encases the entire hair follicle extending down to the level of the papilla where it is only one cell layer thick.[30,31] See diagram.

2.11 *Sebaceous gland*

This gland originates in a proliferation of basal cells in the hair follicle. The fully mature gland is present at birth. The secreted material is derived from decomposition of the entire mature sebaceous gland cell, so-called holocrine secretion. Histologically, the sebaceous gland exhibits a peripheral layer of basal cells that are basophilic. As the cells progress to the center of the gland lipid, droplets appear in the cytoplasm. The fully mature sebaceous cell has a central nucleus surrounded by numerous fat deposits that stain for adipophilin and for androgen.[32–34]

2.12 *Eccrine and apocrine glands*

The eccrine glands are first noted to occur in the basal layer of the epidermis where there are clustered deeply basophilic cells. They do not exhibit the numerous mesenchymal cells at their base that are seen in the developing hair. However, the glands gradually proliferate and form finally a lumen. The cells, as they extend deeply, differentiate either into secretory cells or myoepithelial cells.[27,35]

The apocrine glands take origin as noted above from the upper bulge of hair follicles. However, their development is limited to certain areas of the body such as in the axilla and groin.[36]

2.13 *Basement membrane zone*

The BMZ is a very delicate structure that is present below the epidermis and the adnexal structures separating it from the dermis beneath. This structure is quite complex.[37] The epidermis attaches to the BMZ with hemidesmosomes or attachment plaques in the basal layer.[38] The BMZ adheres to the dermis through anchoring fibrils of collagen Type VII that attach into an anchoring plaque. The BMZ has multiple components, many of which serve as targets for antibodies.[1,37,39,40] These antibodies interact with the different protein products and result in bullous formation with the skin lifting off the dermis. The BMZ is best examined by light microscope with a PAS stain.[41]

2.14 *Dermis*

The dermis is composed principally of connective tissue. The connective tissue contains collagen with elastic fibers, a prominent intracellular ground substance, as well as specific cellular population. The cells include the fibroblasts, dermal denditric cells, macrophages, and mast cells. The fibroblasts produce the extracellular matrix. Also coursing through the dermis are appendages such as the hair follicle, eccrine duct, that are ensheathed in special connective tissue of the dermis and the vasculature of the skin.

2.15 *Extracellular matrix*

The extracellular matrix of the dermis has three specific elements that we have already referred to. These are the collagen fibers that are most abundant in the reticular dermis where they are seen as thick bundles of collagen. Collagen is produced by fibroblasts that are positive for vimentin and CD10.[1] The fibers of the papillary dermis are fine and very discreetly woven as opposed to the very coarse reticular dermal fibers. Type I collagen fibers are most common in the dermis. Reticulum fibers that are laid down by firbroblasts can easily be visualized with silver nitrate. The reticulum fiber is a special type of collagen fiber that is the building block on which the other fibers of collagen are made. Collagen fibers have a very specific cross-striation with periodicity of 68 nm.[42,43] Seven different types of collagen have been antigenicitally described. There is a Type I collagen that is prominent in post-fetal skin.[1,44]

Type III collagen is composed of reticulum fibers. Type III collagen is limited to the subepidermal and periappendageal areas. These correlate with the papillary dermal collagen. BMZ collagen has been designated as Type

IV. Cartilage collagen is Type II. The Type V collagen has been noted to occur in fetal membranes and vascular tissues. Type VI collagen can be seen in muscle. Finally, Type VII occurs in the basement membranes but it has also been found to be a major component of anchoring fibrils.[45] Type III collagen is most common in skin of the human fetus.

2.16 *Elastic fibers*

Elastic fibers appear very early in fetal life, i.e., in the first 3 weeks.[46] To easily visualize them, one must use special elastic tissue stains that contain orcein or resorcin–fuchsin. In the papillary dermis, there are thin fibers that run parallel to the long axis of the epidermis but then these thin fibers also put upward to the papillary dermis and then terminate in the PAS-positive basement membrane. The latter structures have also been called Oxytalan fibers that are membrane associated.[47,48] The remainder of the elastic fibers varies in thickness. They are the thickest and organized along the lines of the position of the collagen fibers in the deeper dermis.

2.17 *Ground substance*

The skin components lie in a very amorphous substance that occupies the spaces between the various fibrils. This material is composed of glycosaminoglycans and/or mucopolysaccharides. In normal skin, these glycans are not abundant.[49] However, it is clear that in certain instances such as chronic thyroid disorder, myxedema and certain connective tissue diseases, there is an extensive increase in ground substance in the dermis. These accumulations are designated as mucin and are best demonstrated with Alcian blue or colloidal iron stains.[1]

2.18 *Dermal vasculature*

Basic structure of vasculature of the dermis includes first, the medium sized arteries that enter into the lower dermis. They become small arteries and form plexuses that extend upward into the dermis or downward into the center of the lobules of the subcutaneous fat. In the dermis, the vessels become progressively smaller and in the superficial portion around the junction of the papillary and reticular dermis form a plexus of arterioles and venules. From this plexus, there is a hairpin-like structure composes of a pre-capillary arteriole limb, a capillary, and a post-capillary venular limb that joins the superficial vascular plexus of venules. This capillary loop is commonly referred to as the capillary-venule.[50,51] The venous plexus connects with small veins and finally to deeper veins mimicking the cascades of the arterial vessels. Among these vessels course the cutaneous nerves. This superficial venular plexus is affected predominantly in superficial type cutaneous disorders in which there are lymphocytes around the superficial plexus that also extends along the plexus of the hair follicles and the eccrine ducts and gland. This superficial plexus is very important in inflammatory diseases of the skin. The epidermis, the papillary dermis and the superficial vascular plexus have been referred to as the superficial reactive unit of the skin. It is the area most affected by disorders limited to the skin, such as psoriasis, seborrheic dermatitis, and contact dermatitis. In the subcutaneous fat, arteries enter into the lobule. They drain into capillary venules found predominately adjacent to the septae.

2.19 *Cellular elements of the dermis*

In the developing embryo, the dermis is composed of predominantly cells and a gelatinous brown substance. Then the cells, as the third month appears, begin to secrete reticulum fibers that stain with silver stain. As these fibers increase, they then organize into bundles which are collagen. The collagen does not stain, particularly with silver stain.

As the fetus develops, the cells of the dermis differentiate and begin to show spindle shaped cells, phagocytic macrophages, as well as the cells that have granules in them. The fibroblast cells become ever more prominent and there is a prominent organization of the collagen bundles into the papillary and reticular dermis.[1,52] There are elastic fibers in the papillary dermis that are very thin and are attached to the basal membrane cells. There are otherwise elastic fibers in the reticular dermis that are quite large and coarse and infiltrate through the dermis itself.

2.20 *Perivascular and interstitial denditric cells and phagocytic macrophages*

Dermal dendritic cells, first described by Headington[53] characteristically express CD1c, HLA-DR and can express factor XIIIa.[1,54] The factor XIIIa dendrocytes are often found in perivascular array and around the sweat glands.[53] They can also be found in certain proliferations such as Kaposi sarcoma and dermatofibroma. CD34 positive dendritic cells are more commonly found in the mid-dermis and deep dermis around the appendages.[55]

Macrophages, otherwise called histiocytes, are also of bone marrow origin. These cells are identified as small round cells as they enter the skin.

When stimulated, they can exhibit phagocytic activity and represent the phagocytic cells of the skin that express lysosomal enzymes such as β-glucronidase, acid phosphatse, lysozyme, and aryl sulfatase.[56] Under constant stimulation, these cells can develop into epithelioid cells and form granulomas. They are considered to be professional phagocytes. The most specific markers of macrophages include CD68 and KP1. However, the CD163[57] appears to be more specifically restrictive to monocytes and tissue macrophages. Macrophages have the ability to ingest large particles and have a characteristic of very delicate nuclear chromatin with an elongated nucleus and ample cytoplasm with vacuoles. These cells ingest Melanin and iron, leading to the designation of melanophage and siderophage respectively.

2.21 *Mast cells*

Mast cells are also bone marrow-derived cells. They are normally present in the dermis where they have two principle configurations. The mast cells are identified because of their numerous granules that stain metachromatically with stains such as Giemsa or toluidine blue.[58] In perivascular disposition, the cells have an elongated form and contain granules of varying number and different degrees of density of staining. In the deep dermis, scattered, especially along elastic fibers, the mast cells are full of numerous granules that are uniformly stained. Granules in the dermis stain for both chymase and tryptase. Granules in the mucosas and in the lamina propria express tryptase but not chymase.[59] The degranulation of mast cells will occur when there is cross-linking of IgE on the cell surface, substance P, or certain mechanical or thermal stimuli.[60] Morphine sulfate has also been shown to cause degranulation of the mast cell. In anaphylactic reactions, that is, immediate hypersensitivity reactions, the granules are extruded rapidly through the peripheral membrane of the cell.[61] In delayed hypersensitivity reactions, the mast cell granules can also be extruded in the early immediate response to the sensitizing antigen. Degranulation results in the entry of lymphocytes into the skin. This lymphocyte response is related to TNFα contained in the cells that is associated with expression of adhesion molecules such as E-selectin on endothelial cells. Lymphocytes adhere to the endothelium and then migrate into the perivascular space and into the dermis depending on the inflammatory disorder.[62,63] There is also a partial degranulation that occurs in the subacute and chronic dermatoses and chronic delayed hypersensitivity reactions. Mast cells have also been shown to have a role in graft-versus-host disease.[64]

2.22 *Other inflammatory cells*

There are numerous other inflammatory cells that are under the skin including neutrophils,[65] eosinophils,[66] and baosphils[67] that enter as a result of various chemotactic factors present in various reactions in the skin or due to infectious agents such as acute neutorphilic response.

2.23 *Lymphocytes*

Lymphocytes, which also arise in the bone marrow, are present as two different types. One type, which migrates from the bone marrow to the thymus where it differentiates and will become a mature lymphocyte, is the so-called T lymphocyte. These T-cells are located in the lymph node, predominately in the interfollicular cortex, also called the paracortex. The B lymphocyte, on the other hand, matures in the bone marrow and hence is called the B lympocyte that migrates to the lymph nodes where they are present in the germinal centers.[1,68] Basically, the T-cell is the responsible agent for cellular immunity and the B-cell, for humoral immunity. The T-cells can be divided into helper cells and cytotoxic/suppressor cells. The T lymphocytes mediate delayed hypersensitivity reactions. Recent studies have shown that there is a large population of resident T cells in the skin. These CD8+ cells provide a role in immunosurveillance.[69,70] The B-cells form antibodies to specific antigens that are designated as the humoral immune response because the antibodies circulate through the blood and serum to the various tissue sites of the antigens. There are both T-cell and B-cell lymphomas that affect the skin as malignant disorders.[71] The T-cell malignancy tends to infiltrate the epidermis whereas the B-cell infiltrates do not have a predisposition to infiltrate the epidermal and follicular structures.

2.24 *Plasma cells*

Plasma cells have a specific histological picture and develop with the aid of T helper cells from B-cells. The nucleus is always round and eccentrically placed along the membrane with coarse chromatin particles peripherally disposed in a manner that resembles a cartwheel. The cytoplasm has a space adjacent to the nucleus where the golgi is present. This obvious visible golgi area is associated with the production of immunoglobulins. Round eosinophilic bodies, called Russell bodies, represent masses of immunoglobulin in the cytoplasm.[72] The plasma cells can show two basic types of light chains. Two-thirds of the cells make a Kappa light chains and one-third makes a

Gamma light chain. In neoplastic disorders of the plasma cells, there is usually restriction to one light chain or the other.[71] Marked infiltration of benign plasma cells is characteristic of certain infections especially syphilis, granuloma inguinale, and rhinoscleroma.

2.25 *Subcutaneous fat*

The subcutaneous fat is composed of lobules of fat cells or adipocytes with an eccentric nucleus that appears spindled and is compressed along the inner plasma membrane of the cell. The lobules are divided by septae of connective tissues, predominantly collagen.[1] The circulation of the subcutaneous fat extends from the dermis into the center of lobule and drains off into the periphery of the lobule along the septae. The vessels and paraseptal array are similar to the vessels of the superficial vascular plexus and are the sites of egress of lymphocytes in inflammatory responses such as erythema nodosum, a common type of panniculitis. The adipocytes in the lobule give an unusual, almost reticulated appearance of the subcutaneous fat that is due to the close disposition of the cell membranes of the different adipocytes.

3. Wound Healing

Advances in molecular biology have allowed for an improved understanding of wound healing. For optimal healing, complex molecular processes must be appropriate and precise. In recent years, the body of literature surrounding normal and abnormal wound healing has grown exponentially. Numerous growth factors have been isolated and tested for their ability to accelerate wound healing (Figure 5).[73,74] Studies using novel tissue culture techniques have elucidated the well-orchestrated roles of various cells in healing.[75] Investigations of the fetal system have led to insights in minimally scarring wound healing.[76–78] The development of transgenic animal models is helping us understand the function of particular proteins in the healing process.[79] The integration of knowledge related to cutaneous repair process can aid in the development of novel treatments.

3.1 *Basic aspects of normal wound healing*

Injury that is restricted to the epidermis is referred to as an erosion (Figure 6); if the injury were to go deeper, involving the dermis, then it is referred to as an ulcer (Figure 7). Ulcers are further categorized into partial thickness, involving up to the mid-dermis with retention of adnexal structure, and

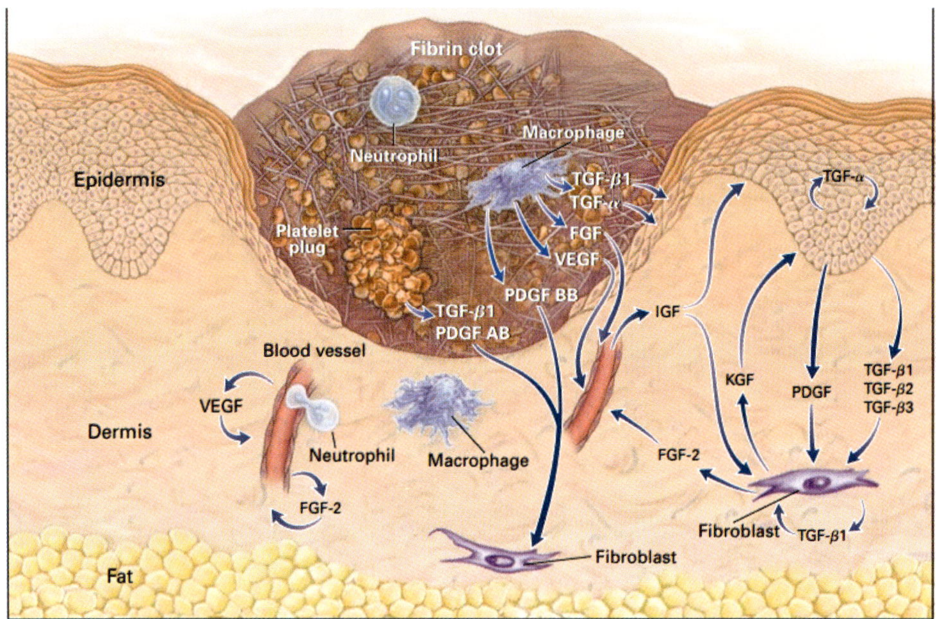

Figure 5. This diagram demonstrates the spectrum of change that occurs in wound healing. The various cells are portrayed, as well as the cytokines that are responsible for the various functions of the different cells and structures. The early neutrophlils release cytokines in cooperation with macrophages. These cells, with fibrin and debris, result in a scale crust (seen in Figure 10). Note the migration of epidermal cells (keratinocytes) toward the base of the wound. Keratinocytes, as well as stromal cells, produce cytokines that affect healing. Some of the fibroblasts are also myofibroblasts. TGF-β1, TGF-β2, and TGF-β3: transforming growth factor β1, β2, and β3, respectively. TGF-α: transforming growth factor α. FGF: fibroblast growth factor. VEGF: vascular endothelial growth factor. PDGF: platelet-derived growth factor. PDGF AB: platelet-derived growth factor AB. PDGF BB: platelet-derived growth factor BB. IGF: insulin-like growth factor. KGF: keratinocyte growth factor. Adapted from Figure 1 of Singer, A. J and Clark R. A. F. Cutaneous wound healing. *N Engl J Med* 341(10), 738–746 (1999).

full-thickness, involving the entire dermis and having loss of adnexal structures (Figure 8).

The depth of injury is of particular interest as it determines if a wound heals with scarring or not.[80] Erosions heal without scar, while dermal injury often heals with scar. Partial thickness wounds have a reservoir of keratinocytes in the hair follicles and in the wound bed, allowing for healing from the wound bed and edges. Full-thickness wounds, on the other hand, heal from the wound edges only and do not have appendages or basal keratinocytes (Figure 9). Further, full-thickness wounds may display contraction, which may result in cosmetic disfigurement. The reason for contraction in

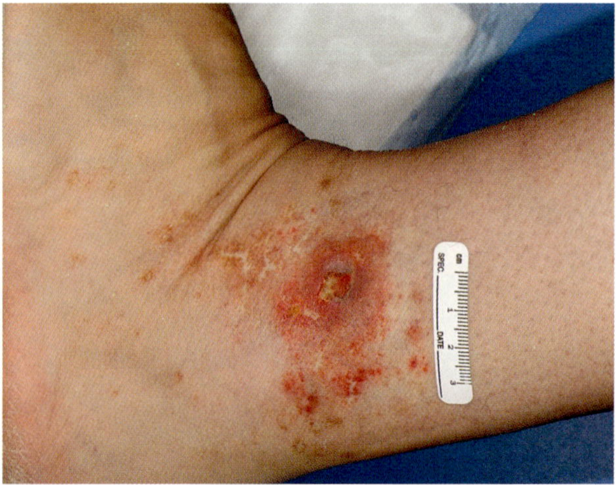

Figure 6.   A superficial erosion in the background of vasculitis reveals loss of epidermis with fibrin in the base.

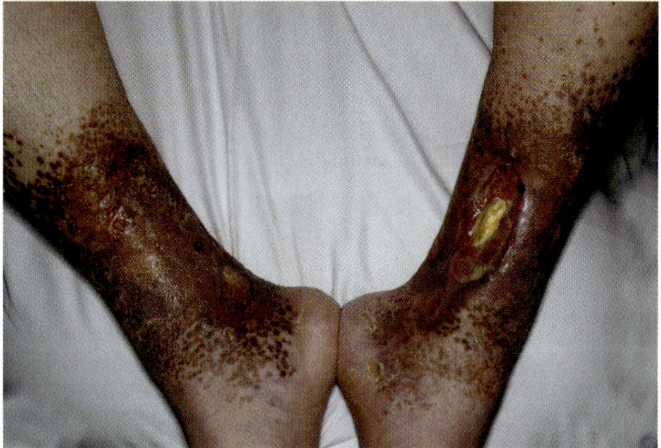

Figure 7.   The large ulcer extends into the subcutaneous fat with loss of the entire skin in the setting of vasculitis that frequently results in ischemia and subsequent ulceration.

full-thickness wounds is not clear, but is thought to be due to fibroblasts–myofibroblasts (Figure 5).[81]

3.2 Primary and secondary intention healing

Primary intention healing is when a wound's edges are approximated and closed; this is often done with the assistance of sutures or staples. Side-to-side closures, grafts and flaps are common surgical techniques applied for primary closure.

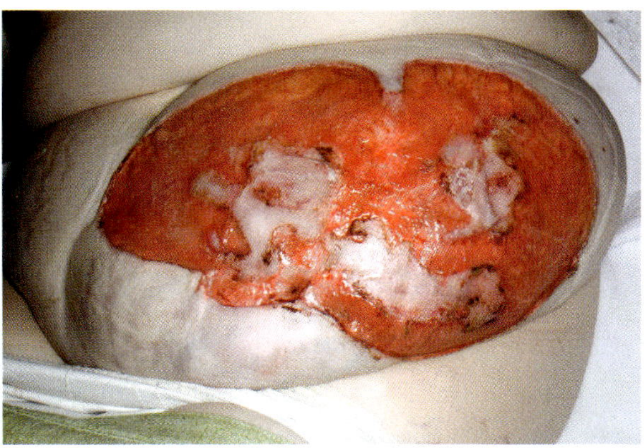

Figure 8. This large abdominal ulcer is caused by massive neutrophilic infiltration of the skin, pyoderma gangrenosum, often associated with inflammatory bowel disease. There are islands of skin in the midst of the ulcer that represent healing that comes from remnants of appendageal keratinocytes.

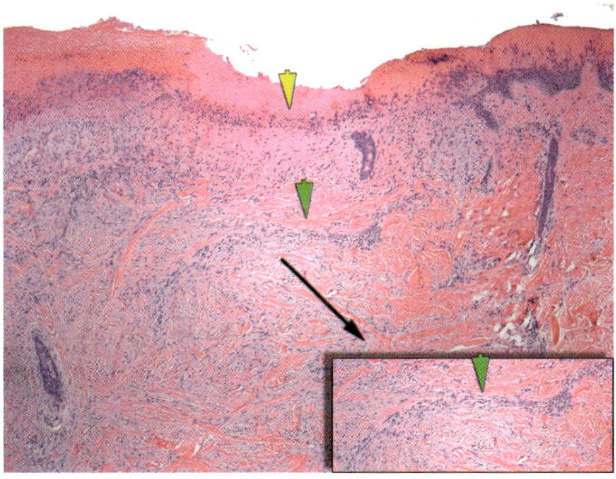

Figure 9. Healing Wound: In the center there is a scale crust fibrin with neutrophils at the base (yellow arrow). Also there is a layer of increased vascularity and fibroblasts overlying fibrosis (green arrow). H&E 45× power.

Secondary intention healing is when a wound is allowed to heal on its own. Many factors influence the time needed for a wound to heal by secondary intention: wound depth (i.e. deeper wounds take longer than shallow wounds), anatomic location (e.g. concave surfaces do better than convex surfaces), vascular supply, shape (i.e. the narrower the diameter, the faster the closure) (Figure 10), underlying comorbidities (e.g. Diabetes, tobacco use) and secondary infection.[82]

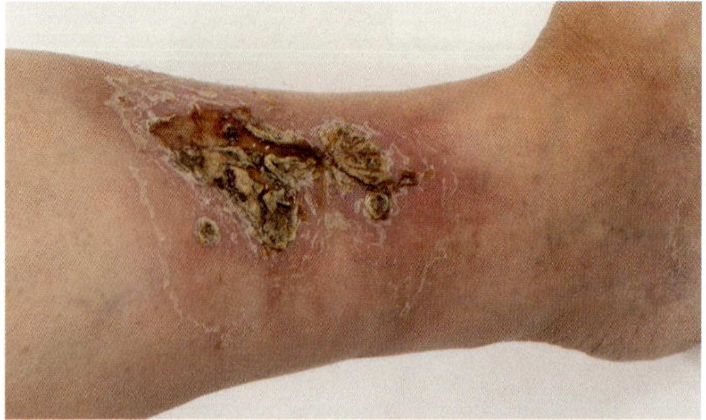

Figure 10. There is variable approximation of the healing wound with the wide scale crust at a site of large ulceration. In the lower portion of the lesion, one sees areas of approximation of the epidermis at either side of the ulcer with a narrow scale crust.

Regardless of differences, both primary and second intention forms of wound healing undergo the four phases of healing, as described below.

3.3 Events of wound healing

Given the coordinated manner by which healing is conducted, the fundamental molecular and biological processes involved in normal healing overlap and cannot easily be divided into clear-cut steps. Nevertheless, it has become easier to divide the repair process into four overlapping steps: 1. Hemostasis, 2. Inflammation, 3. Proliferation, and 4. Remodeling.

Step 1: Hemostasis

In the hemostasis phase of wound healing, the goal is to stop the outflow of blood and lymphatic fluid caused by the initial injury. The hemostatic response must be quick, localized, and carefully regulated. Vasoconstriction is a rapid response to initial injury, wherein the disruption of the vessel endothelium releases inflammatory factors like thromboxanes and prostaglandins that cause the vessel to spasm.[83]

Platelets are considered the main cell type involved in this first step of wound healing. Activation of platelets is mediated by multiple stimuli; pertinent to skin wounds, collagen and thrombin are the most potent platelet activators. The platelets are activated at the site of vascular disruption, forming a platelet plug as an initial response to stop bleeding. Once activated, the platelets undergo significant structural changes, making them extremely adhesive. Extrinsic and intrinsic clotting pathways are also

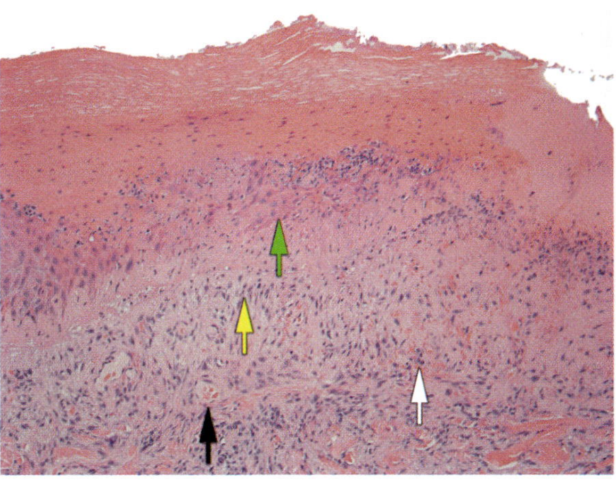

Figure 11.   Healing wound: High power of scale crust. Note at left lateral edge the keratino-cytes migrate toward the center to cover the wound (green arrow). Below these cells there is a plaque of spindle shaped fibroblasts in linear array with prominent mononuclear cells (yellow arrow). Note prominent vascularity with dilated vessels and numerous capillary-venules (black arrow). The newly proliferated vessels have diminished wall integrity so that there are scattered red blood cells extravasated into the tissue (white arrow). At the base, there are macrophages that recruit fibroblasts from the surrounding tissue, leading to the proliferative phase of wound healing. H&E 100× power.

activated by factors released from the wound, resulting in the creation of fibrin cross-linked clot.[84,85]

The platelets that are present in the clot also release growth factors and proinflammatory cytokines, which promote the migration of inflammatory cells and fibroblasts into the wound site (Figure 11).[86]

Step 2: Inflammation

In the inflammatory phase of wound healing, the goal is to bring inflammatory cells into the wound microenvironment to combat invading microbes and release necessary growth factors. The early stage of the inflammatory phase is predominated by polymorphonuclear leukocytes (PMN) and the later stage predominated by macrophages.

Within 6 h of injury, PMNs are readily found within the wound, and their maximum number is attained in 24–48 h. The PMNs play a major role in debridement of the tissue and in controlling infectious agents. They release a variety of antimicrobial substances, including reactive oxygen species (ROS) and proteases. Activated PMNs also release proinflammatory cytokines such as IL-8 and TNF.[87]

In 2–3 days after injury, monocytes enter the wound site from the surrounding blood vessels. Once they leave the vessel, monocytes are termed as macrophages. During the early inflammatory phase, the macrophages work alongside PMNs (Figure 5). These macrophages are termed as M1 macrophages and exhibit inflammatory properties, which include the production of IL-1, nitric oxide, and the clearance of invading infectious agents. As the inflammatory phase enters its later stages (days 3–4), macrophages adopt an anti-inflammatory, profibrotic phenotype (Figure 11). The population of macrophages dominant in this part of the inflammatory phase is termed M2 macrophages. M2 macrophages produce IL-10 and are involved in debris scavenging, angiogenesis, and tissue remodeling. The M2 macrophages also produce growth factors, such as TGF-β, which aids in recruiting fibroblasts from surrounding uninjured tissue to the wound site.[88–90] The recruitment of fibroblasts leads to the next stage of wound healing, the Proliferative Phase (Figure 11).

Step 3: Proliferation

In the proliferative phase of wound healing, the goal is to create healthy granulation tissue at the site of injury. Three cells are central in the formation of granulation tissue: macrophages, fibroblasts and endothelial cells. In days 5–7 after injury, fibroblasts that have been recruited by macrophages help lay down new collagen, predominantly Type III collagen with resulting fibrosis (Figure 12). In addition to collagen, fibroblasts secrete glycosaminoglycans, fibronectin and tenascin — these substances are components of the extracellular wound matrix and help facilitate cell adhesion, migration and proliferation.[91]

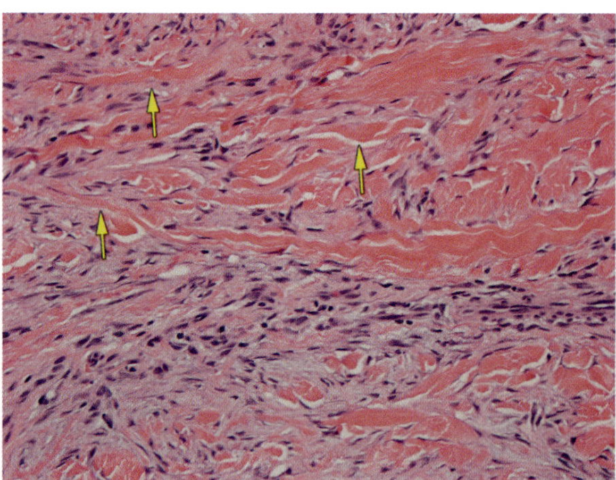

Figure 12.   Healing wound: At base of wound there is fibrosis and strands of fibroblast intersecting the fibrin and reticulin and laying down Type III collagen (yellow arrows). There is residual prominent vascularity. H&E 400× power.

Macrophages help orchestrate the sprouting of new blood vessels (i.e. angiogenesis) via the production of VEGF, among other factors.[92] VEGF and proteins found in the extracellular wound matrix help activate and modulate endothelial cells.[93] The new blood vessels formed by the endothelial cells help deliver the nutrients required by the metabolically active wound tissue as well as reverse the hypoxic and acidotic environment. Healthy granulation tissue is highly vascular, giving it a pink-red color.

With a good vascular-nutrient supply as well as a framework provided from the extracellular matrix, keratinocytes are guided into granulation tissue, leading to re-epithelialization. Depending on the depth of the injury, the keratinocytes may enter the wound site from periphery and/or adnexal structures. The migration and proliferation of keratinocytes, stimulated by keratinocyte growth factor, results in a thin epithelial layer, which bridges the wound (Figure 13).[94]

In a clean and uncontaminated wound, the proliferative phase of wound healing may last for 4 weeks, after which we enter the fourth stage of healing, remodeling.

Step 4: Remodeling

In the remodeling phase of wound healing, the wound undergoes constant alterations, which can last for years after the initial injury occurred. The mechanisms underlying in transformation of granulation tissue into scar tissue are largely undetermined. The scar matrix formed during the proliferative phase is degraded and new components are synthesized. The balance between

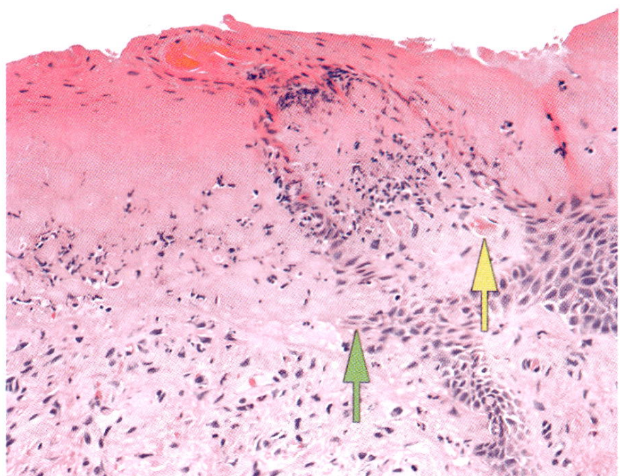

Figure 13. Healing wound: at the edge of the wound, keratinocytes extend into the base of the lesion (green arrow). Note the vessels have focal thrombosis (yellow arrow) and dilation with red blood cells filling the lumens in areas of fibrin deposition. H&E 400× power.

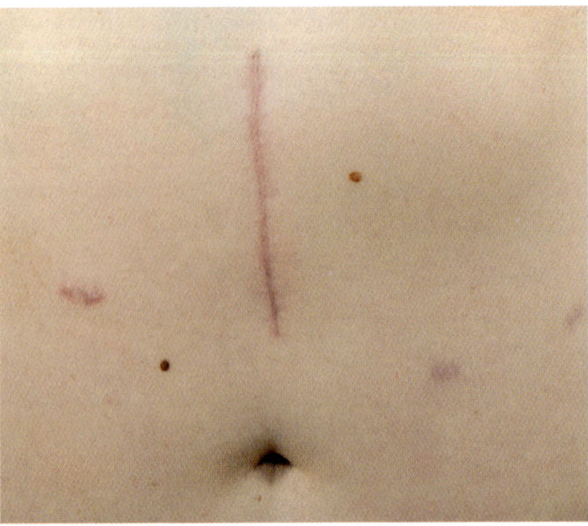

Figure 14. Well healed linear scar showing approximation of normal skin to the healed wound. Both sides of the wound are equidistant and there is no asymmetry.

degradation and synthesis determines if a scar will be normal or abnormal (e.g. atrophic scar, hypertrophic scar, keloid).[95] In the remodeling phase, fibroblasts are transformed into myofibroblasts, which help contract the wound. There is also substitution of the provisional extracellular matrix into permanent collagenous matrix. Maximal contraction is achieved by the 12th week and the ultimate scar strength is 80% of the tensile strength of the original skin that it replaced (Figure 14).[96] Finally, in remodeling, vascular structures regress and there is apoptosis or efflux of any residual inflammatory cells.

3.4 *Abnormal wound healing*

The mature scar formed from normal wound healing is not the equivalent to normal skin, but it achieves the goal of restoring the barrier. A normotrophic scar represents the midpoint in the spectrum of wound-healing responses. In hypertrophic scars and keloids, the healing process is overactive, producing excess scar tissue. In chronic wounds and atrophic scars, the healing process is underactive and, many times, incomplete. Dysfunctional wound healing is a major clinical and socioeconomic problem.

3.5 *Chronic wounds*

A chronic wound is a non-healing wound that persists for over 6 weeks. They are often a symptom of an underlying disease, rather than a disease

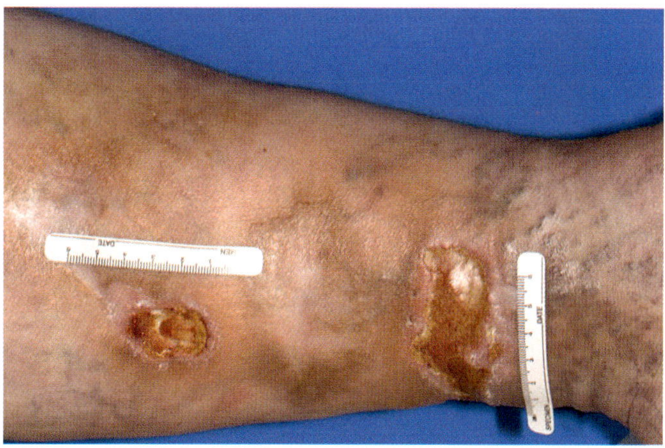

Figure 15. Two ulcers in an edematous leg that shows prominent venous stasis with blue-brown discoloration of the skin.

entity in itself. Commons factors leading to incomplete healing include venous insufficiency (Figure 15), ischemia (Figure 7), diabetes; less common causes include: vasculitis (Figure 7), hypercoagulability, malnutrition, and certain therapies (e.g. hydroxyurea) (Figure 16).[97–101] Poor blood supply, dysregulated inflammatory response (Figure 8), wound-site infection and growth factor deficiency can also contribute to a chronic wound. A chronic wound, regardless of cause, has increased proteolysis and impairments in cell proliferation and migration. Inflammatory cells, primarily PMNs and M1 macrophages, may be present in high numbers within a chronic wound, leading to the production of ROS that damage the extracellular matrix and surrounding cells.[102] Further, ROS and other factors produced by inflammatory cells induce the production of proteolytic enzymes. These proteolytic enzymes degrade components of the extracellular matrix, impeding the migration of keratinocytes into the wound site and, therefore, preventing re-epithelialization.[103,104] Myofibroblasts are also affected by the chronic wound milieu; a decrease in their number leads to delayed healing.[105]

3.6 *Fibroproliferative disorders*

Hypertrophic and keloidal scarring is defined by the presence of excess scar tissue. In hypertrophic scars, the excess scar tissue is limited to the boundary of the original site of the injury; in keloidal scars, the scar tissue extends beyond the site of injury. Hypertrophic scarring is associated with a prolonged inflammatory phase of wound healing. Furthermore, certain injury types have a propensity for hypertrophic scar formation (e.g. Burn injury

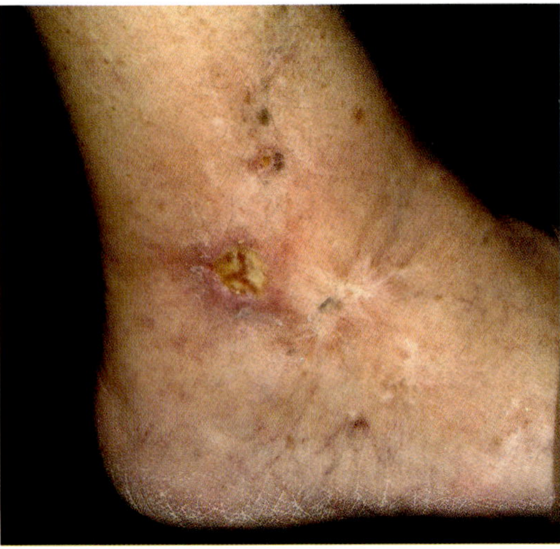

Figure 16. This ulceration occurred in a patient who was on chronic therapy with hydroxurea. Purulent debris covers most of the surface of this ulcer that shows sharply demarcated borders. This represents an unusual side effect of chronic hydroxyurea therapy.

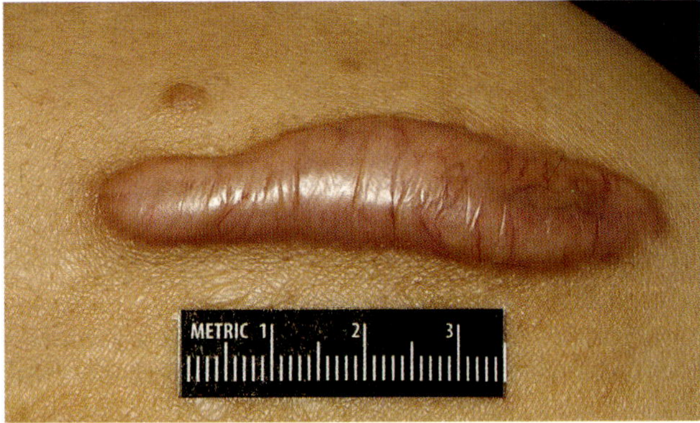

Figure 17. This keloid exhibits a raised tubular excrescence has grown well beyond the original excision site in width. It is always elevated, firm, often uncomfortable, and even painful. These lesions are difficult to treat and often recur.

and trauma to the deep dermis).[106] Keloidal scaring, on the other hand, has a genetic influence that has yet to be ascertained. Keloid formation predominantly occurs in darker skinned individuals and has a propensity to be inherited (Figure 17).

Both hypertrophic and keloid scars may be cosmetically distressing and, secondary to contractures, may be debilitating. Effective treatment, especially for keloids, is limited, given our incomplete understanding of the pathophysiology of these disorders.[107] It has been found that the fibroblasts present in keloids and scars are functionally different from those in normal skin. Instead of undergoing apoptosis or decreasing collagen production, the fibroblasts in keloids and hypertrophic scars remain activated. The fibroblasts in hypertrophic scars express higher levels of TGF-β, a profibrotic cytokine, and lower levels of collagenase, which breaks down collagen.[108] Compared to hypertrophic scars, keloidal fibroblasts are hyperproliferative and produced increased levels of Type I collagen, resulting in a unique increase in the Type I to Type III collagen ratio.[109]

Acknowledgments

The authors would like to acknowledge David DiCaudo, MD at Scottsdale, Arizona for providing the photomicrographs used in Figures 9, 11–13. The authors would also like to acknowledge Mr. Steven Conley for his help in preparing the figures.

References

1. Lian, C. and Murphy, G. Histology of the Skin. In: Elder, D., (ed), *Levers' Histopathology of the Skin*. Philadelphia, PA: Wolters Kluwer, p. 6–75 (2015).
2. Breathnach, A. S. The Herman Beerman lecture: Embryology of human skin, a review of ultrastructural studies. *J Invest Dermatol* 57(3), 133–143 (1971).
3. Holbrook, K. A. and Odland, G. F. The fine structure of developing human epidermis: Light, scanning, and transmission electron microscopy of the periderm. *J Invest Dermatol* **65**(1), 16–38 (1975).
4. Hashimoto, K. and Lever, W. F. The cell surface coat of normal keratinocytes and of acantholytic keratinocytes in pemphigus. An electron microscopic study. *B J Dermatol* 83(2), 282–290 (1970).
5. Wolff, K. and Wolff-Schreiner, E. C. Trends in electron microscopy of skin. *J Invest Dermatol* **67**(1), 39–57 (1976).
6. Amagai, M. Adhesion molecules. I: Keratinocyte-keratinocyte interactions; cadherins and pemphigus. *J Invest Dermatol* **104**(1), 146–152 (1995).
7. Matoltsy, A. G. Desmosomes, filaments, and keratohyaline granules: Their role in the stabilization and keratinization of the epidermis. *J Invest Dermatol* **65**(1), 127–142 (1975).
8. Elias, P. M. Epidermal lipids, barrier function, and desquamation. *J Invest Dermatol* 80(1), 44s–49s (1983).
9. Wolff-Schreiner, E. C. Ultrastructural cytochemistry of the epidermis. *Int J Dermatol* 16(2), 77–102 (1977).
10. Lazarus, G. S. Hatcher, V. B. and Levine, N. Lysosomes and the skin. *J Invest Dermatol* **65**(3), 259–271 (1975).

11. Dale, B. A. Filaggrin, the matrix protein of keratin. *Am J Dermatopathol* 7(1), 65–68 (1985).

12. Murphy, G. F. Flynn, T. C. Rice, R. H. and Pinkus, G. S. Involucrin expression in normal and neoplastic human skin: A marker for keratinocyte differentiation. *J Invest Dermatol* 82(5), 453–457 (1984).

13. Wolff, K. and Schreiner, E. Epidermal lysosomes. Electron microscopic-cytochemical studies. *Arch Dermatol* 101(3), 276–286 (1970).

14. Katz, S. I. Dohi Memorial Lecture. The skin as an immunological organ: Allergic contact dermatitis as a paradigm. *J Dermatol* 20(10), 593–603 (1993).

15. Bos, J. D. and Kapsenberg, M. L. The skin immune system: Progress in cutaneous biology. *Immunol Today* 14(2), 75–78 (1993).

16. Becker, S. W. Jr. and Zimmermann, A. A. Further studies on melanocytes and melanogenesis in the human fetus and newborn. *J Invest Dermatol* 25(2), 103–112 (1955).

17. Mottaz, J. H. and Zelickson, A. S. Melanin transfer: A possible phagocytic process. *J Invest Dermatol* 49(6), 605–610 (1967).

18. Cochran, A. J. The incidence of melanocytes in normal human skin. *J Invest Dermatol* 55(1), 65–70 (1970).

19. Becker, S. W. Jr. Fitzpatrick, T. B. and Montgomery, H. Human melanogenesis; cytology and histology of pigment cells (melanodendrocytes). *AMA Arch Dermatol Syphilol* 65(5), 511–523 (1952).

20. Szeder, V. Grim, M. Halata, Z. and Sieber-Blum, M. Neural crest origin of mammalian Merkel cells. *Dev Bio* 253(2), 258–263 (2003).

21. Moll, I. Roessler, M. Brandner, J. M. Eispert, A. C. Houdek, P. and Moll, R. Human Merkel cells--aspects of cell biology, distribution and functions. *E J Cell Bio* 84(2–3), 259–271 (2005).

22. Moll, R. Moll, I. and Franke, W. W. Identification of Merkel cells in human skin by specific cytokeratin antibodies: Changes of cell density and distribution in fetal and adult plantar epidermis. *Differentiation* 28(2), 136–154 (1984).

23. Wolff, K. The langerhans cell. *Curr Prob Dermatol* 4, 79–145 (1972).

24. Tamaki, K. Stingl, G. and Katz, S. I. The origin of Langerhans cells. *J Invest Dermatol* 74(5), 309–311 (1980).

25. Wolff, K. and Stingl, G. The langerhans cell. *J Invest Dermatol* 80(1), 17s–21s (1983).

26. H. P. The biology of hair growth. In: Montagna W. E. R., (ed.), *Embryology of Hair*. New York, NY: Academic Press, (1958).

27. Hashimoto, K. Gross, B. G. and Lever, W. F. The ultrastructure of human embryo skin. II. The formation of intradermal portion of the eccrine sweat duct and of the secretory segment during the first half of embryonic life. *J Invest Dermatol* 46(6), 513–529 (1966).

28. Kollar, E. J. The induction of hair follicles by embryonic dermal papillae. *J Invest Dermatol* 55(6), 374–378 (1970).

29. Hashimoto, K. The ultrastructure of the skin of human embryos. VII. Formation of the apocrine gland. *Acta Dermato Venereol* 50(4), 241–251 (1970).

30. Pinkus, H. Dermal Pathology. In: Graham, J. H. Johnson, W. C. Helwig, E. B. (ed.), *Anatomy and Histology of the Skin.* Hagerstown MD: Harper & Row, (1972).

31. Yang, C. C. and Cotsarelis, G. Review of hair follicle dermal cells. *J Dermatol Sci* 57(1), 2–11 (2010).

32. Strauss, J. S. and Pochi, P. E. Histology, histochemistry, and electron microscopy of sebaceous glands in man. In: Marchionini A., (ed.), *Handbuch der haut- und Geschlechtskrankheiten, Erganzungswerk.* Vol. 1, Part 1, Berlin, Germany: Springer-Verlag, (1968).

33. Jakobiec, F. A. and Werdich, X. Androgen receptor identification in the diagnosis of eyelid sebaceous carcinomas. *Am J Ophthalmol* 157(3), 687–696, e1–e2 (2014).

34. Plaza, J. A. Mackinnon, A. Carrillo, L. Prieto, V. G. Sangueza, M. and Suster, S. Role of Immunohistochemistry in the diagnosis of sebaceous carcinoma: a clinicopathologic and immunohistochemical study. *Am J Dermatopathol* 37(11), 809–821 (2015).

35. Watanabe, S. Ichikawa, E. Takanashi, S. and Takahashi, H. Immunohistochemical localization of cytokeratins in normal eccrine glands, with monoclonal antibodies in routinely processed, formalin-fixed, paraffin-embedded sections. *J Am Acad Dermatol* 28(2 Pt 1), 203–212 (1993).

36. Inoue, T. Scanning electron microscopic study of the human axillary apocrine glands. *J Dermatol* 6(5), 299–308 (1979).

37. Eady, R. A. The basement membrane. Interface between the epithelium and the dermis: Structural features. *Arch Dermatol* 124(5), 709–712 (1988).

38. Yancey, K. B. Adhesion molecules. II: Interactions of keratinocytes with epidermal basement membrane. *J Invest Dermatol* 104(6), 1008–1014 (1995).

39. Moll, R. and Moll, I. Epidermal adhesion molecules and basement membrane components as target structures of autoimmunity. *Virchows Archiv* 432(6), 487–504 (1998).

40. Vafia, K., Groth, S. Beckmann, T. Hirose, M. Dworschak, J. Recke, A. *et al.* Pathogenicity of autoantibodies in anti-p200 pemphigoid. *PloS one* 7(7), e41769 (2012).

41. Fassihi, H. Wong, T. Wessagowit, V. McGrath, J. A. and Mellerio, J. E. Target proteins in inherited and acquired blistering skin disorders. *Clin Exp Dermatol* 31(2), 252–259 (2006).

42. Lazarus, G. S. Collagen, collagenase and clinicians. *B J Dermatol* 86(2), 193–199 (1972).

43. Grant, M. E. and Prockop, D.J. The biosynthesis of collagen. 1. *N Eng J Med* 286(4), 194–199 (1972).

44. Stenn, K. Collagen heterogeneity of skin. *Am J Dermatopathol* 1(1), 87–88 (1979).

45. Leigh, I. M, Eady, R. A. Heagerty, A. H. Purkis, P. E. Whitehead, P. A. and Burgeson, R. E. Type VII collagen is a normal component of epidermal basement membrane, which shows altered expression in recessive dystrophic epidermolysis bullosa. *J Invest Dermatol* **90**(5), 639–642 (1988).

46. Deutsch, T. A. and Esterly, N. B. Elastic fibers in fetal dermis. *J Invest Dermatol* **65**(3), 320–323 (1975).

47. Frances, C. and Robert, L. Elastin and elastic fibers in normal and pathologic skin. *Int J Dermatol* **23**(3), 166–179 (1984).

48. Bruckner-Tuderman, L. Ruegger, S. Odermatt, B. Mitsuhashi, Y. and Schnyder, U.W. Lack of type VII collagen in unaffected skin of patients with severe recessive dystrophic epidermolysis bullosa. *Dermatologica.* **176**(2), 57–64 (1988).

49. Winand, R. Biosynthesis, organization and degradation of mucopolysaccharides. *Arch Bel Dermatol Syphiligr* **28**(1), 35–40 (1972).

50. Yen, A. and Braverman, I. M. Ultrastructure of the human dermal microcirculation: The horizontal plexus of the papillary dermis. *J Invest Dermatol* **66**(3), 131–42 (1976).

51. Braverman, I. M. and Yen, A. Ultrastructure of the human dermal microcirculation. II. The capillary loops of the dermal papillae. *J Invest Dermatol* **68**(1), 44–52 (1977).

52. Nigra, T. P. Friedland, M. and Martin, G. R. Controls of connective tissue synthesis: Collagen metabolism. *J Invest Dermatol* **59**(1), 44–9 (1972).

53. Headington, J. T. The dermal dendrocyte. *Adv Dermatol* **1**, 159–171 (1986).

54. Cerio, R. Griffiths, C. E. Cooper, K. D. Nickoloff, B. J. and Headington J. T. Characterization of factor XIIIa positive dermal dendritic cells in normal and inflamed skin. *Br J Dermatol* **121**(4), 421–431 (1989).

55. Nestle, F. O. Zheng, X. G. Thompson, C. B. Turka, L. A. and Nickoloff, B. J. Characterization of dermal dendritic cells obtained from normal human skin reveals phenotypic and functionally distinctive subsets. *J Immunol* **151**(11), 6535–6545 (1993).

56. Eyden, B. The myofibroblast: An assessment of controversial issues and a definition useful in diagnosis and research. *Ultrastruc Pathol* **25**(1), 39–50 (2001).

57. Lau, S. K. Chu, P. G. and Weiss, L. M. CD163: A specific marker of macrophages in paraffin-embedded tissue samples. *Am J Clin Pathol* **122**(5), 794–801 (2004).

58. Dvorak, H. F. and Dvorak, A. M. Basophils, mast cells, and cellular immunity in animals and man. *Hum Pathol* **3**(4), 454–456 (1972).

59. Irani, A. M. Bradford, T. R. Kepley, C. L. Schechter, N. M. and Schwartz, L. B. Detection of MCT and MCTC types of human mast cells by immunohistochemistry using new monoclonal anti-tryptase and anti-chymase antibodies. *J Histochem Cytochem* **37**(10), 1509–1515 (1989).

60. Kobayasi, T. and Asboe-Hansen, G. Degranulation and regranulation of human mast cells. An electron microscopic study of the whealing reaction in urticaria pigmentosa. *Acta Dermato Venereol* **49**(4), 369–381 (1969).

61. Kaminer, M. S. Lavker, R. M. Walsh, L. J. Whitaker, D. Zweiman, B. and Murphy, G. F. Extracellular localization of human connective tissue mast cell granule contents. *J Invest Dermatol* **96**(6), 857–863 (1991).

62. Klein, L. M. Lavker, R. M. Matis, W. L. and Murphy, G. F. Degranulation of human mast cells induces an endothelial antigen central to leukocyte adhesion. *Pro Natl Acad Sci USA* **86**(22), 8972–896 (1989).

63. Christofidou-Solomidou, M. Murphy, G. F. and Albelda, S. M. Induction of E-selectin-dependent leukocyte recruitment by mast cell degranulation in human skin grafts transplanted on SCID mice. *Am J Pathol* **148**(1), 177–188 (1996).

64. Murphy, G. F, Sueki, H. Teuscher, C. Whitaker, D. and Korngold, R. Role of mast cells in early epithelial target cell injury in experimental acute graft-versus-host disease. *J Invest Dermatol* **102**(4), 451–461 (1994).

65. Jones, D. A. Abbassi, O. McIntire, L. V. McEver, R. P and Smith, C. W. P-selectin mediates neutrophil rolling on histamine-stimulated endothelial cells. *Biophys J* **65**(4), 1560–1569 (1993).

66. Dvorak, A. M. Mihm, M. C. Jr. Osage, J. E. Kwan, T. H. Austen, K. F. and Wintroub, B. U. Bullous pemphigoid, an ultrastructural study of the inflammatory response: Eosinophil, basophil and mast cell granule changes in multiple biopsies from one patient. *J Invest Dermatol* **78**(2), 91–101 (1982).

67. Dvorak, H. F. and Mihm, M. C., Jr. Basophilic leukocytes in allergic contact dermatitis. *J Exp Med* **135**(2), 235–254 (1972).

68. Weissman, I. L. Warnke, R. Butcher, E. C. Rouse, R. and Levy, R. The lymphoid system. Its normal architecture and the potential for understanding the system through the study of lymphoproliferative diseases. *Hum Pathol* **9**(1), 25–45 (1978).

69. Clark, R. A. Chong, B. Mirchandani, N. Brinster, N. K. Yamanaka, K. Dowgiert, R. K. *et al*. The vast majority of CLA+ T cells are resident in normal skin. *J Immunol* **176**(7), 4431–4439 (2006).

70. Jiang, X. Clark, R. A. Liu, L. Wagers, A. J. Fuhlbrigge, R. C. and Kupper, T. S. Skin infection generates non-migratory memory CD8+ T(RM) cells providing global skin immunity. *Nature* **483**(7388), 227–231 (2012).

71. Murphy, G. and Mihm, M. J. Benign, dysplastic, and malignant lymphoid infiltrates of the skin: An approach based on pattern analysis. In: Murphy G, Mihm MJ, (ed.), Lympho-proliferative disorders of the skin. Boston, MA: Butterworths, p. 123–141 (1986).

72. Erlach, E. Gebhart, W. and Niebauer, G. Ultrastructural investigations on the morphogenesis of Russell bodies. *J Cutan Path* (3), 145 (1976).

73. Werner, S. and Grose, R. Regulation of wound healing by growth factors and cytokines. *Physiol Rev* **83**(3), 835–870 (2003).

74. Clark, R. *The Molecular and Cellular Biology of Wound Repair*: Springer Science & Business Media, (2013).

75. Bell, E. Ehrlich, H. P. Buttle, D. J. and Nakatsuji, T. Living tissue formed *in vitro* and accepted as skin-equivalent tissue of full thickness. *Science* **211**(4486), 1052–1054 (1981).

76. Longaker, M. T. Chiu, E. S. Adzick, N. S. Stern, M. Harrison, M. R. and Stern, R. Studies in fetal wound healing. V. A prolonged presence of hyaluronic acid characterizes fetal wound fluid. *Ann Surg* **213**(4), 292–296 (1991).

77. Longaker, M. T. Whitby, D. J. Ferguson, M. W. Lorenz, H. P. Harrison, M. R. and Adzick, N.S. Adult skin wounds in the fetal environment heal with scar formation. *Ann Surg* **219**(1), 65–72 (1994).

78. Mackool, R. J. Gittes, G. K. and Longaker, M. T. Scarless healing. The fetal wound. *Clin Plast Surg* **25**(3), 357–365 (1998).

79. Martin, P. Wound healing--aiming for perfect skin regeneration. *Science.* **276**(5309), 75–81 (1997).

80. Dunkin, C. S. Pleat, J. M. Gillespie, P. H. Tyler, M. P. Roberts, A. H. and McGrouther, D. A. Scarring occurs at a critical depth of skin injury: Precise measurement in a graduated dermal scratch in human volunteers. *Plast Reconstr Surg* **119**(6), 1722–1732 (2007); discussion 33–34.

81. Berry, D. P. Harding, K. G. Stanton, M. R. Jasani, B. and Ehrlich, H. P. Human wound contraction: Collagen organization, fibroblasts, and myofibroblasts. *Plast Reconstr Surg* **102**(1), 124–131 (1998); discussion 32–34.

82. Zitelli, J. A. Secondary intention healing: An alternative to surgical repair. *Clin Dermatol* **2**(3), 92–106 (1984).

83. Baum, C. L. and Arpey, C. J. Normal cutaneous wound healing: Clinical correlation with cellular and molecular events. *Derm Surg* **31**(6), 674–86 (2005); discussion 86.

84. Broughton, G. 2nd, Janis, J. E. and Attinger, C. E. The basic science of wound healing. *Plast Reconstr Surg* **117**(7), 12S–34S (2006).

85. Packham, M. A. Role of platelets in thrombosis and hemostasis. *Can J Physiol Pharmacol* **72**(3), 278–84 (1994).

86. Szpaderska, A. M. Egozi, E. I. Gamelli, R. L. and DiPietro, L. A. The effect of thrombocytopenia on dermal wound healing. *J Invest Dermatol* **120**(6), 1130–1137 (2003).

87. Hart, J. Inflammation. 1: Its role in the healing of acute wounds. *J Wound Care* **11**(6), 205–209 (2002).

88. Koh, T. J. and DiPietro, L. A. Inflammation and wound healing: The role of the macrophage. *Expert Rev Molecul Med* **13**, e23 (2011).

89. Mosser, D. M. and Edwards, J. P. Exploring the full spectrum of macrophage activation. *Nat Rev Immunol* **8**(12), 958–969 (2008).

90. Mahdavian Delavary, B. van der Veer, W. M. van Egmond, M. Niessen, F. B. and Beelen, R. H. Macrophages in skin injury and repair. *Immunobiol* **216**(7), 753–762 (2011).

91. Darby, I. A. Laverdet, B. Bonte, F. and Desmouliere, A. Fibroblasts and myofibroblasts in wound healing. *Clin Cosmet Invest Dermatol* **7**, 301–311 (2014).

92. Jetten, N. Verbruggen, S. Gijbels, M. J. Post, M. J. De Winther, M. P. and Donners, M. M. Anti-inflammatory M2, but not pro-inflammatory M1 macrophages promote angiogenesis *in vivo*. *Angiogenesis.* **17**(1), 109–118 (2014).

93. Wilgus, T. A. and DiPietro, L. A. Complex roles for VEGF in dermal wound healing. *J Invest Dermatol* **132**(2), 493–494 (2012).

94. Guo, S. and Dipietro, L. A. Factors affecting wound healing. *J Dent Res* **89**(3), 219–229 (2010).

95. Bayat, A. McGrouther, D. A. and Ferguson, M. W. Skin scarring. *Br Med J.* **326**(7380), 88–92 (2003).

96. Diegelmann, R. F. and Evans, M. C. Wound healing: An overview of acute, fibrotic and delayed healing. *Front Biosci* **9**, 283–289 (2004).

97. Stadelmann, W. K. Digenis, A. G. and Tobin, G. R. Physiology and healing dynamics of chronic cutaneous wounds. *Am J Surg* **176**(2A), 26S–38S (1998).

98. Nwomeh, B. C. Yager, D. R. and Cohen, I. K. Physiology of the chronic wound. *Clin Plast Surg* **25**(3), 341–356 (1998).

99. Falanga, V. The chronic wound: Impaired healing and solutions in the context of wound bed preparation. *Blood Cells Mol Dis* **32**(1), 88–94 (2004).

100. Goldman, R. Growth factors and chronic wound healing: Past, present, and future. *Adv Skin Wound Care* **17**(1), 24–35 (2004).

101. Velnar, T. Bailey, T. and Smrkolj, V. The wound healing process: An overview of the cellular and molecular mechanisms. *J Inter Med Res* **37**(5), 1528–1542 (2009).

102. Brancato, S. K. and Albina, J. E. Wound macrophages as key regulators of repair: Origin, phenotype, and function. *Am J Pathol* **178**(1), 19–25 (2011).

103. Yager, D. R. and Nwomeh, B. C. The proteolytic environment of chronic wounds. *Wound Repair Regen* **7**(6), 433–441 (1999).

104. Trengove, N. J. Stacey, M. C. MacAuley, S. Bennett, N. Gibson, J. Burslem, F. *et al.* Analysis of the acute and chronic wound environments: The role of proteases and their inhibitors. *Wound Repair Regen* **7**(6), 442–452 (1999).

105. Van De Water, L. Varney, S. and Tomasek, J. J. Mechanoregulation of the myofibroblast in wound contraction, scarring, and fibrosis: Opportunities for new therapeutic intervention. *Adv Wound Care* **2**(4), 122–141 (2013).

106. Zhu, Z. Ding, J. Shankowsky, H. A. and Tredget, E. E. The molecular mechanism of hypertrophic scar. *J Cell Commun Signal* **7**(4), 239–252 (2013).

107. Al-Attar, A. Mess, S. Thomassen, J. M. Kauffman, C. L. and Davison, S. P. Keloid pathogenesis and treatment. *Plast Reconstr Surg* **117**(1), 286–300 (2006).

108. Tuan, T. L. and Nichter, L. S. The molecular basis of keloid and hypertrophic scar formation. *Mol Med Today* **4**(1), 19–24 (1998).

109. Uitto, J. Perejda, A. J. Abergel, R. P. Chu, M. L. and Ramirez, F. Altered steady-state ratio of type I/III procollagen mRNAs correlates with selectively increased type I procollagen biosynthesis in cultured keloid fibroblasts. *Proc Natl Acad Sci USA* **82**(17), 5935–5939 (1985).

3. Deprived and Enriched Environments: How Sensory Stimulation Affects Wound Healing

Jonathan G. Fricchione* and John B. Levine*,†

*Massachusetts General Hospital,
55 Fruit St, Boston, MA 02114, USA
†Harvard Medical School,
25 Shattuck St, Boston, MA 02115, USA

Abstract

Tissue damage that follows an injury results in a cascade of biological events that in ideal conditions result in activation of local and systemic responses that activate wound healing. Optimal wound healing becomes impaired in a variety of clinical conditions such as when other chronic illnesses are present (e.g. wounds in a patient with pre-existing diabetes) or when environmental factors negatively impact the healing response (e.g. wounds in a person with limited physical activity). In this chapter, we review a wide body of research on two models that have elucidated the mechanisms involved in impaired wound healing due to environmental conditions. The two conditions examined, termed deprived environment (DE) and enriched environments (EE) are reviewed by examining how the systemic response to sensory inputs in these two conditions are altered. To examine the complex mechanisms involved in the wound healing response to these two conditions, we divide them into three components: (a) how the sensory regions of the brain are changed in response to these conditions, (b) how the sensory changes in the brain result in altered brain outputs to injured tissue during wound healing, and (c) how such outputs lead to impaired healing in the DE condition and improved healing in the EE condition. For each of these components of the wound healing response in DE and EE conditions, we identify abnormal (DE conditions) and normal responses (EE conditions) that can potentially be altered with therapy aimed at improving wound healing.

1. Overview

Over the past two decades, a growing body of evidence from both clinical and basic science research suggests that differential sensory input from the external environment affects wound healing. Results of this growing literature suggest that suboptimal environmental stimulation or exposure to deprived environments (DEs) impairs wound healing, while enhancing environmental stimulation through enriched environments (EEs) can facilitate wound healing. The pathways by which DE and EE impact wound healing and other physical illnesses is sometimes referred to as the "brain-skin" axis.[1–3] To emphasize the focus of this chapter on how afferent[a] environmental stimuli mediate the brain's impact on efferent[b] skin signals during wound healing (and vice versa), we refer to the brain-skin axis as the environment-brain-skin (EBS) circuit.

Because of the many synapses[c] in the EBS circuit, in this chapter we group them into three sub-circuits: (a) afferent input from the environment to neurons[d] in the brain's sensory cortex,[e] as outlined in Figure 1, (b) efferent output from neurons in the brain's sensory cortex to various brain nuclei[f] that, in turn, send efferent signals that affect wound healing, as outlined in Figure 2, and (c) signals from brain efferents to skin cells involved in wound healing, as outlined in Figure 3.

---

[a] Afferent signals are biological inputs *to* a nerve cell *from* other nerve or non-nervous-system cells (e.g. input from *touch receptors* in skin that projects *to nerve cells* in the tactile sensory portion of the nervous system).

[b] Efferent signals are biological outputs *from* a nerve cell *to* other nerve or non-nervous-system signals (e.g. input on changes in pressure or position *from nerve cells* in the tactile portion of nervous system *to other nerve and somatic cells*. In short, afferent signals are inputs to nerve cells and efferent signals are outputs from nerve cells).

[c] Synapse: the connection, whether afferent or efferent, between a given nerve cell and its biological input or output.

[d] Neuron: nerve cells that have their cell bodies in a portion of the central nervous system (versus nerve cells that have their cell bodies in the peripheral nervous system).

[e] Cortex: the outer portion of the brain associated with higher level cognitive functioning (includes the frontal, parietal (motor and sensory), occipital, and temporal lobes).

[f] Nuclei: closely located, groups of neuronal or nerve cell bodies.

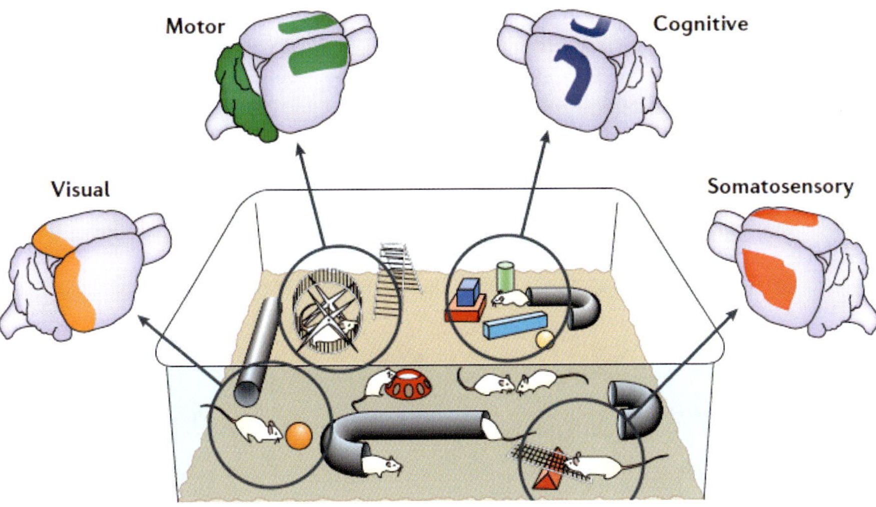

Figure 1.   The components of the EBS circuit linking the environment with brain sensory cortex.[4]

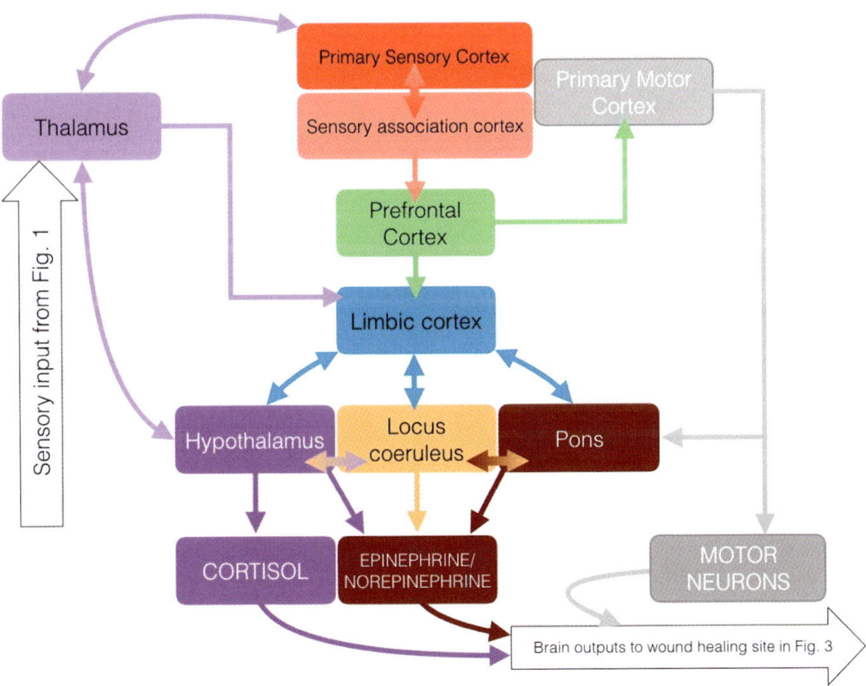

Figure 2.   Proposed component of the EBS circuit linking brain afferent input with brain efferent output that impacts cutaneous function. It can be divided into two components: first (A) efferent output from sensory cortex (red) to limbic cortex (blue) and (B) from limbic cortex (blue) to neuroendocrine regions (hypothalamus — purple, locus coeruleus — orange, pons — maroon) (Fricchione and Levine, unpublished).

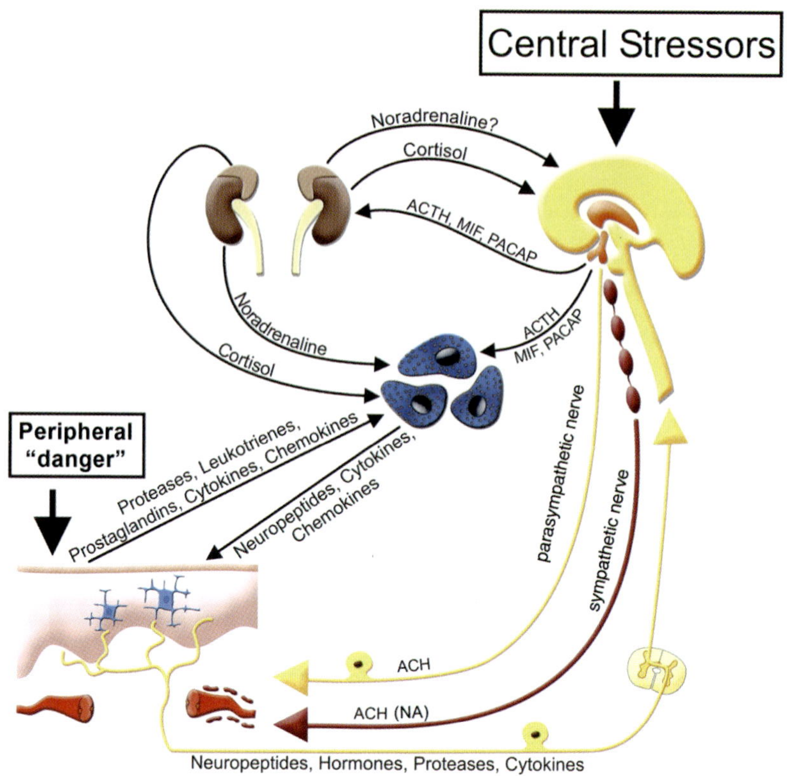

Figure 3.   The components of EBS circuit relaying brain outputs to skin and wound healing.[5]

Not surprisingly, the complexity of this system has led researchers to focus on one of the sub-circuits of the brain–skin system. The sub-circuit linking neurons communicating with the environment to neurons communicating with the skin is the most complicated (Figure 2). As a result, the connections in this pathway are less well understood compared to those in the sub-circuits linking environment to brain and brain to skin (Figures 1 and 3).

2. Observational Data from Human and Animal Studies Supports a Link between Environmental Sensory Inputs and Wound Healing

In this section, we review the observational data on environmental effects on wound healing that led systems biologists to investigate the underlying mechanisms, as outlined in Figures 1–3 and discussed in sections 3–5.

2.1 *Human subjects research*

Much of the research demonstrating the effect of the environment on wound healing in human subjects has involved social stimuli in DE compared to EE. Several studies in this area have examined the effect of relationship stress. For example, studies examining the effects of positive versus negative marital interactions found differential healing of experimentally administered blister wounds.[6,7] In addition, the rate of healing differs with the level of stress associated with marital separation. Specifically, individuals with the highest stress levels during marital separation healed the slowest after experimentally induced damage to the stratum corneum.[g,8] Relationship stress in non-marital relationships was also associated with slower wound healing. In this study, subjects reporting the highest difficulty managing anger had the slowest resolution of blister wounds.[9]

In terms of studies with non-social EE, physical exercise, which has been shown to enhance wound healing in animal models,[10] also appears to facilitate wound healing in human subjects.[11]

While not specific to wound healing, clinical observations on the effect of DE and EE on general health date back to the early 1900s. Spitz[12] coined the terms hospitalism and anaclitic depression to describe the detrimental effects on infants between 0 and 5 months who were raised in an orphanage (the Foundling Home) compared to those raised in a nursery within a jail where their mothers were incarcerated. The nursery was more environmentally stimulating and infants there had regular visits with their mothers. As shown in Figure 4, over the first 5 months of life, while the Foundling Home babies started out with higher cognitive (left panel) and motor (right panel) scores, these fell below (on cognitive assessment) or to the same levels (on motor assessment) as the infants raised in the jail nursery. Of further note, a measles epidemic that occurred during the second year of these children's lives led to a 40% death rate among those raised in the Foundling Home, while less than 1% of the children raised in the jail, succumbed to the infection.

Similar to Spitz's findings, recent findings on transitioning children from orphanage housing to foster care in Romania after the Ceausescu regime was overthrown showed detrimental and beneficial effects of factors related to DE and EE, respectively, on physical and mental health. Specifically, both cognitive and motor delays[13,14] were associated with the duration of institutional rearing. Time reared in institutional environments also had

---

[g]Stratum Corteum: the outermost layer of the epidermis.

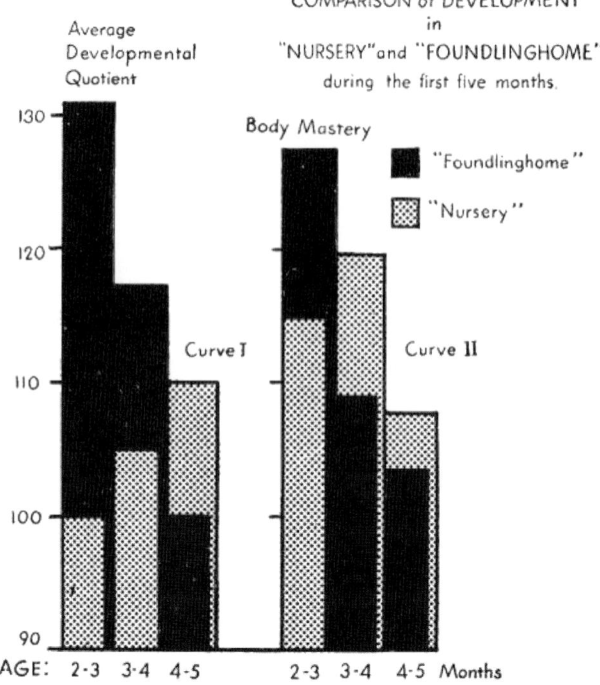

Figure 4. Over the first five months of life, children raised in a jail that had high environmental stimulation and had regular visits with their incarcerated mothers (Nursery) had stable cognitive (left panel) and physical (right panel) development, while the scores of those raised in an orphanage (Foundling Home infants) started out with higher scores that declined.[12]

detrimental impacts on children in China[15] and Guatemala.[16] Of particular interest to the effect of environmental enrichment on wound healing is that for US adopted children from Romania and other eastern European countries, the ability to integrate input from multiple sensory modalities negatively correlated with the time spent in institutions prior to adoption.[17]

2.2 *Animal studies*

Animal studies on DE and EE and wound healing have generally examined the effects of cage differences in social and non-social stimuli. Specially, in DE conditions, rodents are typically singly housed (referred to in the literature as social isolation or isolation rearing) and/or given limited inanimate objects to manipulate (reduced "cage complexity"). In contrast, in EE conditions animals are typically group housed and/or provided increased cage complexity. Examples of objects used for increasing cage complexity in EE conditions are shown in Figure 12.

In reviewing studies on DE and EE in wound healing, it is important for the reader to consider that the optimal amount of cage complexity and number of conspecifics[h] available for interaction is not a matter of simply increasing these variables. For example, if cage size is not increased, increasing the number of rats per cage leads to crowding which is stressful for most animals. Further, the effect of changing such variables is species and developmentally dependent. For example, rats and polygamous prairie voles would typically find increasing the number of conspecifics more enriching than less social mice and monogamous prairie voles. Further, young rodents typically seek more proximity to conspecifics than adult rodents. For these reasons, in designing DE and EE studies, a reasonable understanding of animal husbandry is important.

Despite the variability in the literature noted above regarding the variables used in different experiments to provide DE or EE conditions, the preponderance of the literature does show that DE has a detrimental effect on wound healing,[18–21] while most of the EE literature shows a facilitative impact on wound healing.[22–25]

In our studies, we have used single housing with standard bedding conditions as the DE condition[18,26] and either group housing with three rats per cage[18] or enhanced bedding[22,27] as the EE condition. Differences in activity level and behavior between the single and group housed rats from our studies can be observed in videotapes of these conditions at https://drive.google.com/open?id=0B08QbYOl86SoMkRELXhldXVtWEk (for single housed rats) and at https://drive.google.com/open?id=0B08QbYOl86SoN2d0WmlLSzRxUjA (for group housed rats). For our inanimate EE condition, we enhanced the standard cage bedding with either Nestlets® (Figure 5A) or Carefresh® (Figure 5C) bedding. Nestlets are chemically inert, odorless, non-ingestible 1×1 inch squares of sterilized pulped cotton. Carefresh is a soft, grayish paper pulp that forms a loose, low-density pile.

In our studies, we almost always found dramatic differences with social DE compared to social EE. The effect was less consistent but evident

---

[h] Conspecific: an animal of the same species (i.e. in an EE rodent study with rats, housing with increased numbers of conspecifics refers to increasing the number of rats per cage).

(A)   (B)

(C)   (D)

Figure 5. Inanimate EE using bedding with condition Nestlets (A and B) and Carefresh® (C and D).[27]

to a significant degree with inanimate (non-social) DE versus EE. Specially, the rate and quality of healing was much worse among singly compared to group housed rats in almost all of our experiments. The rate and quality of healing were also significantly worse for singly housed rats without enhanced bedding (inanimate DE condition) compared to those with enhanced bedding (inanimate EE condition), but the effect was less consistent compared to the difference between rats exposed to a socially DE versus EE. This is shown qualitatively and quantitatively in Figures 6 and 7, respectively.

3. Brain Cortical[i] Changes Result from ED and EE

Having reviewed the observational data that DE and ED conditions alter wound healing in Section 2, below, we review our current understanding of the mechanisms underlying this effect. In this section, we review the

---

[i] Cortical: of or pertaining to the cortex.

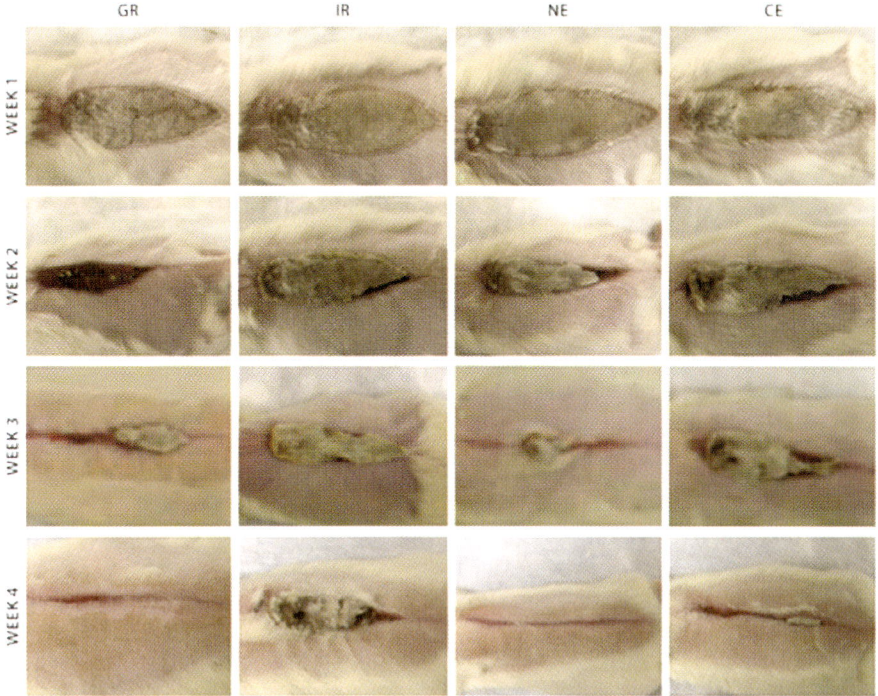

Figure 6.   Typical examples of healing in social and inanimate DE compared to EE. Four weeks following a 40% total body surface area burn, wound closure is much better for rats in social EE compared to social DE (compare columns 1 and 2 along row 4). We find a similar difference for rats with inanimate EE compare to inanimate DE (compare column 2 to columns 1, 3 and 4 along row 4). GR — group housed, IR — singly housed, NE — singly housed rats with Nestlet enriched bedding, CE — singly housed rats with Carefresh enriched bedding.[27]

first of the sub-circuits (Figure 1) linking changes in sensory receptors due to DE or EE to cortical changes. We review this sub-circuit for three sensory systems impacted by DE and EE: the visual, auditory, and tactile systems. The latter sensory system, as noted in Section 3.3, is most impacted during wound injury and repair.

3.1 *DE and EE lead to changes in the visual cortex*

Hubel and Wiesel's classic studies on the effect of external visual sensory input (for which they earned the 1981 Nobel Prize in physiology) established that DE in the form of monocular visual deprivation results in both atrophy of the

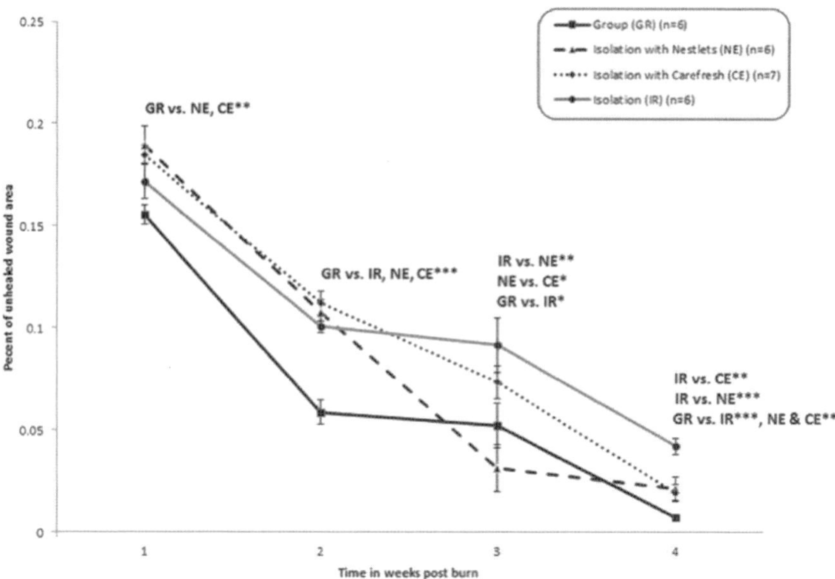

Figure 7.   Quantification of differences in rate of healing in socially DE compared to socially EE (single versus group housing — compare lines labeled group versus isolation) and inanimate DE compared to inanimate EE (single housing with standard versus enriched bedding — compare lines labeled isolation to lines labeled isolation with Nestlets and Isolation with Carefresh).[27]

thalamus[j,28] and altered firing of cells in the visual cortex.[29] In these studies, the effect of DE on visual sensory receptors in the retina was assessed by measuring electrical signals in the visual cortex of cats in which stimulation of visual sensory receptors was blocked in the retina of one eye (due to delayed opening of one eye or closing one eye with sutures.) Of interest, the effect was greater when the DE was monoocular than when both eyes were blocked.

The greater impact of monocular compared to binocular visual deprivation was found to be due to altered timing of sensory input from the sensory cells to the cortex when one eye was blocked. This indicated that the impact of the monocular DE was not due to disuse alone, but rather to a change in

---

[j]Thalamus: the brain region lying below the cortex that links incoming sensory signals to the cortex.

the ratio of ipsilateral[k] and contralateral[l] neuronal stimulation of the cortical receptive fields of each eye.[30,31] The impact of mismatched input from the retina to the visual cortex suggested by Hubel and Weiss' studies is consistent with our current understanding of the clinical disorder termed amblyopia, which is a syndrome involving distorted visual perception that arises from various conditions that cause asynchronous binocular visual input (e.g. strabismus, astigmatisms, or congenital cataracts).

As with the studies cited in Section 2 that indicate EE conditions reverse impaired wound healing resulting from DE, the abnormal changes to the visual cortex resulting from the experimentally induced DE can be reversed with application of EE conditions.[32,33] Furthermore, clinical treatments for visual distortions associated with amblyopia involve EE features. For example, patching the functional eye and increasing visual input to the impaired eye often leads to improved vision. Furthermore, a technique termed "perceptual learning" (PL), that has shown promise for the treatment of amblyopia[33,34] involves an EE component. Specifically, the strategy relies on increasing visual sensory input by increasing the focus on small visual sensory changes in particular patterns of sensory input such as depth and direction, as shown in Figure 8.[35]

3.1.1 Changes in excitatory transmission underlie the effect of DE and EE in the visual cortex

The studies reviewed above examining visual system sensory inputs to the visual cortex provide some insight into potential mechanisms underlying the resulting cortical changes. In the studies on amblyopia, the evidence suggests that an important component is the relative degree of excitatory versus inhibitory communication between neurons in this region. During a condition of DE, such as reduced monocular vision that leads to conditions such as amblyopia, the ratio of excitatory post-synaptic potentials (EPSPs)[m]

---

[k] Ipsilateral neuronal stimulation: activation of brain nerve cells (neurons) on the same side of the brain as the eye being stimulated.

[l] Contralateral neuronal stimulation — activation of brain nerve cells (neurons) on the opposite side of the brain as the eye being stimulated.

[m] EPSP: changes that result in a more positive neuronal cell membrane potential (i.e. a more positive voltage develops between the neuron's extracellular and intracellular fluid). When the voltage becomes more positive it disinhibits (or "excites") the neuron. The chemical glutamate is the most common molecule in the nervous system that leads to EPSPs, though there are many other chemicals have this effect.

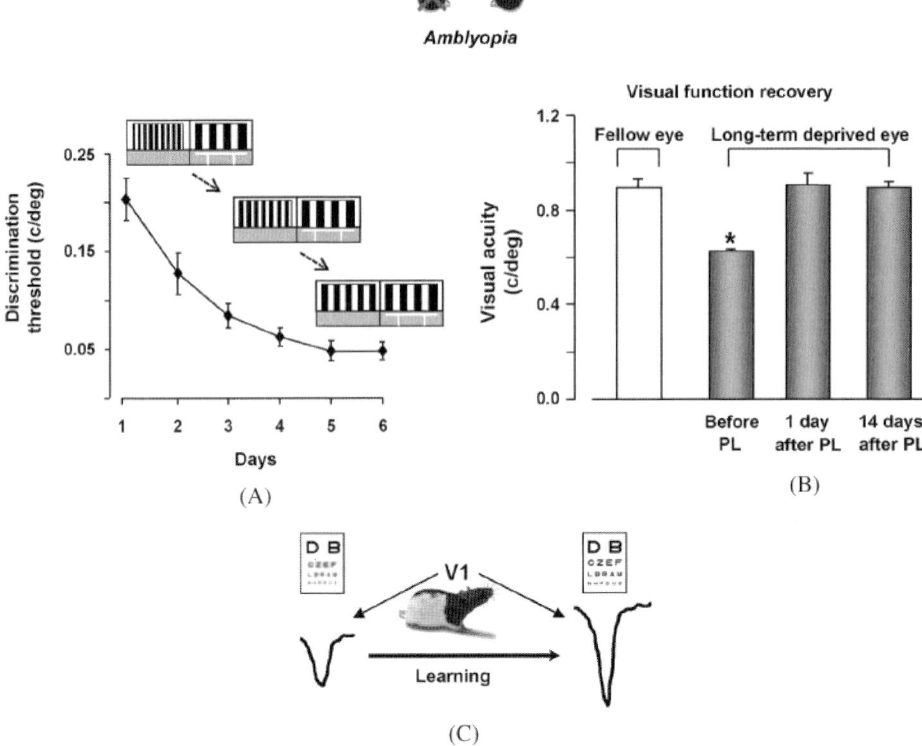

Figure 8. Increasing the difficulty of pattern discrimination shown to the abnormally functioning eye (panel A) resulted in improved visual acuity (panel B) and was associated with long-term changes in electrical signaling from neurons in the visual cortex (panel C).[35]

to inhibitory post-synaptic potentials (IPSPs)[n] transmission decreases, such that there are relatively more IPSPs and abnormal inhibition in visual cortical neurons that receive aberrant input from the retina. In contrast, the ratio of EPSPs to IPSPs changes in the opposite direction during an EE condition that increases stimulation to retina of the abnormal eye, and reverses the visual deficits associated with amblyopia, such with PL. In this case, the ratio of EPSPs to IPSPs transmission increases, such that there are

---

[n] IPSP: changes that result in the neuron's cell membrane potential becoming more negative (i.e. a more negative voltage develops between the neuron's extracellular and intracellular fluid). When the voltage becomes more negative it inhibits the neuron. The chemical gamma butyric acid (GABA) is the most common molecule in the nervous system that leads to IPSPs, though many other chemicals have this effect.

relatively more EPSPs and greater dis-inhibition (excitation) in visual cortical neurons that are overly inhibited after an ocular DE that results in
amblyopia.[4,36,37]

Mechanisms related to the relative changes in EPSPs and IPSPs in the visual
cortex have been identified. For example, in the DE condition of monocular
deprivation that leads to amblyopia, an increase in brain glucocorticoids[o] may
indirectly contribute to the relative increase in IPSPs.[38] This is because brain
glucocorticoids down-regulate amino-hydroxy-methyl-isoxazolepropionic
(AMPA) receptors, which are receptors for glutamate that increase EPSPs.[38]
Further, the beneficial effects of EE on amblyopia can be reversed if GABA, the
major inhibitory neurotransmitter in the brain, is delivered to the visual cortex
during the procedure.[32] In addition, following treatment of amblyopia, long-
term potentiation and brain-derived growth factor (BDNF) increase.[32] Both
increased BDNF and LTP are associated with an increased ratio of EPSPs to
IPSPs in cortical regions that respond to retinal input in amblyopia.

3.2 *DE and EE lead to changes in the auditory cortex*

The receptive fields of cells in the auditory cortex also shift their pattern of
response after exposure to sensory input that is altered based on DE and EE
conditions. Some of these studies used an experimental paradigm similar to
studies reviewed in Section 3.1 using monocular vision as the DE condition.
Specifically, this research examined monaural sensory deprivation by reducing
auditory sensory input to one vestibulocholear[p] nerve. When this is done, the
DE condition of decreased sensory input from the vestibulocholear nerve to one
ear results in an abnormal balance of excitatory versus inhibitory communication in the auditory cortex,[39,40] similar to the effect of monocular sensory
stimulation on this balance in the visual cortex as reviewed in Section 3.1. As
shown in Figures 9 and 10, in response to decreased auditory stimulation from
one ear, cortical neurons that respond to stimuli from this ear produce an abnormal amount of inhibitory stimuli (IPSPs) relative to excitatory stimuli (EPSPs).[40]
An additional example of how the auditory cortex is modified in response to
altered auditory sensory input is observed when cells become increasingly
responsive (have increased EPSPs relative to IPSPs) after exposure to a particular
frequency from the sound spectrum,[41] as shown in Figure 11 (when the exposure occurs during a critical developmental period as noted in the Figure legend).

---

[o] Glucocorticoids: hormones released in the body and brain in response to stress.
[p] Vestibulocochlear nerve: the 8th cranial nerve; it transmits auditory input from the ear's
inner sensory organ (the cochlea) to the auditory cortex.

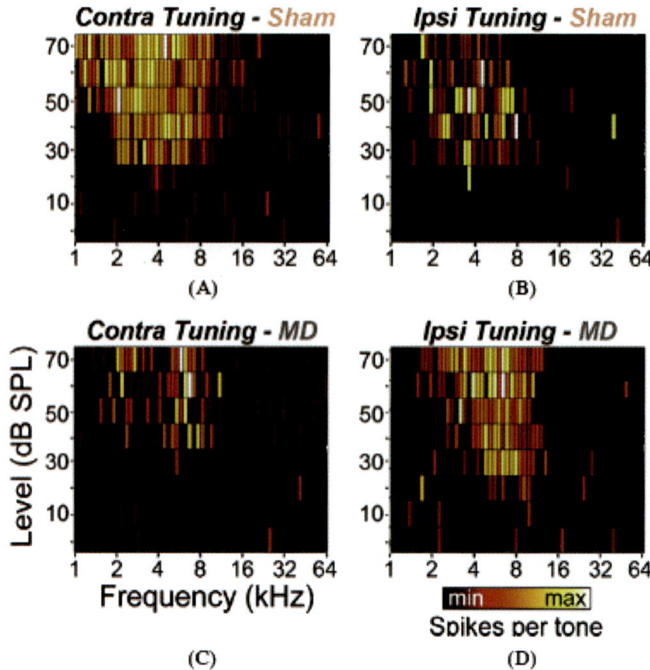

Figure 9. Compared to rats in the sham condition (A and B), in rats in the monaural stimulation condition (C and D), cells in the auditory cortex that typically respond to auditory sensation A and C — (labeled "Contra Tuning") were much less responsive (compare bright lines in A to C), while stimuli that typically evoke a weak or inhibitory effect in the same cells (B and D, labeled "Ipsi Tuning") became more active (compare bright lines in B to D). Thus, as with the monocular experiments in amblyopia where there is a shift in the ratio of EPSPs to IPSPs in neurons in the visual cortex cells, with monaural deprivation, a similar shift occurs in the auditory cortex. From Ref. 40.

Altered auditory receptive field responses to atypical sensory inputs discussed above and shown in Figures 9–11, can be reversed if this is followed by EE,[42] particularly during critical developmental periods. The EE condition involves increasing cage size, number of conspecifics, and inanimate objects, as shown in Figure 12. When placed in this EE condition, the abnormal auditory cortical effects noted above are reduced in both the primary (A1) and secondary (AII) auditory cortices, with the effect more predominant in AII.[43]

The finding that both aberrant and enriched sound exposure affect the response of auditory cortical cells predominantly during critical periods is consistent with the finding that during such critical developmental periods, the EPSP to IPSP ratio is greater than during periods of maturation.[44,45] Thus, changes to the auditory cortex induced by EE replicate changes during

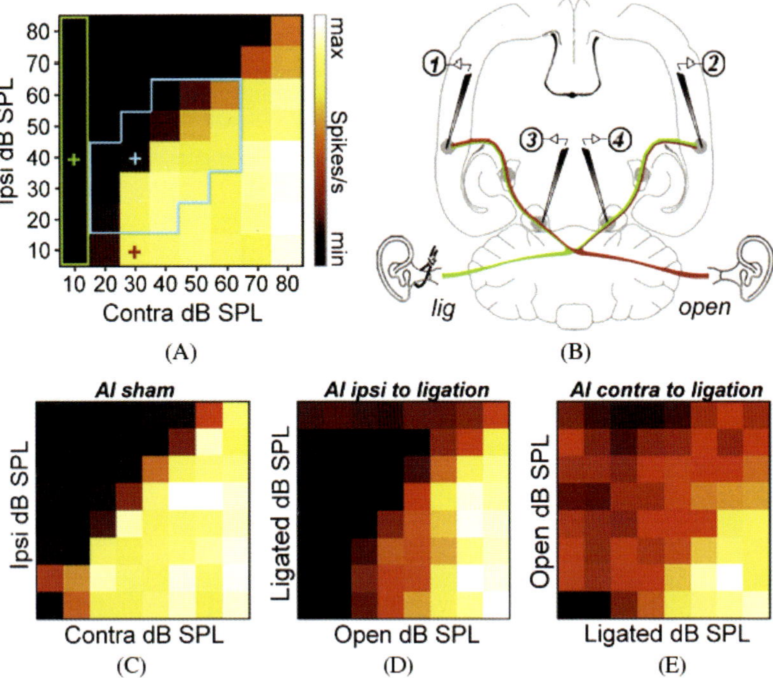

Figure 10. With monaural stimulation, auditory cells that typically have an activated response (an EPSP) in response to auditory stimulation (yellow boxes in panels (A and C) are reduced (compare yellow boxes in panels A and C to yellow boxes in panels C–E. In contrast, cells that normally have an inhibitory response (an IPSP) to the same auditory stimulation (dark colored boxes in panels) are increased (compare dark boxes in panels A and C to panels D and E).[40]

normal development. On the other hand, as noted above, in DE (such as monaural auditory deprivation), the ratio of EPSP to IPSP responses in cortical neurons becomes negative and inhibitory. A predominance of inhibition over activation of neurons is not seen in either normal developmental critical periods or in the mature cortex.[45] Thus, DE induces an aberrant cortical state associated with negative functional effects on vision when it occurs in the visual cortex (Section 3.1) and hearing when it occurs in the auditory cortex.

3.3 *DE and EE lead to changes in the somatosensory cortex*

The changes to visual and auditory cortex neurons due to altered sensory input in DE and EE conditions, as discussed in Sections 3.1 and 3.2 also occur in the somatosensory cortex. This is particularly relevant for wound healing,

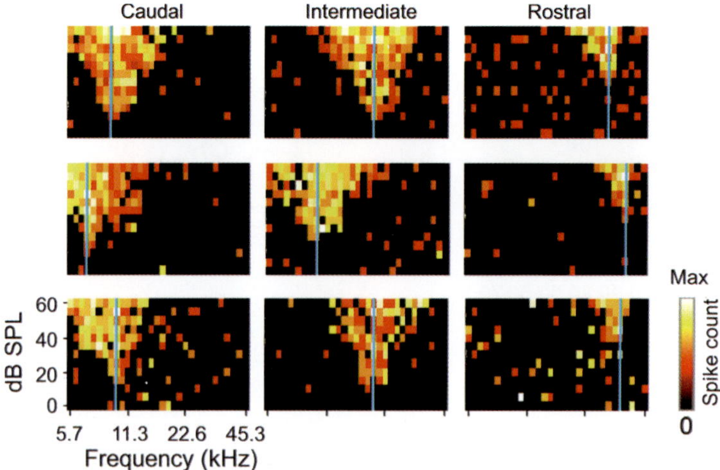

Figure 11. Cells that respond to fast frequency sound waves when listening to the normal sound spectrum (middle column, top row) become entrained to respond to slower frequencies when they hear frequencies in this range more than other sound frequencies during post-natal days 11–15 (middle column, middle row) but not when the exposure occurred during post-natal days 16–20 (middle column, bottom row).[41]

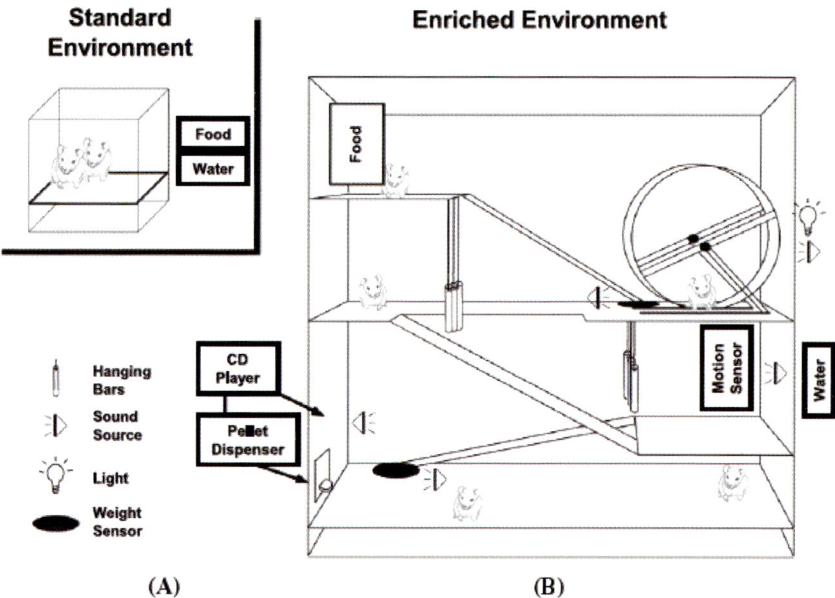

Figure 12. Visual depiction of an EE condition that reverses the effect on the cortex of abnormal auditory sensory stimulation.[42]

as the somatosensory cortex receives critical sensory input about wound healing. Namely, cells in the somatosensory cortex exhibit an altered pattern of responding when tactile sensory input from the skin is reduced or increased in DE and EE conditions during wound injury and repair.

Studies of peripheral nerve damage provide good examples of the way DE and EE conditions involving the skin affect the somatosensory cortex. In particular, stimulation of skin regions innervated by the damaged peripheral sensory nerves result in activation of an abnormally large, but more fragmented, region of the relevant somatosensory cortex adult (though not in very young) animals.[46,47]

Further, as noted with EE during monocular and monaural DE conditions, provision of an EE condition after peripheral nerve damage reverses abnormal activation of the somatosensory cortex. For example, following median nerve damage in the hand, when EE was provided, the abnormal activation of non-specific regions of the somatosensory cortex became refined and specific. Namely, neuronal activation in the somatosensory cortex became limited to cells that typically react to cutaneous stimulation relayed from the median sensory nerve.[48] While the amount of cellular activation was normalized by this EE intervention, it did not correct the abnormal ordering of the skin surfaces represented in this area of the cortex (referred to as the cortical somatotopic map of this skin region).[48]

With a more substantial DE condition that occurs with limb amputation, the area of somatosensory cortex that previously received input from the amputated limb undergoes a different type of cortical remodeling. In this case, the reduced activation of the somatosensory cortex in the area that had received input from the amputated limb becomes innervated over time by neurons that receive input from other skin regions.[49,50] After this occurs, regions of the somatosensory cortex that had previously responded to input from the amputated limb, respond to signals from other skin regions. This may contribute to distorted perceptions of the amputated limb (so-called "phantom limb" sensations[q]).

The aberrant activation of cells that had not previously responded to activation of the limb that was amputated can be reduced with an EE type intervention termed constraint inhibition (CI). CI therapy involves constraining use of the intact limb and encouraging sensory stimulation of the

---

[q]Phantom limb sensation: experiencing abnormal, often painful, sensations that are experienced as coming from the amputated limb.

amputation stump.[51] Improved motor function of the residual limb, improved mobility with prosthetic devices, and decreased phantom limb pain has been associated with CI.

The improvement with this approach is thought to result from decreasing "learned disuse" of the amputated limb (by forcing its use through constraint of the intact limb). The resulting increased use of the residual portions of the amputated limb has been associated with less abnormal activation of the somatosensory cortex after amputation. This seems to involve recruitment of pathways to the cortical receptive fields of the amputated limb that can compensate for the lost function and limit the activation of this territory by irrelevant somatosensory cortical neurons.[52,53] While first applied in animal models, CI therapy is currently used with patients who have had amputations or loss of limb functions due to other conditions (see Figure 13).

Interestingly, some components of CI therapy overlap with the EE strategy of PL that has been effective for the visual deficits in amblyopia discussed above in Section 3.1. As reviewed in that section, PL[35] is a kind of EE condition that improves vision and reduces abnormal neuronal activity

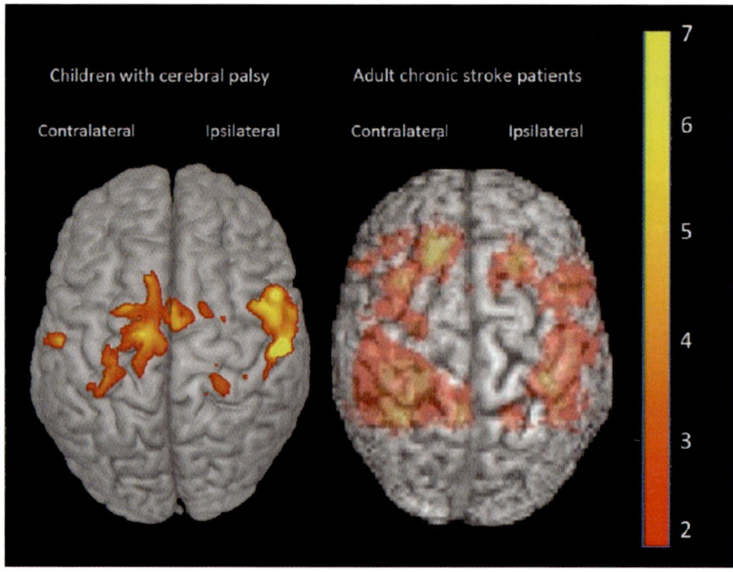

Figure 13.   Regions of somatosensory cortex that become more active following CI therapy for children with cerebral palsy (left side) and adults who have motor impairments due to a stroke (right side).[53]

in the visual cortex after monocular deprivation results in amblyopia. It involves increasing sensory input from the deprived eye by exposing it to increasingly subtle visual discrimination tasks. Likewise, the EE component of CI involves increasing sensory input from sensory nerves receiving input from damaged regions of the musculoskeletal system.

3.4 *DE and EE lead to changes in the thalamus before affecting the cortex*

Changes in sensory input, whether visual (Section 3.1), auditory (Section 3.2), or tactile (Section 3.3) due to DE or EE conditions, are transmitted to neurons that have their cell bodies in the thalamus, a subcortical region in the middle of the brain before connecting with the cortex. The thalamus (see Figure 14)[56] serves as a relay station for connecting different types of sensory input with the relevant sensory cortex (e.g. one section will receive mostly auditory, another mostly visual, etc.). Thus, in addition to impacting the cortex, as reviewed above, DE and EE conditions also impact the thalamus, though the degree of change in cortical compared to thalamic neurons varies with the type of DE or EE condition.[48,54] In our model, we found that activity in both the thalamus and frontal cortex was reduced with a social form of DE; and that the largest changes were in the thalamus, as shown in Figure 16.[55]

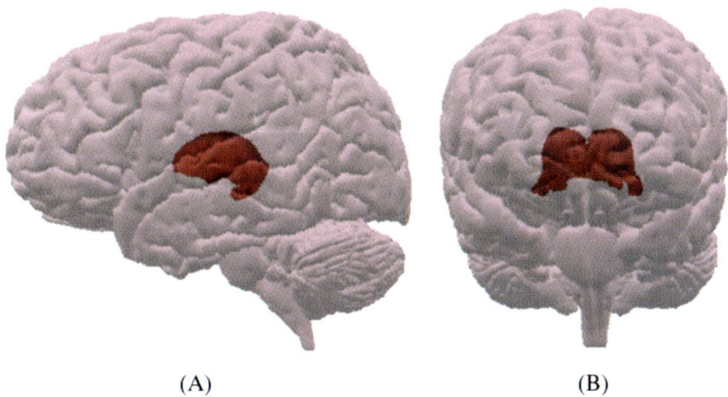

(A)   (B)

Figure 14.   The thalamus is located in the middle of the brain below the cerebrum (shown in A from the side/axial view and in B from the front/"coronal" view). From Ref. 56, Figure 16-5, panels A and B.

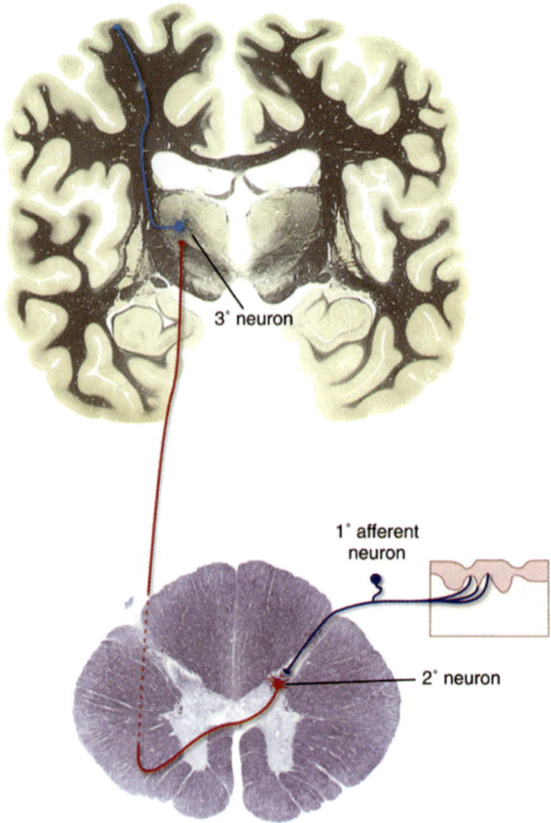

Figure 15.   The role of the thalamus in relaying sensory information. Information about a specific sensation (in this case touch) reaches the cortex through three nerves that travel, as shown in the figure from: (a) skin (in the case of the sense of touch) to spinal cord, (b) from spinal cord to the thalamus, and (c) from the thalamus to the appropriate sensory cortex (in the case of touch the somatosensory cortex discussed in Section 3.3). From Ref. 56, adapted from Figure 3-29.

4. Changes in Wound Healing due to DE and EE Conditions Involve Multiple Brain Circuits that Link Cortical Changes with Brain Output to Skin

In this section, we review sub-circuit 2 (see Section 1, Figure 2) which links sensory inputs as reviewed in Section 3 with brain outputs that affect wound healing, as discussed in Section 5. Though studied less than the other sub-circuits that comprise the overall effect of DE and EE on wound healing, there is increasing understanding of the pathways that are related to this sub-circuit, as reviewed below.

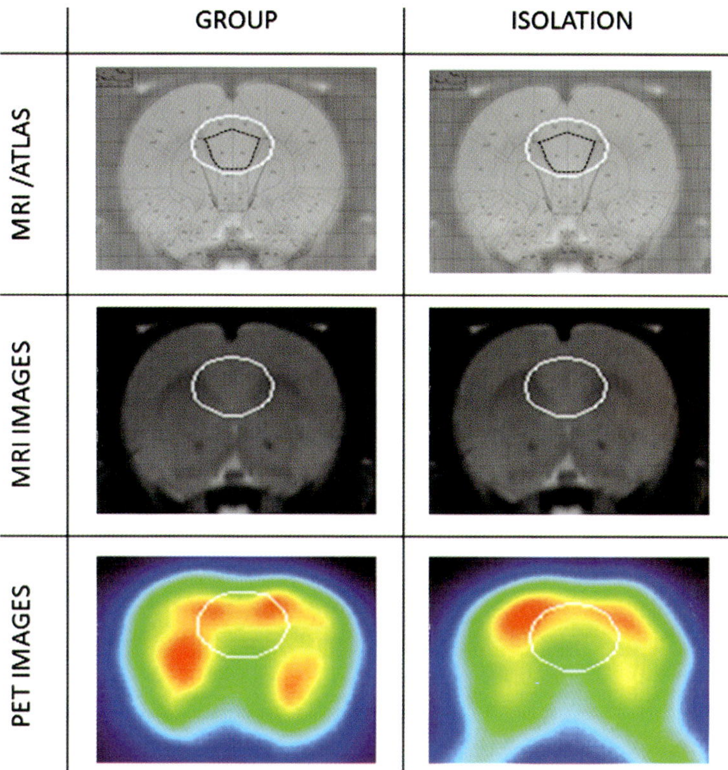

Figure 16. Coronal (from front) view of the rat thalamus (encircled regions columns 1 and 2, rows 1 and 2). These regions had reduced glucose uptake (a measure of metabolic activity in this region) in the DE condition of isolation reared (single housed) compared to group reared rats as seen when comparing the color of the circled regions of the images in row 3 (red color reflects the highest level of neural activity as measured by glucose metabolism).[55]

While somewhat oversimplified, the current understanding is that input that reaches the sensory cortex through the thalamus as reviewed in Sections 3.1–3.4 at the level of the "primary sensory cortex" (S1 in Figure 17), is passed to "association cortex" (SA in Figure 17) where basic chunks of sensory input (e.g. the pressure of a particular finger) is integrated with other pieces of sensory input from S1 (e.g. combining the pressure sensation from S1 with the position sensation in S1) to develop a more complex sensory representation of a particular sensory experience.

Information in the sensory association areas is then conveyed to motor association cortex (PM in Figure 17) and then to primary motor cortex (M1 in Figure 17). M1 cells are activated by input from association areas and then transmitted through the long corticospinal tract (green line in

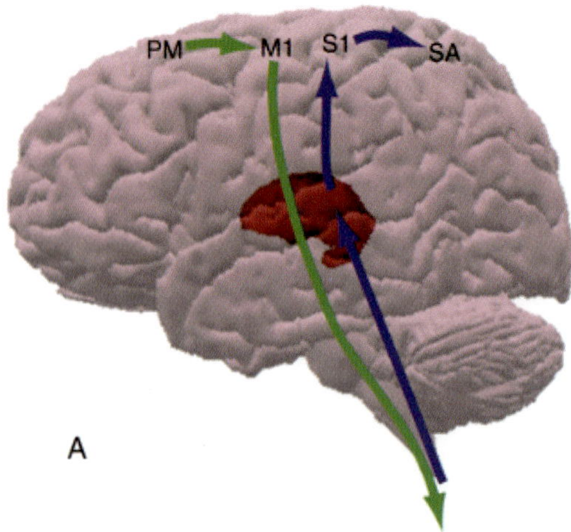

Figure 17.   Sensory input from the periphery travels through the spinothalamic tract (blue) to the primary sensory cortex (S1) and then this information is conveyed in a more integrated way in the SA cortex. SA then relays information to the premotor cortex (PM); this is an area of cortex that integrates sensory and motor information) that is then conveyed to the M1 cortex and conveyed to the body through the corticospinal tract (green line). From Ref. 56, adapted from Figure 22-14, panel A.

Figure 17) to nerves from the spinal cord that travel to organs in the rest of the body to cause motor changes in response to sensory input (e.g. turning the eyes or head in the right direction in response to visual or auditory inputs or moving a part of the body in response to sensory input from the skin).

Other brain regions receive sensory input from the thalamus that is further integrated with input from sensory and motor associational regions to inform the motor output from M1. These include the cerebellum, the basal ganglion, and input from other sensory cortices. These other inputs that impact the motor output of M1 are shown in Figure 18.[56] The eventual impact on musculoskeletal function of M1 output through the corticospinal tract has downstream effects on several aspects of local tissue during wound healing reviewed in Section 5, and could explain some of the ways sensory changes with DE and EE impact wound healing.

In addition to the impact on the musculoskeletal system, changes to MI resulting from altered sensory cortex activity in DE and EE conditions also impact the autonomic nervous system (nerves that lead to a stress response by activating the "fight flight" response) and the endocrine system. Through an offshoot of the corticospinal tract, the corticopontine tract alters the

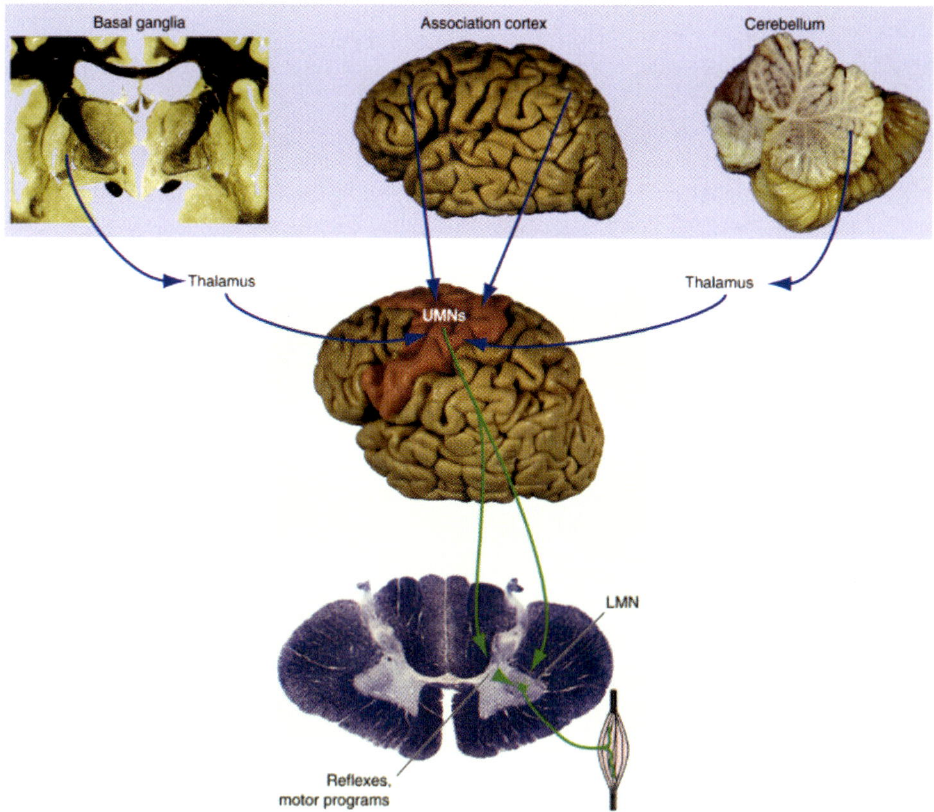

Figure 18. Information from the basal ganglia, cerebellum, and association cortex that integrates sensory input from one modality with other modalities is conveyed to M1 to inform how cells in this region are activated to initiate a motor response appropriate to the incoming sensory input. From Ref. 56, adapted from Figure 18-7.

functioning of cells in the pons[r] associated with alertness and the functioning of the sympathetic nervous system (SNS), particularly by impacting the reticular formation cells in the pons. Figure 19 shows that many regions of the brain impact different regions of the pons in addition to the corticopontine tract.[56]

Furthermore, endocrine outputs that affect wound healing are altered by sensory changes due to DE and EE conditions and their impact on the sensory cortex. Specifically, premotor sensory–motor association areas

---

[r]Pons: a region of the brainstem associated with the (SNS) that projects afferent fibers to the thalamus and efferent fibers to the cerebellum and medulla.

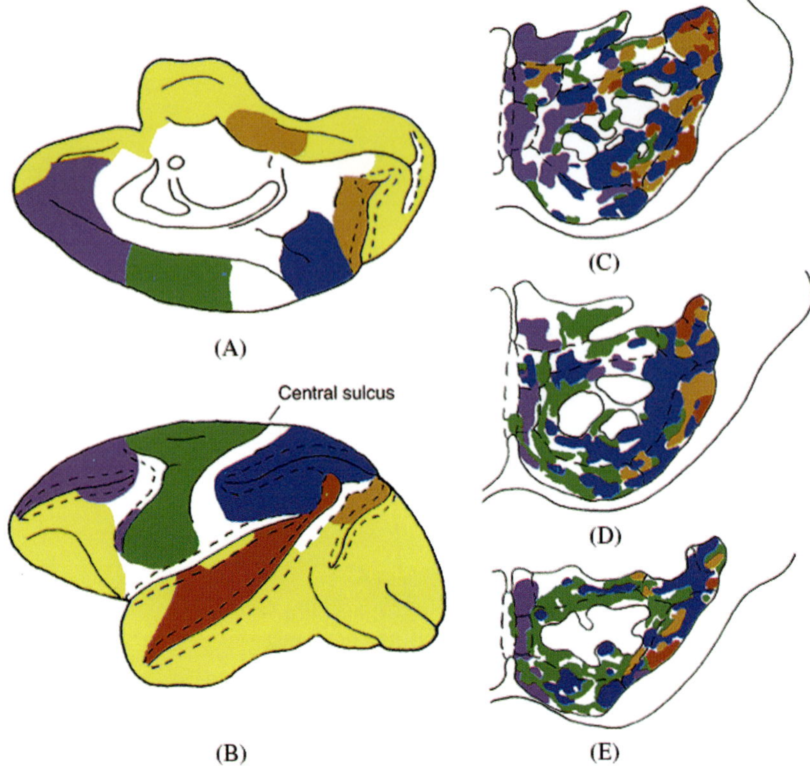

Figure 19.   Changes to the sensory cortex due to DE and EE conditions alter an offshoot of the corticospinal tract termed the corticopontine tract that terminates in the different regions of the pons (C–E). As with the additional brain regions that affect the corticospinal tract output (Figure 18), neurons from other brain association regions send direct processes to the pons. These association areas are from the mid and lateral regions of the brain images shown in A and B. Neurons from these association cortices synapse on portions of the pons that are identified by color in C–E, as shown in prefrontal (purple), posterior parietal (blue), temporal (red), parastriate/parahippocampal (orange) regions as well as motor, premotor, and supplementary motor areas (green). From Ref. 56, adapted from Figure 20-19.

affected by sensory cortical changes, as reviewed above and shown in Figure 17, affect limbic structures and the region of the diencephalon (the hypothalamus) that regulates the pituitary gland and adrenal corticotropin hormone (ACTH) release, as shown in Figure 20.[56] Levels of ACTH regulate systemic cortisol levels, that as discussed in Section 2, are abnormal in many of the studies of the wound healing effects of sensory changes due to DE and EE conditions.

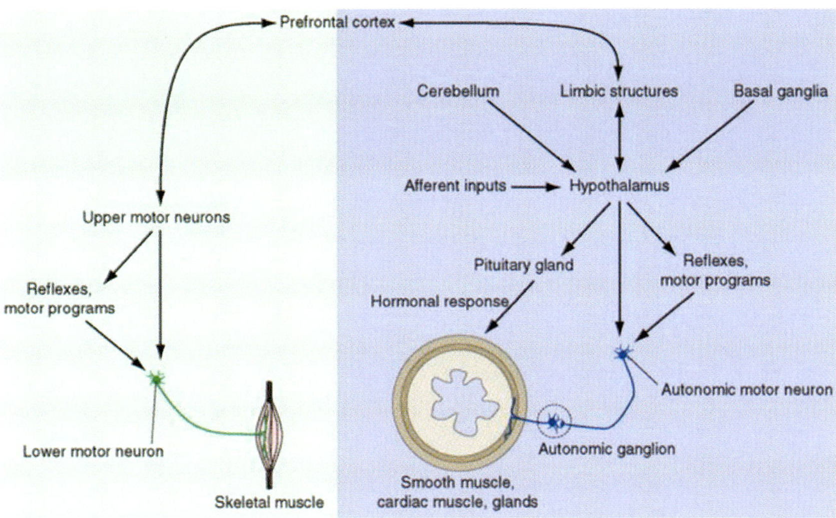

Figure 20. The prefrontal cortex (particularly PM, as illustrated in Figure 17) receives input on sensory changes from the sensory association cortex (SA in Figure 17), which affects the musculoskeletal system by its input to the primary primary motor cortex (M1 in Figure 17), as shown on the green left panel. It also communicates information about sensory changes to the hypothalamus through limbic structures as shown on the purple right side. From Ref. 56, adapted from Figure 23-2.

5. DE and EE Conditions That Alter Wound Healing, Also Change Brain Outputs to the Skin

Having reviewed how DE and EE conditions affect the brain sensory cortex (Figure 1, as reviewed in Section 3), as well as how this alters signals from sensory cortex to brain regions that affect wound healing (Figure 2, as reviewed in Section 4), in the current section, we review how changes in brain outputs to skin lead to differential wound healing in DE and EE conditions (Figure 3).

The brain outputs, as reviewed in Section 4, involved in altered wound healing due to changes in environmental sensory input, as occurs in DE and EE conditions are: (1) from the corticospinal tract to the spinal cord (green lines from M1 in Figures 17 and UMN in Figure 18), (2) from the corticobulbar tracts (an offshoot of the corticospinal tract) to the cranial nerve nuclei, (3) from the corticopontine tract (another offshoot of the corticospinal tract) to various nuclei in the pons (outlined in Figure 19), (4) from the prefrontal cortex to the hormonal output of the hypothalamus (shown on the right side of Figure 20), (5) from the prefrontal cortex to the autonomic outputs (shown on the right side of Figure 20), and (6) from the prefrontal cortex to spinal cord outputs to muscle cells (green side of Figure 20). These six brain outputs affect wound healing by changes to: (1) the endocrine system, through the pathways that change output from the hypothalamus (right side of Figure 20), (2) the autonomic nervous system, through effects of the corticopontine tract (Figure 19) and downstream effects of altered hypothalamic output (right side of Figure 20), and (3) the peripheral motor system, through effects on the corticospinal tract (Figures 17 and 18) and corticobulbar tracts. In Sections 5.1–5.3, we review these three effects of altered brain output on wound healing due to changes in environmental sensory input.

5.1 *Levels of hypothalamic regulated hormones change in DE and EE conditions associated with altered wound healing*

The human and animal studies reviewed in Section 2[6,7,9,11,19,25] showed that altered release of hormones by the hypothalamus and the pituitary gland (through the hypothalamic–pituitary axis or HPA) in DE and EE conditions lead to wound healing changes. The results show that HPA reactivity becomes "blunted" in response to DE conditions and normalized with EE conditions.[57–59] This leads to a reduction in the secretion of adrenocortictropic (ACTH) hormone from the pituitary gland and cortisol (corticosterone in rats) from the adrenal gland to the bloodstream, as shown in Figure 21.[60]

Reduced cortisol in DE conditions (and its normalization during EE conditions) impacts wound healing through interactions with cellular changes during the inflammatory phase of wound healing.[60,61] Specifically, cortisol modulates the release of pro-inflammatory proteins from cells responding to tissue damage (e.g. neutrophils) at the site of tissue injury,[62,63] as shown in Figure 21. This occurs during the inflammatory and early proliferative phases of wound healing (Figure 22A and B).[64]

Based on this normal role of cortisol in modulating the pro-inflammatory phase of wound healing, a blunted cortisol response due to DE conditions as reviewed in Section 2[6,7,9,11,19,25] will result in an abnormal inflammatory response in healing tissue that likely contributes to the impaired healing observed in DE conditions. In contrast, in EE conditions, normalization of blood cortisol levels will allow appropriate dampening and modulation of the pro-inflammatory phase of wound healing.

A useful model for understanding how DE conditions result in a dampened cortisol response to tissue damage involves allostasis and allo-

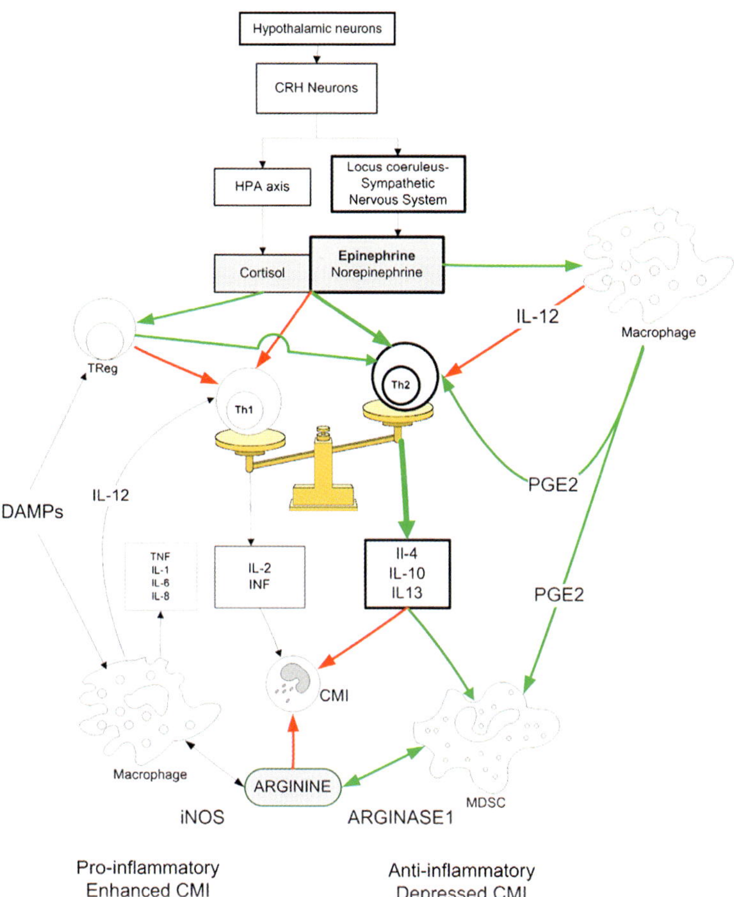

Figure 21. Hypothalamic effects on hormonal output (left side of figure) and the autonomic nervous system (right side of figure). On the left side, altered corticotropin releasing hormone (CRH in the figure) from neurons in the hypothalamus leads to changes in the pituitary gland (HPA axis in the figure) that result in changes in cortisol released from the adrenal gland that then results in complex changes to cytokines released by immune cells in tissue undergoing wound healing. On the right side, altered hypothalamic output from CRH neurons leads to changes in the sympathetic component of the autonomic nervous system, that results in increased epinephrine and norepinephrine levels, that then has similar effects to cortisol on cytokines released from immune cells in tissue undergoing wound healing (see overlap of arrows to immune cells from the right and left sides of the figure).[60]

static when applied in a biological context as proposed by Sterling and Eyer in 1988.[65,66] As discussed by McEwen,[67,68] when applied in biology, allostasis is the body's ability to generate an effective chemical response (such as a therapeutic increase in cortisol from the HPA) to an acute event,

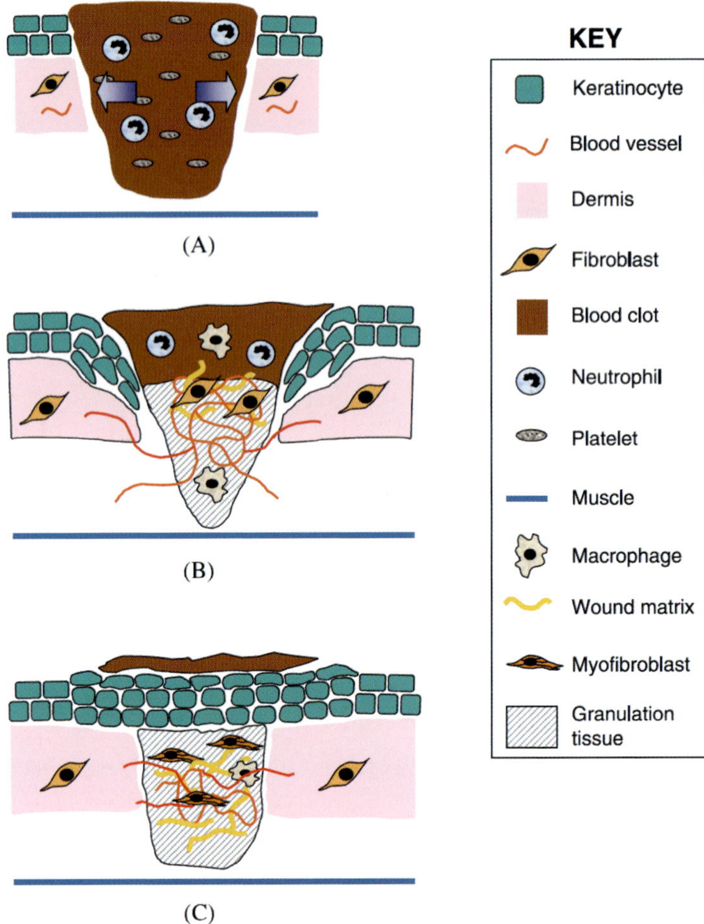

Figure 22. The phases of wound healing can be divided into: an initial inflammatory response during days 1–2 after a wound (A) that involves increased neutrophils at the site of tissue damage which release factors involved in recruiting other pro-inflammatory cells and in protecting the tissue from further damage, as well as in "cleaning up" (through phagocytosis) cellular debris of the damaged tissue, a proliferative phase during days 3–7 after a wound (B) that involves recruitment of macrophages involved in building new blood vessels and fibroblasts that produce collagen, and the remodeling phase between days 7 and 14 after a wound (C) in which fibroblasts shift to myofibroblasts involved in wound contraction, granulation tissue, and formation of a mature scar.[64]

such as after injuries resulting in burns or other wounds. The ability to generate an effective response to an acute event is limited when it occurs in the context of a highly stressful environment (a high allostatic load) and leads to impaired allostasis that can be potentially be quantified.[69] Thus, the blunted cortisol response in response to a wound in DE compared to EE conditions is an example of impaired allostasis due to high allostatic load.

5.1.1 *Epigenetic changes may lead to the blunted cortisol response in DE conditions*

Elegant mechanistic studies from Michael Meaney's lab at McGill[70–72] indicate that the transcription of genes involved in the expression of cortisol is altered in an animal model comparing DE to EE conditions. These studies examined gene expression in rat pups born to dams that expressed either high (the EE condition) or low (the DE condition) social contact based on the degree of licking and grooming. Differences between the two groups were found in the binding of the immediate early gene (IEG) and transcription factor, nerve growth factor A (NGFI-A), on an exon of the promoter region for the glucocorticoid (GC) receptor gene in hippocampal neurons. This difference was found to result from epigenetic modification of the NGFI-A binding exon. Further evidence that the differential NGFI-A binding and GC receptor expression resulted from an epigenetic mechanism was that when rat pups from low licking rat dams were cross-fostered with dams providing high licking and grooming, NGFI-A binding to the GC promoter exon normalized.

While we did not investigate GC expression in our work, we did find decreased expression of IEG transcription factors including one related to NGFI-A[18,22] though for this IEG, as shown in Figure 23, the difference was not significant. Further, in these studies, we found that both impaired wound healing and decreased IEG gene expression normalized to the control group (socially reared rats), when isolation reared rats were provided non-social EE condition of augmenting their cage with Nestlets©.[22]

5.2 *Output from the sympathetic nervous system (SNS) changes in DE and EE conditions associated with altered wound healing*

While changes in outputs effecting cortisol due to DE conditions result in a blunted and reduced response, the release of adrenergic molecules (epinephrine and noradrenaline) from the sympathetic nervous system (SNS), as shown on the right side of Figure 21, is increased[73,74] via output from the hypothalamus

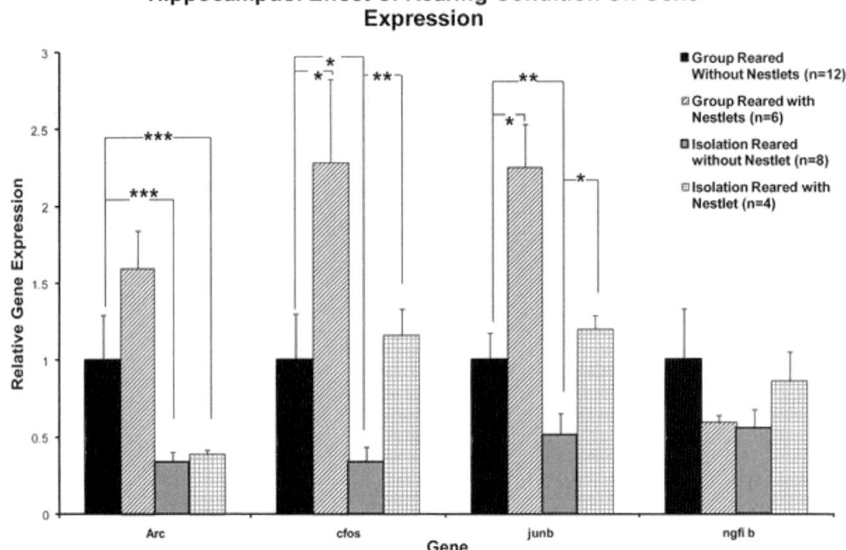

Figure 23. IEG transcription factors were reduced in our DE condition of isolation rearing and returned to normal levels found in group reared rats in our EE condition of providing Nestlets to the isolation reared rats (the change was significant for the IEGs Arc, cFos, and junb; there was a non-significant trend in the same direction for NGFI).[22]

(through the pathway shown in Figure 20) and the pons (through the pathway shown in Figure 19).

At the wound site, the increased SNS activity and secretion of adrenergic molecules has a similar effect on the inflammatory phase as blunted cortisol response, namely reduced modulation of the pro-inflammatory phase, as reviewed in Section 5.1. In addition, increased sympathetic output impairs the proliferative phase of wound healing through effects on blood vessels that are formed in the damaged tissue during this phase,[64] as shown in panel B, Figure 22. In contrast, in EE conditions, normalization of the SNS and its adrenergic output has a facilitative effect on vascular growth during wound healing.

The effect of abnormal SNS activity on blood vessel proliferation and growth factor release from these vessels in wound healing tissue has been shown in an animal model[75] of spinal cord injury (SCI), a condition known to result in aberrant SNS regulation of blood vessels. As shown in Figure 24, rats with spinal cord damage had more congested blood vessels in healing dermal layers (arrowheads in main panel D, arrows in panel D inset, Figure 24) compared to non-SCI rats (arrows panel B, Figure 24).

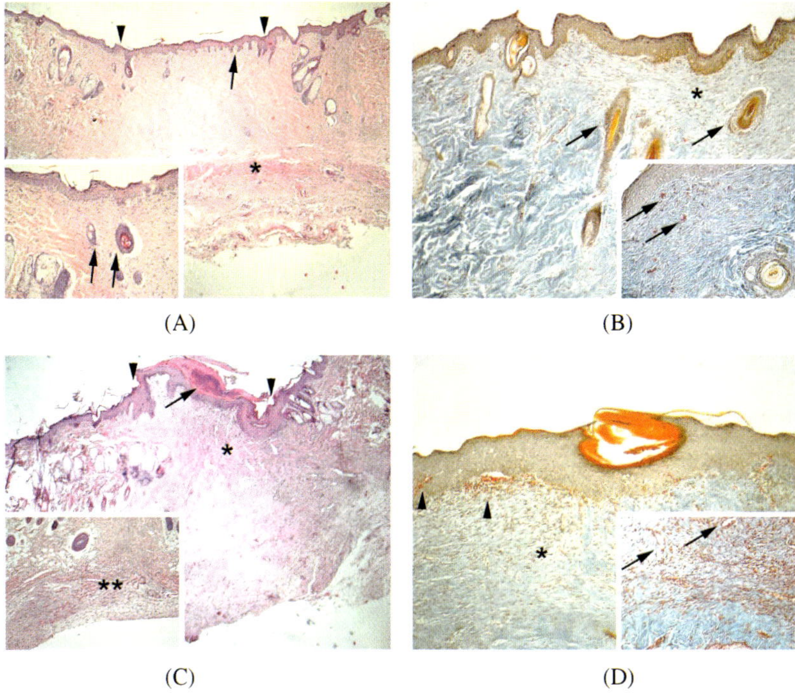

(A)   (B)

(C)   (D)

Figure 24.   Thicker dermal keratinization (arrows) and less organized dermal collagen deposition (*) in skin healing after SCI (panel C) compared to wounds without SCI (panel A). Less prominent and more congested vascularity of dermal layer of cutaneous wound healing of SCI rats (arrows and arrowheads in panel D) compared to wound in animals without SCI (arrows in panel B).[75]

Growth factors are altered by the abnormal SNS activity in rats with SCI.[75] Staining for transforming growth factor (TGF) was reduced in both the dermal and epidermal layers of wound healing skin of SCI compared to non-SCI rats (compare arrows in panels A and B, Figure 25). Models of DE involving social isolation have found suppressed expression of other growth factors, including keratinocyte growth factor and vascular endothelial growth factor (VEGF) gene expression,[19] which in some cases involved increased micro-RNA expression.[20] Higher micro-RNA expression, associated with poor wound healing seems to impact the proliferative phase more than the inflammatory phase of wound healing,[76] and can be activated by SNS ouput.[77]

Growth factors altered by abnormal SNS activity during DE conditions impact all three phases of wound healing. In the inflammatory phase, they act in conjunction with cytokines to modulate immune cell trafficking.

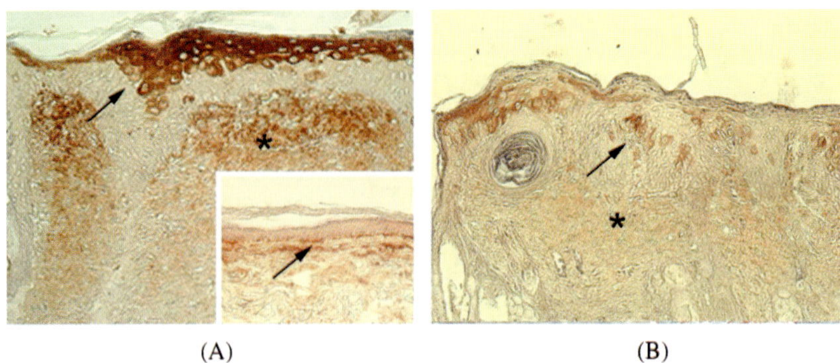

(A)　　　　　　　　(B)

Figure 25.　Decreased transforming growth factor-beta1 (TGF-b1) in the epidermal (arrows) and dermal (*) layers of wounds of rats with SCI (panel (B)) compared to wound healing of control rats without SCI (panel A).[75]

In the proliferative phase, their release from newly forming blood vessels is regulated by SNS control of vascular tone, and in the remodeling phase (panel C, Figure 22), they induce transition of fibroblasts to myofibroblasts.

5.3　*Output from the motor system to wound healing tissue changes in DE and EE conditions associated with altered wound healing*

The effect of DE and EE conditions on the sensory and M1 cortex as reviewed in Sections 3 and 4, is conveyed through outputs to the skin through the corticospinal tract (see Figures 17 and 18). While this output system has been less investigated than the effect of cortisol and SNS changes on wound healing, it may play a role in remodeling phase of wound healing. Specifically, recent literature[77] indicates that output from the corticospinal tract to peripheral motor nerves impacts the transition of fibroblasts to myofibroblasts during the remodeling phase of wound healing. Overactive myofibroblast activity during the remodeling phase impairs effective wound healing. Thus, as DE conditions result in abnormal corticospinal tract output, it is possible that this impacts the wound healing impairments associated with DE conditions through increasing myofibroblast activity. Consistent with this hypothesis, in the model of SCI reviewed above,[75] a condition that leads to abnormal corticospinal tract output, the extracellular matrix after wound healing was more disorganized in the SCI condition (compare panes A and C in Figure 24).

6. Summary and Implications

In this chapter, we examined the large body of research on the effect of environmental sensory input on wound healing by examining deprived environments (DE) and enriched environments (EE). As reviewed, the complex path that results in altered wound healing in DE and EE has three main components: (a) altered activation of the sensory cortex (Figure 1, reviewed in Section 3), (b) altered inhibitory (IPSP) versus excitatory (EPSP) post-synaptic potentials in synapses linking sensory cortex with brain outputs to skin (Figure 2, reviewed in Section 4), and (c) changes in brain outputs to skin (Figure 3, as discussed in Section 5).

While most patients undergoing wound healing are not subjected to DE or EE conditions, these extreme environmental conditions provide model systems for examining how the sensory system is involved in normal wound healing following tissue injury. Furthermore, insights gleaned on how differential sensory input in DE and EE modulates wound healing suggests potential interventions to improve wound healing impairments that result from altered sensory input, including from the tissue damage itself.

References

1. Hunter, H. J. Momen, S. E. and Kleyn, C. E. The impact of psychosocial stress on healthy skin. *Clin Exp Dermatol* **40**, 540–546 (2015).
2. Bowe, W. P. and Logan, A. C. Clinical implications of lipid peroxidation in acne vulgaris: Old wine in new bottles. *Lipids Health Dis* **9**, 141 (2010).
3. Kleyn, C. E. Schneider, L. Saraceno, R. Mantovani, C. Richards, H. L. Fortune, D. G. Cumberbatch, M. Dearman, R. J. Terenghi, G. Kimber, I. and Griffiths, C. E. The effects of acute social stress on epidermal Langerhans' cell frequency and expression of cutaneous neuropeptides. *J Invest Dermatol* **128**, 1273–1279 (2008).
4. Nithiananantharajah, J. and Hannan, A. J. Enriched environments, experience-dependent plasticity and disorders of the nervous system. *Nat Rev Neurosci* **7**, 697–709 (2006).
5. Roosterman, D. Goerge, T. Schneider, S. W. Bunnett, N. W. and Steinhoff, M. Neuronal control of skin function: the skin as a neuroimmunoendocrine organ. *Physiol Rev* **86**, 1309–1379 (2006).
6. Gouin, J. P. Carter, C. S. Pournajafi-Nazarloo, H. Glaser, R. Malarkey, W. B. Loving, T. J. Stowell, J. and Kiecolt-Glaser, J. K. Marital behavior, oxytocin,

vasopressin, and wound healing. *Psychoneuroendocrinology* 35, 1082–1090 (2010).

7. Kiecolt-Glaser, J. K. Loving, T. J. Stowell, J. R. Malarkey, W. B. Lemeshow, S. Dickinson, S. L. and Glaser, R. Hostile marital interactions, proinflammatory cytokine production, and wound healing. *Arch Gen Psychiatry* 62, 1377–1384 (2005).

8. Muizzuddin, N. Matsui, M. S. Marenus, K. D. and Maes, D. H. Impact of stress of marital dissolution on skin barrier recovery: Tape stripping and measurement of trans-epidermal water loss (TEWL). *Skin Res Technol* 9, 34–38 (2003).

9. Gouin, J. P. Kiecolt-Glaser, J. K. Malarkey, W. B. and Glaser, R. The influence of anger expression on wound healing. *Brain Behav Immun* 22, 699–708 (2008).

10. Keylock, K. T. Vieira, V. J. Wallig, M. A. DiPietro, L. A. Schrementi, M. and Woods, J. A. Exercise accelerates cutaneous wound healing and decreases wound inflammation in aged mice. *Am J Physiol Regul Integr Comp Physiol* 294, R179–R184 (2008).

11. Emery, C. F. Kiecolt-Glaser, J. K. Glaser, R. Malarkey, W. B. and Frid, D. J. Exercise accelerates wound healing among healthy older adults: a preliminary investigation. *J Gerontol Ser A Biol Sci Med Sci* 60, 1432–1436 (2005).

12. Spitz, R. A. Hospitalism; an inquiry into the genesis of psychiatric conditions in early childhood. *Psychoanal Study Child* 1, 53–74 (1945).

13. Johnson, D. E. Guthrie, D. Smyke, A. T. Koga, S. F. Fox, N. A. Zeanah, C. H. and Nelson, C. A. 3rd. Growth and associations between auxology, caregiving environment, and cognition in socially deprived Romanian children randomized to foster vs ongoing institutional care. *Arch Pediatr Adolesc Med* 164, 507–516 (2010).

14. Levin, A. R. Zeanah, C. H. Jr. Fox, N. A. and Nelson, C. A. Motor outcomes in children exposed to early psychosocial deprivation. *J Pediatr* 164, 123–129 e121 (2014).

15. Miller, L. C. and Hendrie, N. W. Health of children adopted from China. *Pediatrics* 105, E76 (2000).

16. Miller, L. Chan, W. Comfort, K. and Tirella, L. Health of children adopted from Guatemala: Comparison of orphanage and foster care. *Pediatrics* 115, e710–e717 (2005).

17. Lin, S. H. Cermak, S. Coster, W. J. and Miller, L. The relation between length of institutionalization and sensory integration in children adopted from Eastern Europe. *Am J Occup Ther* 59, 139–147 (2005).

18. Levine, J. B. Leeder, A. D. Parekkadan, B. Berdichevsky, Y. Rauch, S. L. Smoller, J. W. Konradi, C. Berthialime, F. and Yarmush, M. L. Isolation rearing impairs wound healing and is associated with increased locomotion and decreased immediate early gene expression in the medial prefrontal cortex of juvenile rats. *Neuroscience* 151, 589–603 (2008).

19. Pyter, L. M. Yang, L. da Rocha, J. M. and Engeland, C. G. The effects of social isolation on wound healing mechanisms in female mice. *Physiol Behav* **127**, 64–70 (2014).

20. Yang, L. Engeland, C. G. and Cheng, B. Social isolation impairs oral palatal wound healing in sprague-dawley rats: A role for miR-29 and miR-203 via VEGF suppression. *PLoS One* **8**, e72359 (2013).

21. Martin, L. B. 2nd, Glasper, E. R. Nelson, R. J. and Devries, A. C. Prolonged separation delays wound healing in monogamous California mice, Peromyscus californicus, but not in polygynous white-footed mice, *P. leucopus. Physiol Behav* **87**, 837–841 (2006).

22. Vitalo, A. Fricchione, J. Casali, M. Berdichevsky, Y. Hoge, E. A. Rauch, S. L. Berthiaume, F. Yarmush, M. L. Benson, H. Fricchione, G. L. and Levine, J. B. Nest making and oxytocin comparably promote wound healing in isolation reared rats. *Plos One* **4** (2009).

23. Detillion, C. E. Craft, T. K. Glasper, E. R. Prendergast, B. J. and DeVries, A. C. Social facilitation of wound healing. *Psychoneuroendocrinology* **29**, 1004–1011 (2004).

24. Glasper, E. R. and Devries, A. C. Social structure influences effects of pair-housing on wound healing. *Brain Behav Immun* **19**, 61–68 (2005).

25. DeVries, A. C. Craft, T. K. Glasper, E. R. Neigh, G. N. and Alexander, J. K. 2006 Curt P. Richter award winner: Social influences on stress responses and health. *Psychoneuroendocrinology* **32**, 587–603 (2007).

26. Levine, J. B. Youngs, R. M. MacDonald, M. L. Chu, M. Leeder, A. D. Berthiaume, F. and Konradi, C. Isolation rearing and hyperlocomotion are associated with reduced immediate early gene expression levels in the medial prefrontal cortex. *Neuroscience* **145**, 42–55 (2007).

27. Vitalo, A. G. Gorantla, S. Fricchione, J. G. Scichilone, J. M. Camacho, J. Niemi, S. M. Denninger, J. W. Benson, H. Yarmush, M. L. and Levine, J. B. Environmental enrichment with nesting material accelerates wound healing in isolation-reared rats. *Behav Brain Res* **226**, 606–612 (2012).

28. Wiesel, T. N. and Hubel, D. H. Effects of visual deprivation on morphology and physiology of cells in the cats lateral geniculate body. *J Neurophysiol* **26**, 978–993 (1963).

29. Wiesel, T. N. and Hubel, D. H. Single-cell responses in striate cortex of kittens deprived of vision in one eye. *J Neurophysiol* **26**, 1003–1017 (1963).

30. Hubel, D. H. and Wiesel, T. N. Binocular interaction in striate cortex of kittens reared with artificial squint. *J Neurophysiol* **28**, 1041–1059 (1965).

31. Hubel, D. H. and Wiesel, T. N. Early exploration of the visual cortex. *Neuron* **20**, 401–412 (1998).

32. Sale, A. Maya Vetencourt, J. F. Medini, P. Cenni, M. C. Baroncelli, L. De Pasquale, R. and Maffei, L. Environmental enrichment in adulthood promotes amblyopia recovery through a reduction of intracortical inhibition. *Nat Neurosci* **10**, 679–681 (2007).

33. Sale, A. De Pasquale, R. Bonaccorsi, J. Pietra, G. Olivieri, D. Berardi, N. and Maffei, L. Visual perceptual learning induces long-term potentiation in the visual cortex. *Neuroscience* **172**, 219–225 (2011).

34. Zhou, Y. Huang, C. Xu, P. Tao, L. Qiu, Z. Li, X. and Lu, Z. L. Perceptual learning improves contrast sensitivity and visual acuity in adults with anisometropic amblyopia. *Vision Res* **46**, 739–750 (2006).

35. Bonaccorsi, J. Berardi, N. and Sale, A. Treatment of amblyopia in the adult: Insights from a new rodent model of visual perceptual learning. *Front Neural Circuits* **8**, 82 (2014).

36. Sale, A. Berardi, N. and Maffei, L. Environment and brain plasticity: Towards an endogenous pharmacotherapy. *Physiol Rev* **94**, 189–234 (2014).

37. Maffei, A. Enriching the environment to disinhibit the brain and improve cognition. *Front Cell Neurosci* **6**, 53 (2012).

38. Miyazaki, T. Takase, K. Nakajima, W. Tada, H. Ohya, D. Sano, A. Goto, T. Hirase, H. Malinow, R. and Takahashi, T. Disrupted cortical function underlies behavior dysfunction due to social isolation. *J Clin Invest* **122**, 2690–2701 (2012).

39. Froemke, R. C. and Jones, B. J. Development of auditory cortical synaptic receptive fields. *Neurosci Biobehav Rev* **35**, 2105–2113 (2011).

40. Popescu, M. V. and Polley, D. B. Monaural deprivation disrupts development of binaural selectivity in auditory midbrain and cortex. *Neuron* **65**, 718–731 (2010).

41. Barkat, T. R. Polley, D. B. and Hensch, T. K. A critical period for auditory thalamocortical connectivity. *Nat Neurosci* **14**, 1189–1194 (2011).

42. Jiang, C. Xu, X. Yu, L. Xu, J. and Zhang, J. Environmental enrichment rescues the degraded auditory temporal resolution of cortical neurons induced by early noise exposure. *Eur J Neurosci* **42**, 2144–2154 (2015).

43. Jakkamsetti, V. Chang, K. Q. and Kilgard, M. P. Reorganization in processing of spectral and temporal input in the rat posterior auditory field induced by environmental enrichment. *J Neurophysiol* **107**, 1457–1475 (2012).

44. Sun, Y. J. Wu, G. K. Liu, B. H. Li, P. Zhou, M. Xiao, Z. Tao, H. W. and Zhang, L. I. Fine-tuning of pre-balanced excitation and inhibition during auditory cortical development. *Nature* **465**, 927–931 (2010).

45. Zhao, Y. Zhang, Z. Liu, X. Xiong, C. Xiao, Z. and Yan, J. Imbalance of excitation and inhibition at threshold level in the auditory cortex. *Front Neural Circuits* **9**, 11 (2015).

46. Florence, S. L. Garraghty, P. E. Wall, J. T. and Kaas, J. H. Sensory afferent projections and area 3b somatotopy following median nerve cut and repair in macaque monkeys. *Cereb Cortex* **4**, 391–407 (1994).

47. Florence, S. L. Jain, N. Pospichal, M. W. Beck, P. D. Sly, D. L. and Kaas, J. H. Central reorganization of sensory pathways following peripheral nerve regeneration in fetal monkeys. *Nature* **381**, 69–71 (1996).

48. Florence, S. L. Boydston, L. A. Hackett, T. A. Lachoff, H. T. Strata, F. and Niblock, M. M. Sensory enrichment after peripheral nerve injury restores cortical, not thalamic, receptive field organization. *Eur J Neurosci* **13**, 1755–1766 (2001).

49. Florence, S. L. Taub, H. B. and Kaas, J. H. Large-scale sprouting of cortical connections after peripheral injury in adult macaque monkeys. *Science* **282**, 1117–1121 (1998).

50. Merzenich, M. M. Nelson, R. J. Stryker, M. P. Cynader, M. S. Schoppmann, A. and Zook, J. M. Somatosensory cortical map changes following digit amputation in adult monkeys. *J Comp Neurol* **224**, 591–605 (1984).

51. Taub, E. Uswatte, G. and Mark, V. W. The functional significance of cortical reorganization and the parallel development of CI therapy. *Front Hum Neurosci* **8**, 396 (2014).

52. Mark, V. W. Taub, E. and Morris, D. M. Neuroplasticity and constraint-induced movement therapy. *Eura Medicophys* **42**, 269–284 (2006).

53. Uswatte, G. and Taub, E. Constraint-induced movement therapy: A method for harnessing neuroplasticity to treat motor disorders. *Prog Brain Res* **207**, 379–401 (2013).

54. Florence, S. L. Hackett, T. A. and Strata, F. Thalamic and cortical contributions to neural plasticity after limb amputation. *J Neurophysiol* **83**, 3154–3159 (2000).

55. Bonab, A. A. Fricchione, J. G. Gorantla, S. Vitalo, A. G. Auster, M. E. Levine, S. J. Scichilone, J. M. Hegde, M. Foote, W. Fricchione, G. L. Denninger, J. W. Yarmush, D. M. Fischman, A. J. Yarmush, M. L. and Levine, J. B. Isolation rearing significantly perturbs brain metabolism in the thalamus and hippocampus. *Neuroscience* **223**, 457–464 (2012).

56. Vanderah, T. W. *Nolte's The Human Brain*, 7th Edition, Elsevier, Philadelphia, PA: Elsevier, (2016).

57. McLaughlin, K. A. Sheridan, M. A. Tibu, F. Fox, N. A. Zeanah, C. H. and Nelson, C. A. 3rd. Causal effects of the early caregiving environment on development of stress response systems in children. *Proc Natl Acad Sci USA* **112**, 5637–5642 (2015).

58. McEwen, B. S. Protective and damaging effects of stress mediators. *N Engl J Med* **338**, 171–179 (1998).

59. McEwen, B. S. The brain on stress: Toward an integrative approach to brain, body, and behavior. *Perspect Psychol Sci* **8**, 673–675 (2013).

60. Marik, P. E. and Flemmer, M. The immune response to surgery and trauma: Implications for treatment. *J Trauma Acute Care Surg* **73**, 801–808 (2012).

61. Watt, D. G. Horgan, P. G. and McMillan, D. C. Routine clinical markers of the magnitude of the systemic inflammatory response after elective operation: A systematic review. *Surgery* **157**, 362–380 (2015).

62. Chapman, K. E. Coutinho, A. E. Zhang, Z. Kipari, T. Savill, J. S. and Seckl, J. R. Changing glucocorticoid action: 11beta-hydroxysteroid dehydrogenase

type 1 in acute and chronic inflammation. *J Steroid Biochem Mol Biol* **137**, 82–92 (2013).

63. Coutinho, A. E. and Chapman, K. E. The anti-inflammatory and immunosuppressive effects of glucocorticoids, recent developments and mechanistic insights. *Mol Cell Endocrinol* **335**, 2–13 (2011).

64. Werner, S. and Grose, R. Regulation of wound healing by growth factors and cytokines. *Physiol Rev* **83**, 835–870 (2003).

65. Sterling, P. Allostasis: A model of predictive regulation. *Physiol Behav* **106**, 5–15 (2012).

66. Sterling, P. E. J. Allostasis: A new paradigm to explain arousal pathology. In: Fisher, S. R. J. T. (ed.), *Handbook of Life Stress, Cognition, and Health*. Chicester, NY: Wiley, (1988).

67. Androulakis, I. P. Systems engineering meets quantitative systems pharmacology: From low-level targets to engaging the host defenses. *Wiley Interdiscip Rev Syst Biol Med* **7**, 101–112 (2015).

68. McEwen, B. S. and Gianaros, P. J. Stress- and allostasis-induced brain plasticity. *Annu Rev Med* **62**, 431–445 (2011).

69. Bizik, G. Picard, M. Nijjar, R. Tourjman, V. McEwen, B. S. Lupien, S. J. and Juster, R. P. Allostatic load as a tool for monitoring physiological dysregulations and comorbidities in patients with severe mental illnesses. *Harv Rev Psychiatry* **21**, 296–313 (2013).

70. Champagne, F. and Meaney, M. J. Like mother, like daughter: Evisdence for non-genomic transmission of parental behavior and stress responsivity. *Prog Brain Res* **133**, 287–302 (2001).

71. Francis, D. Diorio, J. Liu, D. and Meaney, M. J. Nongenomic transmission across generations of maternal behavior and stress responses in the rat. *Science* **286**, 1155–1158 (1999).

72. Weaver, I. C. Cervoni, N. Champagne, F. A. D'Alessio, A. C. Sharma, S. Seckl, J. R. Dymov, S. Szyf, M. and Meaney, M. J. Epigenetic programming by maternal behavior. *Nat Neurosci* **7**, 847–854 (2004).

73. Romana-Souza, B. Otranto, M. Vieira, A. M. Filgueiras, C. C. Fierro, I. M. and Monte-Alto-Costa, A. Rotational stress-induced increase in epinephrine levels delays cutaneous wound healing in mice. *Brain Behav Immun* **24**, 427–437 (2010).

74. Romana-Souza, B. Porto, L. C. and Monte-Alto-Costa, A. Cutaneous wound healing of chronically stressed mice is improved through catecholamines blockade. *Exp Dermatol* **19**, 821–829 (2010).

75. Konya, D. Gercek, A. Akakin, A. Akakin, D. Tural, S. Cetinel, S. Ozgen, S. and Pamir, M. N. The effects of inflammatory response associated with traumatic spinal cord injury in cutaneous wound healing and on expression of transforming growth factor-beta1 (TGF-beta1) and platelet-derived growth factor (PDGF)-A at the wound site in rats. *Growth Factors* **26**, 74–79 (2008).

76. Jin, Y. Tymen, S. D. Chen, D. Fang, Z. J. Zhao, Y. Dragas, D. Dai, Y. Marucha, P. T. and Zhou, X. MicroRNA-99 family targets AKT/mTOR signaling pathway in dermal wound healing. *PLoS One* 8, e64434 (2013).

77. Pu, J. Bai, D. Yang, X. Lu, X. Xu, L. and Lu, J. Adrenaline promotes cell proliferation and increases chemoresistance in colon cancer HT29 cells through induction of miR-155. *Biochem Biophys Res Commun* **428**, 210–215 (2012).

4. Models of Ischemic and Vascular Wounds

Michael T. Watkins and Hassan Albadawi

Division of Vascular and Endovascular Surgery
Department of Surgery
Massachusetts General Hospital
Boston, MA 02114

Abstract

Delayed wound healing is a prevalent clinical problem with substantial economic burden on the healthcare system. This is due to an ageing population and increasing comorbidities such as diabetes and peripheral vascular disease. Several elements including chronic tissue ischemia or repeated episodes of ischemia reperfusion injury contribute to the development of skin ulcers and wound healing delay in the lower limbs. One of the challenges in studying wound healing is the lack of accurate models that fully recapitulate the human condition. Despite these differences, several models contributed to our understanding of the role of disease conditions associated with aberrant wound healing and provided a biological system for designing and testing novel therapies. In this chapter, we described the factors involved in wound healing impairment and the different experimental models which may simulate ischemic wounds in humans.

1. Introduction

Wound healing occurs naturally through an advancement of highly integrated biological processes. These processes include hemostasis, inflammation, proliferation, and tissue remodeling. Chronic wounds or ulcers are frequently associated with diabetes mellitus and peripheral vascular disease, which result in tissue ischemia. There is a considerable evidence indicating that ischemic, non-healing wounds in diabetic patients with peripheral arterial occlusive disease (PAD) involve tissue hypoxia which is associated with metabolic dysfunction, altered inflammation, and decreased protein synthesis. Significant insight for understanding the role of diabetes and ischemia in the pathophysiology of delayed wound healing has been gained using numerous *in vitro* and *in vivo* animal models. These models have simulated various aspects of wound healing impairments associated with underlying disease conditions, including ischemia and diabetes. Also, these models may provide valuable tools for designing and testing novel therapies aimed at enhancing wound healing in humans. However, due to differences in biology between human and other animal species as well as experimental design, these models may have several disadvantages which are important to understand when designing novel treatment strategies. In this chapter, we describe wound healing impairment in humans and experimental models used to simulate ischemic wounds in humans.

2. The Clinical Problem

Delayed healing of both acute and chronic wounds signals that the wound healing process is ceasing to progress in a normal synchronized fashion, causing a shift towards a chronic non-healing wound with a pathologic inflammatory state. The clinical and economic burden of chronic non-healing wounds is estimated to be more than three billion dollars annually in the US, and it is projected to grow as a result of an estimated 2–3-fold increase in the prevalence of type-II diabetes and its associated comorbidities by 2050.[1] Chronic wounds or ulcers are commonly associated with peripheral vascular disease, diabetes mellitus, or prolonged pressure.[2] It is estimated that the lifetime risk of developing a foot ulcer is 25% among diabetic patients.[3] An estimated 7–10% of individuals suffering from diabetic neuropathy will develop a chronic foot ulcer. Whereas, the rate for patients with additional risk factors, such as PAD, foot deformity, previous ulcers, or past amputation, increases by 25–30%.[4–7] Other conditions that can complicate wound healing are inflammatory abnormalities, such as systemic lupus erythematosus, kidney failure, and lymphedema. Approximately 80% of chronic wounds localized on the lower legs are the result of chronic venous insufficiency.[8–10] Ischemic wound ulcers may appear in multiple locations, such as feet, heels, tips of the toes, or anywhere near boney projections. There are local and systemic factors that may interfere with one or more wound healing processes; repeatedly the etiology of wound healing delay can be heterogeneous, thereby complicating clinical interventions aimed at reversing the impairment.[2]

3. Pathophysiology of Impaired Wound Healing

The pathophysiology of ischemic wound healing impairment has not been fully described due to the lack of understanding of the molecular factors that cause normal wound repair to cease or delay at a certain stage of the wounded tissue regeneration process. It is well known that successful wound healing is dependent on sufficient systemic oxygenation, intact local vascular network, diffusion distance, and the degree of oxygen consumption.[11,12] Disruption of the vasculature and increased metabolic demand are the two main factors that lead to local tissue hypoxia at the wound site. Local hypoxia triggers *de novo* vascularization at the wound site, which ensues via two parallel processes. The first process, termed angiogenesis, involves new blood vessel formation from local endothelial cells; and the second process is vasculogenesis, which involves mobilization and recruitment of bone marrow endothelial progenitor cells (EPCs) to the wound site. Transient hypoxia appears to be important for initiation of the wound healing process by signaling through the hypoxia-inducible factor-1α (HIF-1α) pathway and promoting vasculogenesis through chemokine-induced recruitment of bone marrow-derived EPCs. In contrast, chronic hypoxia can be detrimental to wound healing. HIF-1α is a master regulator of oxygen homeostasis that mediates the adaptive cellular responses to hypoxia by regulating the expression of genes involved in angiogenesis, metabolic demand, and cell proliferation, migration, and survival.[13] Increased proliferation during wound healing requires higher energy substrates, oxygen and nutrients to fuel keratinocytes, endothelial cells and fibroblast activities needed for extracellular matrix (ECM) synthesis and deposition, angiogenesis, re-epithelialization, and remodeling. Angiogenesis occurs concurrently with fibroblast proliferation, which is important for fibroblast migration in the wound in response to several cytokine and growth factor mediators. These mediators include transforming growth factor beta (TGF-β), vascular endothelial growth factor (VEGF), and platelet-derived growth factor (PDGF).

Compared to acute wounds, human chronic wounds contain markedly elevated levels of pro-inflammatory cytokines and matrix metalloproteinases, while matrix metalloproteinase inhibitors and growth factor activities are diminished. Studies indicate that chronic, non-healing wounds in diabetic patients with PAD are associated with tissue hypoxia. This situation is complicated by the ensuing metabolic dysfunction and decreased capacity for pro-angiogenic mediator synthesis and release, which contributes to impaired wound healing.[14] Chronic wounds display excessive inflammation, fibroblast senescence, and altered wound bed flora.[15–17] Additionally, diabetic patients have diminished EPC mobilization and homing capabilities.[18–21] Furthermore, hyperglycemia is known to disrupt HIF-1α signaling by destabilizing its dimerization, thereby, bolstering functional inhibition.[21] The presence of ischemia can limit the EPCs' ability to reach the distantly located and poorly perfused wound area. On the other hand, oxygen-dependent enzymes, known as oxygenases, which are key regulators of several inflammatory and metabolic pathways, play an important role in wound healing, including collagen release.[22,23] Therefore, sufficient wound perfusion and oxygenation are not only vital to satisfy the increase in metabolic demand and EPC recruitment, but they are also essential for oxidative-mediated elimination of invading pathogens by active oxygenases. Studies suggest that this process is compromised by ischemic conditions due to poor circulation and oxygen insufficiency,[24–27] particularly in diabetic patients who are at an increased risk for infection and tissue loss or amputation.[28,29]

4. Therapeutic Interventions for Chronic Wounds

The current treatment strategy for lower extremity chronic wounds focuses on debridement of necrotic tissue, pressure offloading, negative-pressure therapy, wound dressing, skin grafting, and alleviation of limb ischemia.[30,31] More specific treatments are lacking due to our incomplete understanding of the pathophysiology of chronic wounds.[32–34] Investigators have focused on enhancing survival and homing capabilities of EPCs by augmenting the local levels of selected pro-angiogenic growth factors which are known to be deficient in diabetic wounds, such as VEGF[35–38] and PDGF.[39] The challenges of treating non-healing wounds include alleviating tissue ischemia, augmenting angiogenesis, and avoiding treatment side effects.[40,41]

Several approaches have been developed to overcome low oxygen levels in ischemic wounds using systemic hyperbaric oxygen therapy (HBOT). HBOT has been used to facilitate wound healing in patients with complex wounds that are not amenable to surgical revascularization.[4,42–44] HBOT seems to benefit diabetic patients with chronic ulcers by accelerating wound healing, decreasing amputation rates, and enhancing complete wound closure.[45–48] Animal studies and clinical data have demonstrated the benefits of HBOT for ischemic and chronic wounds that did not respond to other conventional treatments.[4,43,47,49] Several hypotheses have been tested, with the aim of understanding the mechanisms by which HBOT augments wound healing, including the enhancement of neovascularization and increased fibroblast proliferation.[50–52] Other studies have suggested that oxygen delivered via HBOT might act as a signal transducer in addition to being a metabolic substrate due of the synergistic effect that was demonstrated when ischemic wounds were treated with PDGF and HBOT.[53,54] The drawback of HBOT is that it adds a substantial financial burden, requires specialized clinical resources, and may carry the risk of oxygen toxicity.[55–57] Alternatively, regional or topical oxygen therapy has been proposed as an alternative therapeutic adjunct for ischemic wounds.[13] Published reports demonstrated enhanced diabetic foot ulcer healing using local delivery of pressurized oxygen to the wound area.[58] The lack of appropriate wound healing models that recapitulate the complexities of wound healing in humans hinders our ability to develop targeted therapies for chronic wounds in humans. The following summarizes the ischemic wound healing models that have been developed and used by investigators to study specific elements of wound healing and develop or test new treatments.

5. Ischemic Wound Healing Models

The etiology of chronic ischemic wounds can be heterogeneous, which can further complicate the clinical manifestation. Wound healing in general is a dynamic, well-orchestrated, and complex process. A better understanding of the temporal dynamics of the mechanisms underlying chronic wounds in humans is challenged by impractical and ethical factors, which hinders our ability to study the pathophysiology of delayed wound healing in humans, including the need for repeated biopsies. Therefore several wound healing models have been developed using *in vitro* and *in vivo* animal systems.

5.1 In vitro *models of ischemic wounds*

The physiologic parameters which are known to affect wound healing include temperature and pH levels, circulation, local oxygen tension, and adequate nutrient supply. These parameters can be easily manipulated in an *in vitro* tissue culture setting. Altered parameters can be applied such as hypoxia, high glucose levels, increased acidity, and variations in temperature. However, the tissue culture environment is utterly artificial because it lacks the complex ECM composition, immune cell activities, and natural *in vivo* multi-cellular interactions. Therefore, *in vitro* cell culture models have been designed to involve individual or multiple cell types to understand the complex signaling between cells in response to altered physiologic parameters known to be involved in the wound healing process. Several types of 2D culture systems have been used to study wound healing. They consist of engineered constructs composed of stratified squamous epithelial cells grown at an air–liquid interface, above a collagen matrix seeded with dermal fibroblasts. Cellular activity such as proliferation, migration, and apoptosis can also be monitored. *In vitro* wounding of cells can be performed using micro indenters. The influence of hypoxia on migration rates of different cell types involved in wound healing was indentified using this system. However, it was determined that signaling pathways that regulate the relationship between cell growth and differentiation can be optimally synchronized if cells are spatially stratified to mimic the architectural characteristics of the *in vivo* environment. Therefore, investigators developed 3D *in vitro* tissue culture systems.[59] These systems, known as human skin equivalents, have helped advance our understanding of skin biology and provided an *in vivo*-like environment to study epithelial cell proliferation, differentiation, and morphology that better simulates the *in vivo* skin environment.[60]

More recently, hydrogel-based culture systems have emerged as useful tools to examine cellular responses to ECM signaling molecules that are present during wound-healing processes and aspects of the extracellular environment.[61,62] These systems have been used to examine fibroblast functional activities, including adhesion, migration, and activation. Several parameters can be altered in these systems to investigate specific effects, such as protein binding and cytokine signaling. Hypoxia-sensitive hydrogels have recently been described in which oxygen levels and gradients within the hydrogels can be accurately controlled and precisely predicted. Using the 3D hydrogel system, studies demonstrated vascular morphogenesis mediated by HIFs which were activated by matrix metalloproteinases to promote rapid angiogenesis.[61,62]

5.2 In vivo *models of ischemic wounds*

Although *in vitro* systems can be valuable tools to simulate or examine mechanisms involved in wound healing at the molecular level, they cannot fully replicate the complexity of the *in vivo* wound healing environment because they lack key elements of immune responses and pathologic conditions associated with impaired wound healing in humans. Thus, the need to develop reproducible, reliable, and practical animal models that mimic chronic wound healing in humans is justifiable. Many investigators have dedicated significant effort to develop ischemic wounds in animal models because chronic ischemia or ischemia reperfusion injuries are prevalent in chronic wounds. Despite numerous differences that exist in wound development or skin biology between humans and rodents,[63–65] these models have contributed significantly to our understanding of wound healing in humans. Animal models have been used to study several wound healing processes, including immune responses and angiogenesis, as well as test wound healing adjuncts, such as anti-infectious agents. Animal model studies have demonstrated that ischemia contributes to impaired wound healing.[66,67] The majority of ischemic wound studies have been performed in mice due to lower cost and the possibility of genetic manipulation.

5.2.1 *The ischemic rabbit ear wound model*

The rabbit ear ischemic ulcer model was designed to create a wound inside an ischemic zone. An ischemic zone is surgically induced by interrupting the arterial supply to the ear, followed by making full-thickness punch biopsy wounds down to the cartilage that is dependent on peripheral vasculature for blood supply.[68] This model was described using minimally invasive surgical techniques.[69] A modified version of this model was also created to simplify the surgical procedure.[70] For the modified model, the ischemic wound is induced on one ear by creating three small skin incisions to expose the vascular pedicles at 1–2 cm from the ear base. The central artery is ligated and cut along with the nerve. The whole cranial bundle is cut and ligated, leaving only the caudal branch intact. A circumferential subcutaneous tunnel is made through the incision to cut through the subcutaneous tissue, muscles, nerves, and small vessels. The contralateral ear is used as a non-ischemic control. Following the surgical induction of ischemia, four wounds are made on the ventral aspect of each ear, generating four ischemic and four non-ischemic wounds in one animal for pair-wise comparisons. Following surgery, the ischemic ear exhibits less movement and lacks arterial pulse, while skin temperature becomes 1–10°C lower compared to the control side and this persists for more than 1 month. Ischemia can be further confirmed with laser Doppler scanning. Wound healing time is also significantly prolonged and the levels of energy substrate are markedly lower in the ischemic ear compared to the control ear. Several advantages of the rabbit ear model include relatively inexpensive animals, breeding, and housing. However, availability of genetic tools is limited.

This model has been used by investigators to test the therapeutic benefit of various growth factors on chronic wounds, including PDGF-B,[71,72] and to study the synergistic effects of PDGF treatment in combination with hyperbaric oxygen on wound healing enhancement.[53,54] Investigators have also utilized the rabbit model to establish the effects of biofilms on impaired cutaneous wound healing,[73] and to study the effects of nitric oxide-producing probiotics on infected wounds.[74] Another study verified the effects of diabetic neuroischemic wound healing, whereby investigators created four, 6-mm punch biopsy, full-thickness wounds following ligation of central and rostral arteries in diabetic and non-diabetic rabbits, after which investigators resected the central and rostral nerves.[75] The concurrent ischemic and neurologic deficiency was shown to exacerbate delayed wound healing and was associated with higher M1/M2 macrophages ratio, diminished cytokine response, and lower neuropeptide expression.

The rabbit model has also been used to test the effect of transdermal sustained oxygen delivery on wound healing in the New Zealand rabbit.[76] During the study, investigators created four, 7-mm punch biopsy, full-thickness wounds on each ear. The treated ears received 100% oxygen supply via silicone tubing tunneled subcutaneously under a semi-occlusive dressing at a continuous flow. Oxygen delivery was found to significantly enhance re-epithelialization and wound coverage. Under ischemic conditions, hypoxia activates HIF, which triggers downstream gene activation by binding to the nuclear hypoxia response element (HRE). In another study, this model was used to monitor hypoxia-related signaling pathway activities by tracking HRE activity using HRE-luciferase-reporter gene plasmid construct.[77] Since impaired re-epithelialization is one of the hallmarks of chronic ischemic wounds, Kloeters *et al.* used the ischemic rabbit ear wound model to study the role of the TGF-β pathway in the normal and hypoxic wound healing response.[78] The investigators increased the expression of the cytoplasmic signaling molecule, Smad-3, which is a signal transducer and transcriptional modulator that is activated by TGF-β and activin Type 1 receptor kinases. The increased expression level of Smad-3 in the ischemic wounds resulted in enhanced re-epithelialization during wound healing. Other investigators have studied the role of keratinocyte growth factor-2 (KGF-2) in wound healing and scar formation using the rabbit ear ischemia model.[79] KGF-2 is a member of the fibroblast growth factor family which is exclusively expressed by epithelial cells in the skin. Investigators have demonstrated that administration of KGF-2 directly to the wound led to significant improvement of epithelial growth and tissue granulation without enhancing scar formation.

A variation of the rabbit ear wound model has also been created in the hairless mouse to study the effect of ischemia on tissue oxygenation and wound healing. Ischemia is induced in the mouse ears by ligating two of the three main vessel bundles resulting in reduced tissue oxygenation and prolonged wound healing. Kamler *et al.* showed that a vasoactive medication enhanced reoxygenation in the hairless mouse model and reversed the adverse effects of ischemia on wound healing.[80]

5.2.2 *Mouse model of wound healing in ischemic cutaneous flaps*

The most frequently used model is the ischemic skin flap model in mice (Figure 1), as reviewed by Wong *et al.*[65] The single-pedicle skin flap model produces more severe tissue ischemia that eventually leads to necrosis of the distal flap area.

This model consists of a three-sided, full-skin thickness flap peninsula that is created on the mouse dorsum and implantation of an impermeable silicone sheet that is placed underneath the skin flap, followed by suturing of the flap borders. The silicone sheet will create a barrier that prevents revascularization originating from the underlying wound bed and allows blood supply only through the flap pedicle side. This ischemic flap model creates a reproducible ischemic gradient extending from the proximal to the distal area of the tissue.

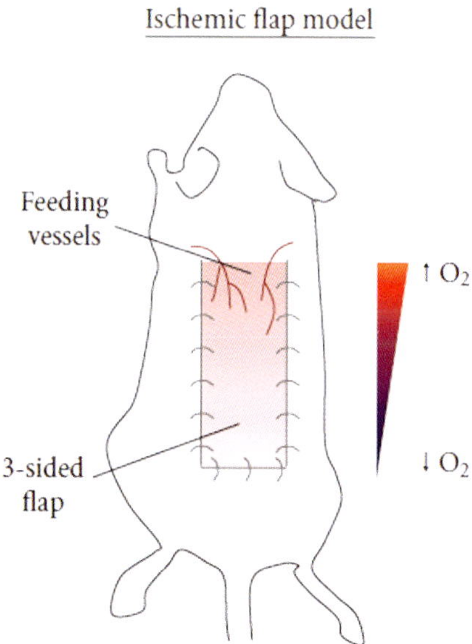

Figure 1.   Single-pedicle ischemic skin flap model in mice. The dorsal skin ischemic flap model produces a controlled gradient of ischemia based on the number and location of feeding blood vessels. Skin regions furthest from the vessel/s are the most ischemic. With permission, from Ref. 65.

Variations of the ischemic flap model have also been developed in rats using a single-pedicle or bipedicle dorsal ischemic skin flap. This model has been important in elucidating novel mechanisms of neovascularization, and describing the role of hypoxia-mediated signaling and chemokine pathways in progenitor cell recruitment to the wound tissue.

5.2.3 *Rat model of bipedicle skin flap with ischemic and non-ischemic wounds*

The bipedicle skin flap design in rats (Figure 2) includes full-thickness excision wounds that are made using a punch biopsy tool inside the flap area which produces transient and moderate ischemia associated with delayed wound healing.[81] The model has been described using Sprague Dawley rats, which were wounded under Isoflurane anesthesia. For this model, a permanent marker is used to outline a 3.0-cm × 10.5-cm rectangular area on the dorsal skin to mark the area where the bipedicle ischemic flap is to be created. Two circular full-thickness excisional wounds are created inside the marked area using a 6-mm, sterile punch biopsy tool, thereby generating ischemic wounds on the back of the rat and inside the flap area. Following creation of the excisional wounds, the bipedicle flap is created by making incisions with a sterile scalpel on each side of the ischemic wounds and along the pre-marked rectangular area lines (10.5 cm in length and 3.0 cm in width). The depth of the two incisions should be down to the level of the paraspinous muscles. Using iris scissors, the panniculus carnosus fascia is separate from the paraspinous muscles without damaging the fascia. A sterile, pre-cut, 10.0-cm × 3.0-cm silicone sheet is inserted between the panniculus carnosus fascia and the paraspinous muscles. The flap borders are closed with eight interrupted, non-absorbable 4-0 silk sutures by anchoring the silicone sheet to the skin on each side along the length of the flap. Two full-thickness circular excisional wounds are created laterally outside the marked skin flap area to obtain non-ischemic wounds on the back of the same rat using a sterile, disposable, punch biopsy tool.

This model results in delayed wound healing by around 30% compared to wounds created in non-ischemic skin. Analysis of biopsies from ischemic wounds in this model have shown increased levels of pro-inflammatory cytokine proteins and mRNA at day 13 after surgery compared to biopsies of wounds created in non-ischemic skin or uninjured normal skin. Levels of matrix metalloproteinases and serine proteases were elevated in ischemic wounds 13 days after injury compared to non-ischemic wounds or normal unwounded skin in this model. Topical treatment of ischemic wounds in the bipedicle skin flap with PDGF has demonstrated accelerated healing. The

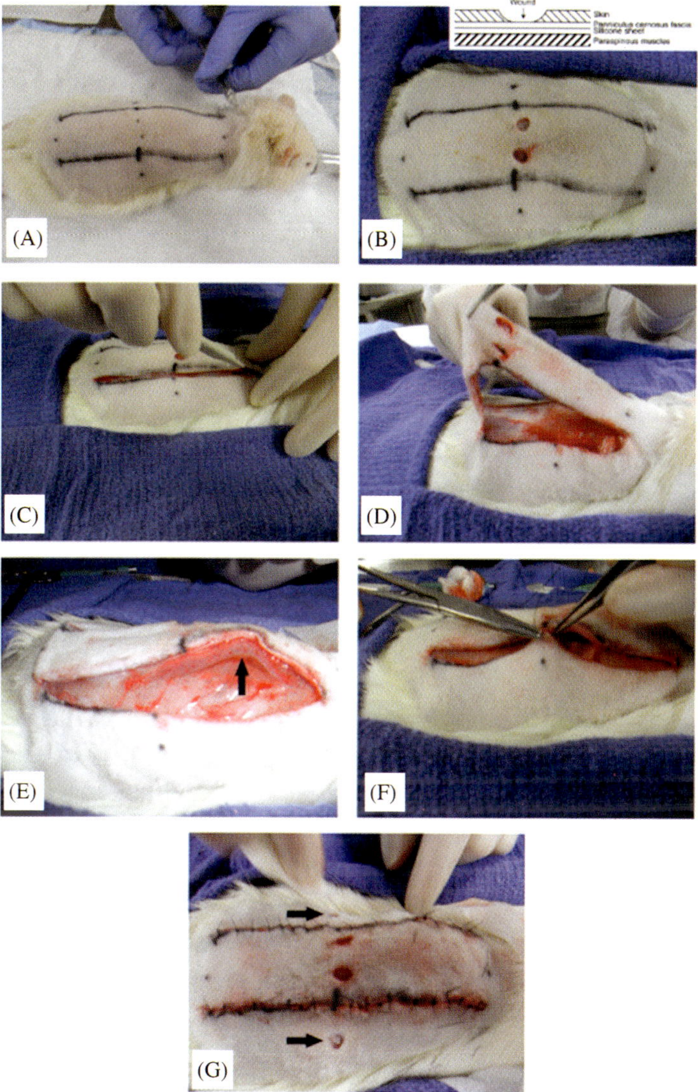

Figure 2. Photographs depicting steps during surgery to create ischemic wounds, the bipedicled flap, and non-ischemic wounds. (A) Pre-surgical hair removal and skin preparation of an anesthetized rat receiving a pre-surgical dose of analgesic (Ketoprofen) subcutaneously for pain management. (B and inset) Full thickness ischemic wounds are created inside the 2 incision markers by utilizing a sterile punch biopsy tool. (C) Incisions are made along the marked lines down to the paraspinous muscle and (D) the bipedicled flap is raised to show the separation of the flap (leaving the panniculus carnosus fascia intact) from the muscle layer below. (E) A sterile silicone sheet (black arrow) is placed between the fascia of the panniculus carnosus and the paraspinous muscle. (F) Black, non-absorbable 4-0 sutures are used to close both incisions by anchoring the silicone sheet to the skin with multiple interrupted stitches along the length of the flap. (G) Two non-ischemic wounds (black arrows) are created using a sterile punch biopsy tool lateral to the bipedicled flap on both sides. With permission from Ref. 81.

ischemic flap models have been used to elucidate mechanisms of wound healing and to evaluate potential wound therapies. Collagen deposition has been studied in the ischemic full-thickness wounds made four days after bipedicle skin flaps were created on the dorsum of rats.[82] This particular study demonstrated significantly lower total collagen content in ischemic wounds, supporting the hypothesis that the delay in wound healing is due to altered collagen metabolism and inhibiting the tissue's ability to maintain appropriate levels of collagen under persistent inflammatory conditions. Other investigators have found use of the bipedicle dorsal flap wound healing model in rats to be helpful when studying human chronic wounds for testing or developing novel therapies for wound care.[83]

One of the major components of the wound healing process that has been studied using these models is wound neovascularization, which involves both angiogenesis manifested by sprouting of new vessels from existing ones and vasculogenesis consisting of *de novo* formation of new vessels from circulating EPCs. Impairment of neovascularization has been recognized to be responsible for delayed wound healing in diabetic conditions and during aging. These models also have demonstrated how systemic stem cell recruitment is mediated by HIF-1α and stromal cell-derived factor-1 pathways. Studies based on these models have reported impairments in HIF-1α-mediated vasculogenesis during aging and established an important role for reactive oxygen species in delayed wound repair in diabetes. Hypoxia is known to induce microRNA-210, which attenuates wound healing and decreases keratinocyte proliferation. Use of these models to study wound healing has elucidated how microRNAs may be implicated in limiting wound re-epithelialization under hypoxia.[84] Ischemia has been evaluated by laser Doppler as well as hyperspectral imaging and shown to result with limited blood flow and lowered tissue oxygen saturation. Electron paramagnetic resonance oximetry has demonstrated that ischemic wound tissue has lower partial oxygen tension.

In summary, these models simulate many of the molecular abnormalities that characterize the environment of ischemic human chronic wounds, including delayed healing, increased levels of inflammatory cytokines, and elevated proteases.

5.2.4 *Rodent models of hind limb ischemia and excisional wound healing*

To study the effect of limb ischemia on bone marrow-derived EPC homing, investigators designed the hind limb ischemic wound model. Unilateral

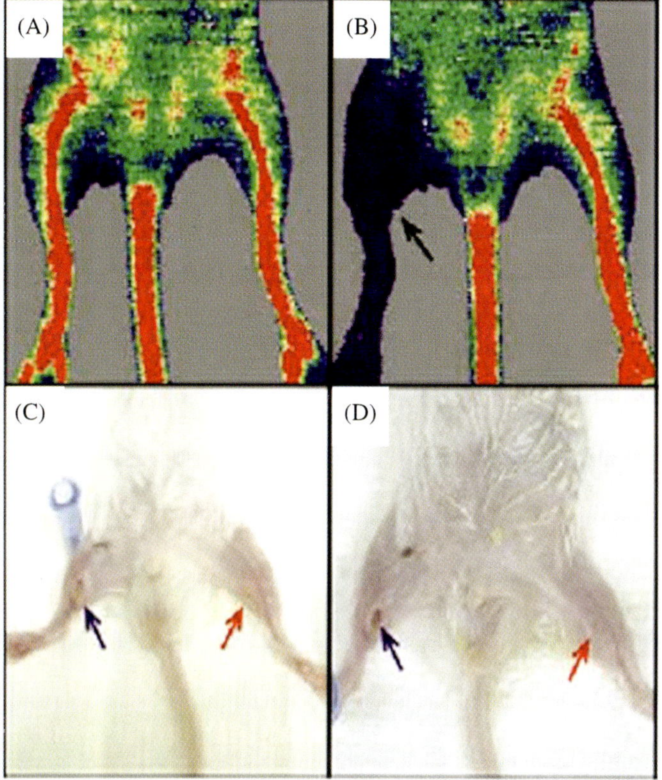

Figure 3. Mouse model of hind limb ischemia and excisional wound healing. In this model femoral artery and vein ligation/excision results in significant hindlimb ischemia and delayed wound healing. (A) Preoperative laser Doppler imaging (LDI) demonstrates good perfusion down each hindlimb. (B) Post-operative LDI demonstrates decreased perfusion in the hindlimb ipsilateral to the femoral ligation (arrow). (C) Representative mouse at post-operative day 5 and (D) day 7 demonstrates delayed healing of hindlimb excisional wounds on the ischemic hindlimbs (blue arrow) compared with the acutely healing wounds on the non-ischemic hindlimb (red arrow). (With permission, Rights Link).

femoral artery and vein ligation and excision surgery was performed to induce hind limb ischemia, followed by the creation of bilateral excisional punch biopsy wounds on the ventral side of each hind limb to create ischemic and non-ischemic wounds distal to the femoral artery ligation access (Figure 3). This model established an essential role of EPCs in vasculogenesis associated with wound healing. These studies demonstrated that granulation tissue has impaired localization of EPCs in ischemic wounds.[85] HBOT has been used as an adjunctive therapy for non-healing ulcers in patients that are not amenable for surgical interventions. Investigators reported HBOT-mediated mobilization of EPCs is associated with increased lower limb spontaneous circulatory recovery after femoral ligation and enhanced closure of ischemic wounds, which is mediated by enhanced nitric oxide (NO) levels.[86] These investigators showed that EPCs, essential for vasculogenesis and wound healing, are diminished systemically and at the wound site in diabetes. Using a similar approach, Gallagher *et al.* identified diminished endothelial nitric oxide synthase (eNOS) activation as being responsible for the defective EPC mobilization in diabetic mice. Furthermore, hyperoxia induced by HBOT reversed the defect in EPCs. They also tested the hypothesis that impaired EPC homing in diabetes is due to decreased stromal cell-derived factor-1 alpha (SDF-1α) expression in cutaneous wounds. They found that HBOT reversed the diabetic defect in EPC mobilization and that increased SDF-1α expression reversed the diabetic defect in EPC homing.[87]

The redox enzyme p66Shc is known to produce hydrogen peroxide and trigger pro-apoptotic signaling. Fadini *et al.* evaluated the role of p66Shc in an animal model of diabetic ischemic wound healing.[88] Skin wounds were created on the hind limbs of wild-type and p66Shc genetically-deficient diabetic and non-diabetic mice with or without hind limb ischemia. Wounds were assessed for collagen content, thickness and vascular density of granulation tissue, apoptosis, re-epithelialization, and expression of c-Myc and beta-catenin. Response to hind limb ischemia was also evaluated. The study demonstrated that genetic deletion of p66Shc improved the rate of wound healing in the setting of diabetes and ischemia.

Another study assessed the effects of injection of human adipose-derived stem cells on enhancing vascularization of hind limb ischemic wounds in diabetic mice.[89] Hind limb ischemic wounds were created in nude diabetic and non-diabetic mice and human adipose tissue-derived stem cells were injected locally. Investigators observed that the stem cell-injected mice had markedly early dense neovascularization and enhanced tissue remodeling associated with lower auto-amputation rates compared to the control mice, suggesting that human adipose-derived stem cells enhance ischemic wound healing of diabetic mice.[89]

Similarly, Alizadeh *et al.* described a new rat model of hind limb ischemia to study the impact of ischemia on wound healing in Wistar rats.[90] The external iliac artery was resected to the femoral arteries level by the knee in one hind limb, and skin wounds were made on both feet. Ischemia was assessed by blood flow measurement, which decreases dramatically in the ischemic limb. A significant delay in wound closure with a decrease in wound contraction was observed in the ischemic limb. Myofibroblast quantification showed a significant delay in appearance as well as a decrease in the number of these cells in the ischemic wound. This was associated with a marked decrease in collagen type-I mRNA in the ischemic granulation tissue after 10 days. The investigator concluded that decreased wound contraction plays an important role in delayed ischemic wound healing due to reduced fibroblast activity.

Guinea pigs have also been used to test the effect of topical hydrogen peroxide treatment on ischemic ulcers.[91,92] Investigators measured vascular perfusion with a laser Doppler velocimeter and gross observations of percentage of non-necrotic wound surface on ischemic wounds in guinea pigs after treatment with either a hydrogen peroxide cream or a placebo cream. Visual evaluations of the percentage of non-necrotic wound surface showed no statistically significant differences among the treatments. In contrast, vascular perfusion measurements resulted in statistically significant differences. Blood flow was significantly higher up to day 15 in ulcers treated with 2% hydrogen peroxide cream compared to those treated with placebo cream. Vascular perfusion was significantly higher in ulcers treated with 3.5% hydrogen peroxide cream compared to ulcers treated with either 1.5% hydrogen peroxide cream or placebo. Adjacent control sites in guinea pigs whose ulcers were treated with hydrogen peroxide cream showed increased vascular perfusion compared to corresponding sites in animals whose ulcers were treated with placebo. Even distant flank control sites of ulcers treated with 3.5% hydrogen peroxide cream showed increased vascular perfusion. Investigators concluded that treatment of ischemia-induced ulcers with hydrogen peroxide cream enhanced cutaneous blood recruitment, not only to ulcers and adjacent sites, but also to distant sites.

5.2.5 *Animal models of ischemia-reperfusion wounds*

The pathogenesis of pressure ulcers is thought to be triggered by repeated episodes of ischemia reperfusion injury induced by cyclic compression of the tissue. This has prompted investigators to develop models to study cellular activities, inflammation, and redox signaling associated with ischemia reperfusion in models that might recapitulate human chronic wounds of pressure ulcers. The reperfusion of ischemic tissue is crucial for tissue survival, but usually results in secondary reperfusion injury associated with inflammatory response and excessive reactive oxygen species generation and cellular apoptosis. One version of skin ischemia reperfusion pressure injury consists of applying cyclical pressure with opposing magnets and inter-positioned skin to mimic ischemia/reperfusion injury which has been implicated in pressure

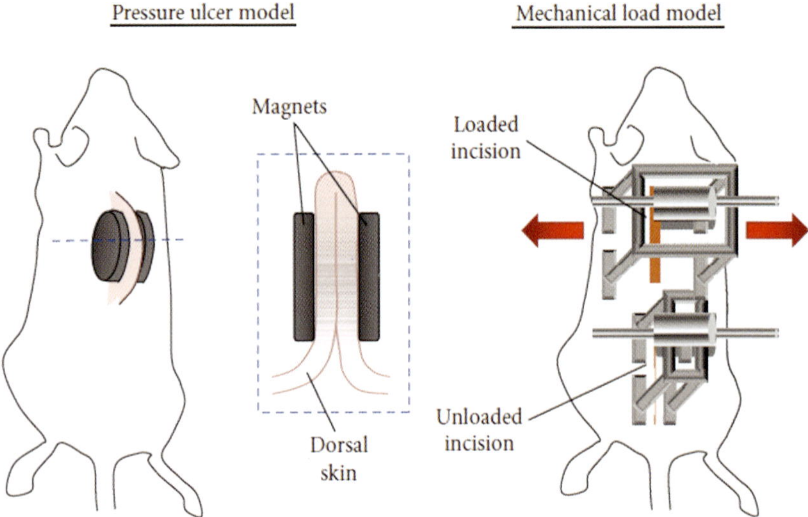

Figure 4. Surgical mouse models of Cutaneous Ischemia Reperfusion. (Left) Cyclical pressure can be applied with opposing magnets and inter-positioned skin to mimic reperfusion/ischemia injury thought to drive pressure ulcer pathophysiology. (Right) The application of exogenous mechanical loading to mouse incisions results in increased wound fibrosis, similar to human hypertrophic scarring.[65] (With permission, Hindawi Publishing Corporation, Wong *et al. J Biomed Biotechnol* 2011.)

ulcer pathophysiology in human (Figure 4, left). Another variation of this approach consists of surgically implanting a metal plate under the skin of a rat, followed by periodic compressions of the skin using an external magnet. The application of exogenous mechanical loading to mouse incisions results in increased wound fibrosis, similar to human hypertrophic scarring (Figure 4, right).[93,94] The injury size can be extended by changing the frequency and duration of applying the compressing magnet, which results in episodes of skin ischemia and tissue hypoxia associated with acute inflammatory response, simulating the human scenario of chronic wounds. Another version of a cyclic ischemia reperfusion model was created using a single pedicle dorsal skin flap in mice that had a single-pedicle of blood supply and which could be clamped in a cyclic fashion to generate episodes of skin ischemia to simulate clinical wound ischemia in humans. Investigators demonstrated that the application of exogenous mechanical loading to mouse incisions results in increased wound fibrosis, similar to human hypertrophic scarring (Figure 4, right).[95,96] These models replicated human pressure ulcer pathophysiology by triggering repetitive episodes of ischemia, followed by reperfusion associated compression, and resulting in skin injury.

These models helped elucidate the cellular and molecular mechanism involved in human pressure ulcers. Cyclic skin ischemia reperfusion was also created using a novel device in the rabbit ear[97] and was used to evaluate ventral ear wound healing. In that model, investigators developed a lightweight clamp apparatus to induce reversible occlusion of the central artery of the ear. Serial analysis for epithelialization and granulation, and gene expression analysis were performed demonstrating significant impairment in epithelial and granulation, and significant up-regulation of heat shock protein-70 and down-regulation of the free radical oxygen scavenger superoxide dismutase-1. This model has several advantages, including minimal skin disruption, longer ischemic time, and higher success rate, compared to other models. This model demonstrated contrasting differences in wound healing characteristics compared to surgically-induced wounds, making it suitable for preclinical research and the wounds are mechanistically similar to pressure ulcers in humans.

5.2.6 *Large animal models of ischemic wounds*

The skin structure and wound healing process in swine resemble more closely the ones in human skin compared to the the majority of smaller animals such as rodents, where repair occurs more predominantly via contraction.[98] This prompted the need to develop preclinical chronic wound models. Recently, the red Duroc pig has extensively been validated as a model for human skin pathology and is increasingly thought of as an ideal, large animal model to study cutaneous disease due to its similarity to human epithelial architecture, nerve density, vascularization, matrix components, and other biological parameters.[99] An ischemic flap wound model has also been developed in swine (Figure 5). The porcine model is widely accepted as an excellent preclinical model for human wounds.[100] A full-thickness bipedicle flap approach was adopted to cause skin ischemia. Closure of excisional wounds placed on ischemic tissue was severely impaired resulting in chronic wounds. Histologically, ischemic wounds suffer from impaired re-epithelialization, delayed macrophage recruitment, and poorer endothelial cell abundance and organization. Compared with the pair-matched, non-ischemic wound, unique aspects of the ischemic wound biology were examined on days 3, 7, 14, and 28 by systematic screening of the wound tissue gene expression using high-density, porcine Gene-Chip analysis.

Investigators reported that ischemia markedly potentiated the expression of arginase-1, a cytosolic enzyme that metabolizes the precursor of NO L-arginine. Ischemia also induced super oxide dismutase-2 in the wound

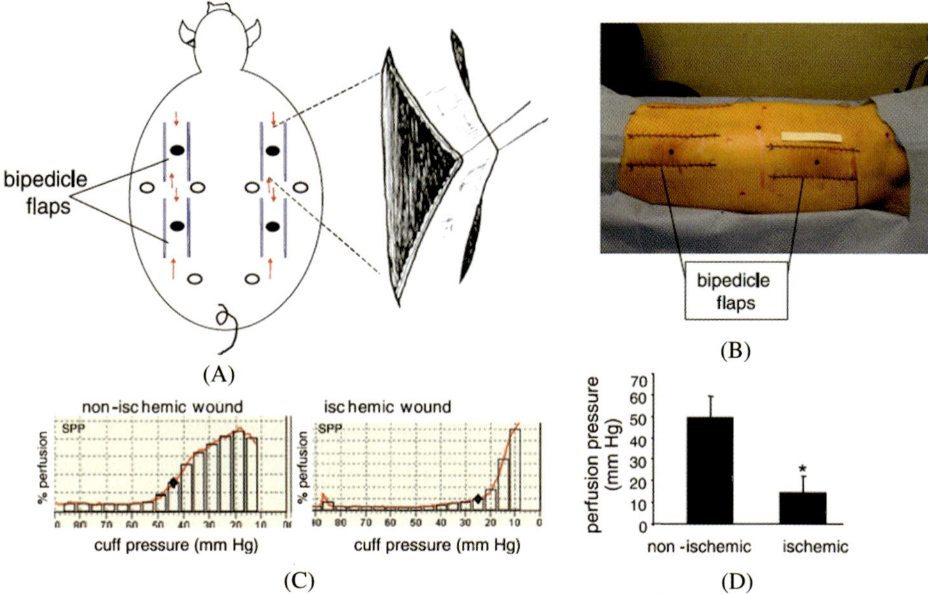

Figure 5. Animal model of preclinical wound healing in swine: wounding approach and skin perfusion pressure (SPP). (A, B) 4 bipedicle flaps (15 cm × 5 cm) were developed on the back of a domestic white swine. The flaps were freed from the underlying muscle and a sterile silicone sheet (15 cm × 5 cm) was placed under it to prevent re-adherence. The incised edges of the flap were sutured to the adjacent skin, incorporating the silicone. Full-thickness excisional wounds were developed in the middle of each flap (filled circles) using an 8-mm sterile, disposable biopsy punch. Six additional paired wounds were developed similarly in the adjacent non-ischemic skin (open circles). Arrows in (A) indicate direction of blood flow. Scale: the paper strip above the left lumbar flap in (B) is 10 cm long. (C) SPP was measured at locations adjacent to the ischemic as well as paired non-ischemic wounds. The cuff pressure was measured where subcutaneous microcirculation was restored. (D) SPP from non-ischemic and ischemic wounds. Data are means ± SD ($n = 3$); $^*p < 0.01$. Permission: Rights Link.

tissue and perhaps is similar to what has been found in chronic wounds in humans. This study demonstrated the usefulness of this model and its ability to provide data that may serve as a valuable tool to test novel hypotheses aimed at elucidating the pathobiology of chronic ischemic wound and tailor novel treatment strategies in a preclinical setting.

An experimental pig model replicated several intrinsic factors clinically relevant to delayed wound healing.[101] The investigators identified that intrinsic factors associated with delayed wound healing in humans are ischemia, infection, the presence of necrotic tissue, and the presence of foreign bodies. Studies were designed to create adaptable intrinsic factors that induce delayed wound healing in the pig, similar to clinical conditions. Ten

4.0 cm × 4.0 cm fresh full-thickness skin wounds were made on the backs of pigs. Double-flanged blocks, made of silicon material that induced foreign body reaction and tissue ischemia, and causes local wound infection, were inserted into half of the wounds and removed at weeks 1, 2, or 3. The other wounds were left open for control purposes. Serial biopsies were obtained for histological evaluation and for assessing inflammatory cytokines levels. Wound healing and perfusion were assessed with digital photography and laser Doppler scanning. Investigators reported that histological findings corresponding to the severity of chronic inflammation also correlated with increased levels of the pro-inflammatory cytokines, interleukin-1β (IL-1β) and tumor necrosis factor-α (TNF-α) in the chronic wounds. Epithelialization was not complete at the end of the fourth week despite removing the silicone blocks in the experimental group. The swine wound model mirrored similar characteristics of chronic ischemic wounds with necrosis and persistent inflammation observed in patients with chronic wounds. The model drawbacks include large expenses for animal acquisition and housing, and, in some instances, the lack of certain molecular tools to perform sample analyses. Therefore, in many instances, the use of large animals for initial large-size wound research projects may be impractical.

5.3. Summary

While important preclinical data has been generated from *in vitro* and *in vivo* models of wound healing, there is no perfect model that can recapitulate all of the relevant factors contributing to poor wound healing in humans. With the admonition of known inherent differences of tissue structure and wound repair process compared to humans, the mouse and rat models remain the most utilized by investigators to study wound repair mechanism or test novel therapies. However, investigators should not be discouraged from using the existing experimental models as long as there is an understanding of their characteristics, advantages, and limitations. Ongoing research in this area is essential to develop more clinically relevant models.

5.4. Acknowledgments

The authors acknowledge support from the Nod and Henry Meyer Fund, and the Division of Vascular and Endovascular Surgery at the Massachusetts General Hospital.

References

1. Boyle, J. P. Thompson, T. J. Gregg, E. W. Barker, L. E. and Williamson, D.F. Projection of the year 2050 burden of diabetes in the US adult population: Dynamic modeling of incidence, mortality, and prediabetes prevalence. *Popul Health Metr* **8**, 29 (2010).

2. Guo, S. and Dipietro, L. A. Factors affecting wound healing. *J Dent Res* **89**(3), 219–229 (2010).

3. Boulton, A. J. Armstrong, D. G. Albert, S. F. Frykberg, R. G. Hellman, R. Kirkman, M. S. Lavery, L. A. Lemaster, J. W. Mills, J. L. Sr., Mueller, M. J. *et al.* Comprehensive foot examination and risk assessment: A report of the task force of the foot care interest group of the American Diabetes Association, with endorsement by the American Association of Clinical Endocrinologists. *Diabetes Care* **31**(8), 1679–1685 (2008).

4. Stoekenbroek, R. M. Santema, T. B. Koelemay, M. J. van Hulst, R. A. Legemate, D. A. Reekers, J. A. and Ubbink, D. T. Is additional hyperbaric oxygen therapy cost-effective for treating ischemic diabetic ulcers? Study protocol for the Dutch DAMOCLES multicenter randomized clinical trial? *Diabetes* **7**(1), 125–132 (2015).

5. Rawles, Z. Assessing the foot in patients with diabetes. *Nursing times* **110**(31), 20–22 (2014).

6. Ramsey, S. D. Newton, K. Blough, D. McCulloch, D. K. Sandhu, N. Reiber, G. E. and Wagner E. H. Incidence, outcomes, and cost of foot ulcers in patients with diabetes. *Diabetes Care* **22**(3), 382–387 (1999).

7. Alexiadou, K. and Doupis, J. Management of diabetic foot ulcers. *Diabetes Ther* **3**(1), 4 (2012).

8. Diabetes Public Health Resource NDSR. Diagnosed and undiagnosed diabetes in the United States, Estimates of diabetes and its burden in the United States. (2014).

9. Situm, M. and Kolic, M. Chronic wounds: Differential diagnosis. *Acta Med Croatica* **67**(1), 11–20 (2013).

10. Situm, M. Kolic, M. Redzepi, G. and Antolic, S. Chronic wounds as a public health problem. *Acta Med Croatica* **68**(1), 5–7 (2014).

11. Gottrup, F. Oxygen, wound healing and the development of infection. Present status. *Eur J Surg* **168**(5), 260–263 (2002).

12. Gottrup, F. Oxygen in wound healing and infection. *World J Surg* **28**(3), 312–315 (2004).

13. Castilla, D. M. Liu, Z. J. and Velazquez O. C. Oxygen: Implications for Wound Healing. *Adv Wound Care* **1**(6), 225–230 (2012).

14. Tandara, A. A. and Mustoe T. A. Oxygen in wound healing — more than a nutrient. *World J Surg* **28**(3), 294–300 (2004).

15. Bosanquet, D. C. and Harding K. G. Wound duration and healing rates: Cause or effect? *Wound Repair Regen* **22**(2), 143–150 (2014).

16. Bosanquet, D. C. Ye, L. Harding, K. G. and Jiang W. G. FERM family proteins and their importance in cellular movements and wound healing (review). *Int J Mol Med* **34**(1), 3–12 (2014).

17. Boateng, J. S. Matthews, K. H. Stevens, H. N. and Eccleston G. M. Wound healing dressings and drug delivery systems: A review. *J Pharm Sci* **97**(8), 2892–2923 (2008).

18. Drela, E. Stankowska, K. Kulwas, A. and Rosc, D. Endothelial progenitor cells in diabetic foot syndrome. *Adv Clin Exp Med* **21**(2), 249–254 (2012).

19. Loomans, C. J. de Koning, E. J. Staal, F. J. Rookmaaker, M. B. Verseyden, C. de Boer, H. C. Verhaar, M. C. Braam, B. Rabelink, T. J. and van Zonneveld, A. J. Endothelial progenitor cell dysfunction: A novel concept in the pathogenesis of vascular complications of type 1 diabetes. *Diabetes* **53**(1), 195–199 (2004).

20. Kim, K. A. Shin, Y. J. Kim, J. H. Lee, H. Noh, S. Y. Jang, S. H. and Bae O. N. Dysfunction of endothelial progenitor cells under diabetic conditions and its underlying mechanisms. *Arch Pharm Res* **35**(2), 223–234 (2012).

21. Catrina, S. B. and Zheng, X. Disturbed hypoxic responses as a pathogenic mechanism of diabetic foot ulcers. *Diabetes Metab Res Rev* (2015).

22. Botusan, I. R. Sunkari, V. G. Savu, O. Catrina, A. I. Grunler, J. Lindberg, S. Pereira, T. Yla-Herttuala, S. Poellinger, L. Brismar, K. *et al.* Stabilization of HIF-1alpha is critical to improve wound healing in diabetic mice. *Proc Natl Acad Sci USA* **105**(49), 19426–19431 (2008).

23. Sander, A. L. Jakob, H. Sommer, K. Sadler, C. Fleming, I. Marzi, I. and Frank, J. Cytochrome P450-derived epoxyeicosatrienoic acids accelerate wound epithelialization and neovascularization in the hairless mouse ear wound model. *Langenbecks Archi Surg / Dtsch Ges Chir* **396**(8), 1245–1253 (2011).

24. Benhaim, P. and Hunt, T. K. Natural resistance to infection: Leukocyte functions. *J Burn Care Rehabil* **13**(2 Pt 2), 287–292 (1992).

25. Nguyen, K. T. Seth, A. K. Hong, S. J. Geringer, M. R. Xie, P. Leung, K. P. Mustoe, T. A. and Galiano, R. D. Deficient cytokine expression and neutrophil oxidative burst contribute to impaired cutaneous wound healing in diabetic, biofilm-containing chronic wounds. *Wound Repair Regen* **21**(6), 833–841 (2013).

26. Allen, R. C. Reduced, radical, and excited state oxygen in leukocyte microbicidal activity. *Front Biol* **48**, 197–233 (1979).

27. Allen, R. C. Yevich, S. J. Orth, R. W. and Steele, R. H. The superoxide anion and singlet molecular oxygen: Their role in the microbicidal activity of the polymorphonuclear leukocyte. *Biochem Biophys Res Commun* **60**(3), 909–917 (1974).

28. Vieru, A. Mazilu, G. and Graur, M. Antibiotic therapy in the infected diabetic foot. *Rev Med Chir Soc Med Nat Lasi* **115**(4), 1091–1096 (2011).

29. Strbova, L. Krahulec, B. Waczulikova, I. Gaspar, L. Ambrozy, E. Bendzala, M. and Dukat, A. Influence of infection on clinical picture of diabetic foot syndrome. *Bratisl Lek Listy* **112**(4), 177–182 (2011).

30. Lepantalo, M. Apelqvist, J. Setacci, C. Ricco, J. B. de Donato, G. Becker, F. Robert-Ebadi, H. Cao, P. Eckstein, H. H. De Rango, P. *et al.* Chapter V: Diabetic foot. *Eur J Vasc Endovasc Surg*, **42**(2), S60–S74 (2011).

31. Mansilha, A. and Brandao, D. Guidelines for treatment of patients with diabetes and infected ulcers. *J Cardiovasc Surg (Torino)* **54**(1), 193–200 (2013).

32. Krakowski, A. C. Diaz, L. Admani, S. Uebelhoer, N. S. and Shumaker, P. R. Healing of chronic wounds with adjunctive ablative fractional laser resurfacing in two pediatric patients. *Lasers Surg Med* (2015).

33. Ribeiro, J. Pereira, T. Amorim, I. Caseiro, A. R. Lopes, M. A. Lima, J. Gartner, A. Santos, J. D. Bartolo, P. J. Rodrigues, J. M. *et al.* Cell therapy with human MSCs isolated from the umbilical cord Wharton jelly associated to a PVA membrane in the treatment of chronic skin wounds. *Int J Med Sci* **11**(10), 979–987 (2014).

34. Liani, M. Trabassi, E. Cusaro, C. Zoppis, E. Maduli, E. Pezzato, R. Piccoli, P. Maraschin, M. Bau, P. Cortese, P. *et al.* Effects of a pulsatile electrostatic field on ischemic injury to the diabetic foot: Evaluation of refractory ulcers. *Prim Care Diabetes* **8**(3), 244–249 (2014).

35. Dhall, S. Do, D. C. Garcia, M. Kim, J. Mirebrahim, S. H. Lyubovitsky, J. Lonardi, S. Nothnagel, E. A. Schiller, N. and Martins-Green, M. Generating and reversing chronic wounds in diabetic mice by manipulating wound redox parameters. *J Diabetes Res* 2014, 562–625 (2014).

36. Jetten, N. Roumans, N. Gijbels, M. J. Romano, A. Post, M. J. de Winther, M. P. van der Hulst, R. R. and Xanthoulea, S. Wound administration of M2-polarized macrophages does not improve murine cutaneous healing responses. *PloS One* **9**(7), e102994 (2014).

37. Park, S. A. Raghunathan, V. K. Shah, N. M. Teixeira, L. Motta, M. J. Covert, J. Dubielzig, R. Schurr, M. Isseroff, R. R. Abbott, N. L. *et al.* PDGF-BB does not accelerate healing in diabetic mice with splinted skin wounds. *PloS One* **9**(8), e104447 (2014).

38. Gainza, G. Bonafonte, D. C. Moreno, B. Aguirre, J. J. Gutierrez, F. B. Villullas, S. Pedraz, J. L. Igartua, M. and Hernandez, R. M. The topical administration of rhEGF-loaded nanostructured lipid carriers (rhEGF-NLC) improves healing in a porcine full-thickness excisional wound model. *J Control Release* **197**, 41–47 (2015).

39. Wieman, T. J. Smiell, J. M. and Su, Y. Efficacy and safely of a topical gel formulation of recombinant human platelet-derived growth factor-BB (becaplermin) in patients with chronic neuropathic diabetic ulcers: A phase III randomized placebo-controlled double-blind study. *Diabetes Care* **21**(5), 822–827 (1998).

40. MacNeil, S. Progress and opportunities for tissue-engineered skin. *Nature* **445**(7130), 874–880 (2007).

41. Zhong, S. P. Zhang, Y. Z. and Lim, C. T. Tissue scaffolds for skin wound healing and dermal reconstruction. *Wiley Interdiscip Rev Nanomed Nanobiotechnol* **2**(5), 510–525 (2010).

42. Heyboer, M. 3rd, Grant, W. D. Byrne, J. Pons, P. Morgan, M. Iqbal, B. and Wojcik, S. M. Hyperbaric oxygen for the treatment of nonhealing arterial insufficiency ulcers. *Wound Repair Regen* **22**(3), 351–355 (2014).

43. Stoekenbroek, R. M. Santema, T. B. Legemate, D. A. Ubbink, D. T. van den Brink, A. and Koelemay, M. J. Hyperbaric oxygen for the treatment of diabetic foot ulcers: A systematic review. *Eur J Vasc Endovasc Surg* **47**(6), 647–655 (2014).

44. Weir, G. R. Smart, H. van Marle, J. Cronje, F. J. and Sibbald, R. G. Arterial disease ulcers, part 2: Treatment. *Adv Skin Wound Care* **27**(10), 462–476; quiz 476–468 (2014).

45. Kalani, M. Jorneskog, G. Naderi, N. Lind, F. and Brismar, K. Hyperbaric oxygen (HBO) therapy in treatment of diabetic foot ulcers. Long-term follow-up. *J Diabetes Complications* **16**(2), 153–158 (2002).

46. Bakker, D. J. Hyperbaric oxygen therapy and the diabetic foot. *Diabetes Metab Res Rev* **16**(1), S55–S58 (2000).

47. Chen, C. E. Ko, J. Y. Fong, C. Y. and John, R. J. Treatment of diabetic foot infection with hyperbaric oxygen therapy. *Foot Ankle Surg* **16**(2), 91–95 (2010).

48. Doctor, N. Pandya, S. and Supe, A. Hyperbaric oxygen therapy in diabetic foot. *J Postgrad Med* **38**(3), 112–114 (1992).

49. Mills, B. J. Wound healing: The evidence for hyperbaric oxygen therapy. *Br J Nurs* **21**(20), 28, 30, 32, 34 (2012).

50. Kang, T. S. Gorti, G. K. Quan, S. Y. Ho, M. and Koch, R. J. Effect of hyperbaric oxygen on the growth factor profile of fibroblasts. *Arch Facial Plast Surg* **6**(1), 31–35 (2004).

51. Brismar, K. Lind, F. and Kratz, G. Dose-dependent hyperbaric oxygen stimulation of human fibroblast proliferation. *Wound Repair Regen* **5**(2), 147–150 (1997).

52. Conconi, M. T. Baiguera, S. Guidolin, D. Furlan, C. Menti, A. M. Vigolo, S. Belloni, A. S. Parnigotto, P. P. and Nussdorfer, G. G. Effects of hyperbaric oxygen on proliferative and apoptotic activities and reactive oxygen species generation in mouse fibroblast 3T3/J2 cell line. *J Investig Med* **51**(4), 227–232 (2003).

53. Bonomo, S. R. Davidson, J. D. Yu, Y. Xia, Y. Lin, X. and Mustoe, T. A. Hyperbaric oxygen as a signal transducer: Upregulation of platelet derived growth factor-beta receptor in the presence of HBO2 and PDGF. *Undersea & hyperbaric medicine* **25**(4), 211–216 (1998).

54. Zhao, L. L. Davidson, J. D. Wee, S. C. Roth, S. I. and Mustoe, T. A. Effect of hyperbaric oxygen and growth factors on rabbit ear ischemic ulcers. *Arch Surg* **129**(10), 1043–1049 (1994).

55. Hinchliffe, R. J. Valk, G. D. Apelqvist, J. Armstrong, D. G. Bakker, K. Game, F. L. Hartemann-Heurtier, A. Londahl, M. Price, P. E. van Houtum, W. H. *et al.* A systematic review of the effectiveness of interventions to enhance the healing of chronic ulcers of the foot in diabetes. *Diabetes/metabolism research and reviews* **24**(1), S119–S144 (2008).

56. Londahl, M. Hyperbaric oxygen therapy as adjunctive treatment of diabetic foot ulcers. *Med Clin North Am* 97(5), 957–980 (2013).

57. Warriner, R. A. 3rd, and Hopf, H. W. The effect of hyperbaric oxygen in the enhancement of healing in selected problem wounds. *Undersea Hyperb Med* 39(5), 923–935 (2012).

58. Blackman, E. Moore, C. Hyatt, J. Railton, R. and Frye, C. Topical wound oxygen therapy in the treatment of severe diabetic foot ulcers: A prospective controlled study. *Ostomy Wound Manage* 56(6), 24–31 (2010).

59. Herman, I. M. and Leung, A. Creation of human skin equivalents for the *in vitro* study of angiogenesis in wound healing. *Methods Mol Biol* 467, 241–248 (2009).

60. Carlson, M. W. Alt-Holland, A. Egles, C. and Garlick, J. A. Three-dimensional tissue models of normal and diseased skin. *Curr Protoc Cell Biol* **Chapter 19**, Unit 19 (2008).

61. Park, K. M. Blatchley, M. R. and Gerecht, S. The design of dextran-based hypoxia-inducible hydrogels via *in situ* oxygen-consuming reaction. *Macromol Rapid Commun* 35(22), 1968–1975 (2014).

62. Park, K. M. and Gerecht, S. Hypoxia-inducible hydrogels. *Nat Commun* 5, 4075 (2014).

63. Chen, J. S. Longaker, M. T. and Gurtner, G. C. Murine models of human wound healing. *Methods Mol Biol* 1037, 265–274 (2013).

64. Nauta, A. C. Gurtner, G. C. and Longaker, M. T. Adult stem cells in small animal wound healing models. *Methods Mol Biol* 1037, 81–98 (2013).

65. Wong, V. W. Sorkin, M. Glotzbach, J. P. Longaker, M. T. and Gurtner, G. C. Surgical approaches to create murine models of human wound healing. *J Biomed Biotechnol* 969618 (2011).

66. Heidenreich, R. Murayama, T. Silver, M. Essl, C. Asahara, T. Rocken, M. and Breier, G. Tracking adult neovascularization during ischemia and inflammation using Vegfr2-LacZ reporter mice. *J Vasc Res* 45(5), 437–444 (2008).

67. Velazquez, O. C. Angiogenesis and vasculogenesis: Inducing the growth of new blood vessels and wound healing by stimulation of bone marrow-derived progenitor cell mobilization and homing. *J Vasc Surg* 45(A), A39–A47 (2007).

68. Ahn, S. T. and Mustoe, T. A. Effects of ischemia on ulcer wound healing: A new model in the rabbit ear. *Ann Plast Surg* 24(1), 17–23 (1990).

69. Chien, S. Ischemic rabbit ear model created by minimally invasive surgery. *Wound Repair Regen* 15(6), 928–935 (2007).

70. Chien, S. and Wilhelmi, B. J. A simplified technique for producing an ischemic wound model. *J Vis Exp* 63, e3341 (2012).

71. Tyrone, J. W. Mogford, J. E. Chandler, L. A. Ma, C. Xia, Y. Pierce, G. F. and Mustoe, T. A. Collagen-embedded platelet-derived growth factor DNA plasmid promotes wound healing in a dermal ulcer model. *J Surg Res* 93(2), 230–236 (2000).

72. Wu, L. Brucker, M. Gruskin, E. Roth, S. I. and Mustoe, T. A. Differential effects of platelet-derived growth factor BB in accelerating wound healing in aged versus young animals: The impact of tissue hypoxia. *Plast Reconstr Surg* **99**(3), 815–822; discussion 823–814 (1997).

73. Gurjala, A. N. Geringer, M. R. Seth, A. K. Hong, S. J. Smeltzer, M. S. Galiano, R. D. Leung, K. P. and Mustoe, T. A. Development of a novel, highly quantitative *in vivo* model for the study of biofilm-impaired cutaneous wound healing. *Wound Repair Regen* **19**(3), 400–410 (2011).

74. Jones, M. Ganopolsky, J. G. Labbe, A. Gilardino, M. Wahl, C. Martoni, C. and Prakash, S. Novel nitric oxide producing probiotic wound healing patch: Preparation and *in vivo* analysis in a New Zealand white rabbit model of ischaemic and infected wounds. *Int Wound J* **9**(3), 330–343 (2012).

75. Pradhan Nabzdyk, L. Kuchibhotla, S. Guthrie, P. Chun, M. Auster, M. E. Nabzdyk, C. Deso, S. Andersen, N. Gnardellis, C. LoGerfo, F. W. *et al.* Expression of neuropeptides and cytokines in a rabbit model of diabetic neuroischemic wound healing. *J Vasc Surg* **58**(3), 766–775, e712 (2013).

76. Said, H. K. Hijjawi, J. Roy, N. Mogford, J. and Mustoe, T. Transdermal sustained-delivery oxygen improves epithelial healing in a rabbit ear wound model. *Arch Surg* **140**(10), 998–1004 (2005).

77. Said, H. K. Roy, N. K. Gurjala, A. N. and Mustoe, T. A. Quantifying tissue level ischemia: Hypoxia response element-luciferase transfection in a rabbit ear model. *Wound Repair Regen* **17**(4), 473–479 (2009).

78. Kloeters, O. Jia, S. X. Roy, N. Schultz, G. S. Leinfellner, G. and Mustoe, T. A. Alteration of Smad3 signaling in ischemic rabbit dermal ulcer wounds. *Wound Repair Regen* **15**(3), 341–349 (2007).

79. Xia, Y. P. Zhao, Y. Marcus, J. Jimenez, P. A. Ruben, S. M. Moore, P. A. Khan, F. and Mustoe, T. A. Effects of keratinocyte growth factor-2 (KGF-2) on wound healing in an ischaemia-impaired rabbit ear model and on scar formation. *J Pathol* **188**(4), 431–438 (1999).

80. Kamler, M. Lehr, H. A. Saetzler, R. K. Galla, T. J. and Messmer, K. Impact of ischemia on tissue oxygenation and wound healing: Improvement by vasoactive medication. *Adv Exp Med Biol* **316**, 419–424 (1992).

81. Trujillo, A. N. Kesl, S. L. Sherwood, J. Wu, M. and Gould, L. J. Demonstration of the rat ischemic skin wound model. *J Vis Exp* **98**, e52637 (2015).

82. Schwarz, D. A. Lindblad, W. J. and Rees, R. R. Altered collagen metabolism and delayed healing in a novel model of ischemic wounds. *Wound Repair Regen* **3**(2), 204–212 (1995).

83. Chen, C. Schultz, G. S. Bloch, M. Edwards, P. D. Tebes, S. and Mast, B. A. Molecular and mechanistic validation of delayed healing rat wounds as a model for human chronic wounds. *Wound Repair Regen* **7**(6), 486–494 (1999).

84. Biswas, S. Roy, S. Banerjee, J. Hussain, S. R. Khanna, S. Meenakshisundaram, G. Kuppusamy, P. Friedman, A. and Sen, C. K. Hypoxia inducible microRNA 210 attenuates keratinocyte proliferation and impairs closure in a murine

model of ischemic wounds. *Proc Natl Acad Sci USA* **107**(15), 6976–6981 (2010).

85. Bauer, S. M. Goldstein, L. J. Bauer, R. J. Chen, H. Putt, M. and Velazquez, O. C. The bone marrow-derived endothelial progenitor cell response is impaired in delayed wound healing from ischemia. *J Vasc Surg* **43**(1), 134–141 (2006).

86. Goldstein, L. J. Gallagher, K. A. Bauer, S. M. Bauer, R. J. Baireddy, V. Liu, Z. J. Buerk, D. G. Thom, S. R. and Velazquez, O. C. Endothelial progenitor cell release into circulation is triggered by hyperoxia-induced increases in bone marrow nitric oxide. *Stem Cells* **24**(10), 2309–2318 (2006).

87. Gallagher, K. A. Liu, Z. J. Xiao, M. Chen, H. Goldstein, L. J. Buerk, D. G. Nedeau, A. Thom, S. R. and Velazquez, O. C. Diabetic impairments in NO-mediated endothelial progenitor cell mobilization and homing are reversed by hyperoxia and SDF-1 alpha. *J Clin Invest* **117**(5), 1249–1259 (2007).

88. Fadini, G. P. Albiero, M. Menegazzo, L. Boscaro, E. Pagnin, E. Iori, E. Cosma, C. Lapolla, A. Pengo, V. Stendardo, M. *et al.* The redox enzyme p66Shc contributes to diabetes and ischemia-induced delay in cutaneous wound healing. *Diabetes* **59**(9), 2306–2314 (2010).

89. Kim, E. K. Li, G. Lee, T. J. and Hong, J. P. The effect of human adipose-derived stem cells on healing of ischemic wounds in a diabetic nude mouse model. *Plast Reconstr Surg* **128**(2), 387–394 (2011).

90. Alizadeh, N. Pepper, M. S. Modarressi, A. Alfo, K. Schlaudraff K, Montandon, D. Gabbiani, G. Bochaton-Piallat, M. L. and Pittet, B. Persistent ischemia impairs myofibroblast development in wound granulation tissue: A new model of delayed wound healing. *Wound Repair Regen* **15**(6), 809–816 (2007).

91. Viders, D. E. Topical hydrogen peroxide treatment of ischemic ulcers in the guinea pig. *J Am Acad Dermatol* **35**(5 Pt 1), 788 (1996).

92. Tur, E. Bolton, L. and Constantine, B. E. Topical hydrogen peroxide treatment of ischemic ulcers in the guinea pig: Blood recruitment in multiple skin sites. *J Am Acad Dermatol* **33**(2 Pt 1), 217–221 (1995).

93. Wassermann, E. van Griensven, M. Gstaltner, K. Oehlinger, W. Schrei, K. and Redl, H. A chronic pressure ulcer model in the nude mouse. *Wound Repair Regen* **17**(4), 480–484 (2009).

94. Peirce, S. M. Skalak, T. C. and Rodeheaver, G. T. Ischemia-reperfusion injury in chronic pressure ulcer formation: A skin model in the rat. *Wound Repair Regen* **8**(1), 68–76 (2000).

95. Tatlidede, S. H. Murphy, A. D. McCormack, M. C. Nguyen, J. T. Eberlin, K. R. Randolph, M. A. Moore, F. D. Jr. and Austen, W. G. Jr. Improved survival of murine island skin flaps by prevention of reperfusion injury. *Plast Reconstr Surg* **123**(5), 1431–1439 (2009).

96. Tatlidede, S. McCormack, M. C. Eberlin, K. R. Nguyen, J. T. Randolph, M. A. and Austen, W. G. Jr. A novel murine island skin flap for ischemic preconditioning. *J Surg Res* **154**(1), 112–117 (2009).

97. Steinberg, J. P. Gurjala, A. N. Jia, S. Hong, S. J. Galiano, R. D. and Mustoe, T. A. Evaluating the effects of subclinical, cyclic ischemia-reperfusion injury on wound healing using a novel device in the rabbit ear. *Ann Plast Surg* **72**(6), 698–705 (2014).

98. Ansell, D. M. Holden, K. A. and Hardman, M. J. Animal models of wound repair: Are they cutting it? *Exp Dermatol* **21**(8), 581–585 (2012).

99. Xie, Y. Zhu, K. Q. Deubner, H. Emerson, D. A. Carrougher, G. J. Gibran, N. S. and Engrav, L. H. The microvasculature in cutaneous wound healing in the female red Duroc pig is similar to that in human hypertrophic scars and different from that in the female Yorkshire pig. *J Burn Care Res* **28**(3), 500–506 (2007).

100. Roy, S. Biswas, S. Khanna, S. Gordillo, G. Bergdall, V. Green, J. Marsh, C. B. Gould, L. J. and Sen, C. K. Characterization of a preclinical model of chronic ischemic wound. *Physiol Genomics* **37**(3), 211–224 (2009).

101. Jung, Y. Son, D. Kwon, S. Kim, J. and Han, K. Experimental pig model of clinically relevant wound healing delay by intrinsic factors. *Int Wound J* **10**(3), 295–305 (2013).

5. Developmental Biology of Skin Wound Healing: On Pathways and Genes Controlling Regeneration Versus Scarring

Sarah Susan Kelangi[*,†] and Marianna Bei[*,†,‡]

*Center for Engineering in Medicine, Massachusetts
General Hospital, Department of Surgery Harvard
Medical School, Boston, Massachusetts 02114, USA
†Shriners Burns Hospital, Boston, Massachusetts, USA
‡Center for Surgery, Innovation and Bioengineering,
Massachusetts General Hospital, Boston,
Massachusetts, USA

Abstract

Cutaneous wound healing depends upon a series of events involving coordination of molecular pathways that regulate the function of specific skin cell types. Key questions in wound healing management involve the identification of the molecular mechanisms by which proteins interact to control wound repair and to contribute to scarless healing. Since there are no sources of stem cells in adult skin to be used for this reason, any attempt to create skin *de novo* or to treat wounds in a way to generate healing without scar formation will likely require the re-programming of other cell types. Thus, the fundamental understanding of the control mechanisms responsible for scarless skin repair is imperative.

I. Introduction

The ultimate goal of research on wound healing is to achieve complete regeneration instead of imperfect wound repair. In nature, some multi-cellular organisms can regenerate the missing part of the body and repair their wounds while others cannot. For example, in earlier branching animals like *Sponges* regeneration is fully achieved and signaling pathways such as Wnt, TGF-β, Notch and Hedgehog play a significant role in this process.[1] Lower vertebrates including fish (*zebra fish*) and amphibians (*axolots* and *xenopus*) also possess the ability to perfectly regenerate their organs. After full thickness excisional wounds in *xenopus* froglets and *axolots* the entire skin including appendages regenerate.[2,3] Even pigmentation pattern of skin can be fully re-established. Zebra fish can recover its striped pigmentation pattern after wounding, and regenerate subcutaneous adipocytes and scales, thus making the regenerated skin almost indistinguishable from the original.[3,4] In contrast, in adult mammals including humans it is difficult to achieve organ regeneration except for liver, bone, and infants' fingertips.[5–7] Wound healing in adult mammals result in scar tissue that prevents the regeneration and function of the affected by the wound organ.

Skin, is the largest organ of our body whose morphogenesis and wound healing after injury are development-or age-dependent. The purpose of this chapter is to discuss the role of genes and pathways that control skin wound healing and to provide some perspectives on the molecular mechanisms that coordinate the process of skin wound healing towards regeneration.

1.1 *Stages of cutaneous wound healing*

Mammalian epidermis is a stratified squamous epithelium whose maintenance relies on proliferation and differentiation of the basal layer of the epidermis.[8] As basal epidermal cells differentiate and move towards the surface, they give rise to suprabasal cells and the granular layer that will differentiate eventually into enucleated corneocytes composing the stratum corneum.[8] Following injury, rapid closure of wound site takes place by migration and proliferation of epithelial cells. Several progenitor populations located in hair follicle bulge region, in hair follicle junctional zone next to the sebaceous gland and in the infundibulum have shown to contribute to regenerating the skin.[9,10]

The process of skin wound healing in humans can be divided into four cellular phases: (i) Hemostasis, (ii) Inflammation, (iii) Proliferation and, (iv) Remodeling.[11,12] The first two phases of hemostasis and inflammation are characterized by fibrin clot formation followed by neutrophil recruitment.[13] Neutrophils arrive within the first 1–2 days of injury.[13,14] Their function is to prevent bacterial infection and to activate keratinocytes, fibroblasts, and other immune cells.[13,14] Circulating monocytes arrive within 2–3 days after injury and differentiate into macrophages.[11,15] Macrophages along with neutrophils function to prevent infection and wound debridement of necrotic tissue.[11,15] Tissue debris and invading microorganisms are cleared in this phase.[11,12] During the third phase known as proliferation, granulation tissue is formed replacing the fibrin clot.[11,12] Macrophages in this stage act as anti-inflammatory agents contributing to pro-fibrotic phenotype of the wound.[13,15] Growth factors like TGF–β and VEGF recruit fibroblasts and stimulate migration of endothelial cells into the wound bed resulting in the formation of blood vessels. Re-epithelialization is achieved by fibroblasts and myofibroblasts that are responsible for *extracellular matrix (ECM)* synthesis that is mainly composed by immature Type III collagen different from the ECM of the unwounded skin that is composed by collagen Type I.[13] The re-organization of the granulation tissue matrix results in the reduction of the wound size, known as contraction of the wound.[16] In addition, hair follicle stem cells are activated and participate in re-epithelialization along with an epithelial to mesenchymal transition of the epithelial cells.[17,18] It is obvious from the above that the proliferative phase is characterized by a series of interactions between different types of cells and their surrounding ECM in a spatially and temporally controlled manner.[17,18] The fourth phase of remodeling is long, begins in the third week after injury and lasts for months even years, depending on the size, location and type of wound.[17,18] Apoptosis of myofibroblasts and overall decrease in fibroblast collagen content occurs. Remodeling occurs by transitioning from Type III to Type I collagen and its cross-linking.[17,18] Matrix metalloproteinases (MMPs) and their inhibitors work to re-organize the collagen into a stronger network.

Thus, during these four phases, complex biological processes such as migration, proliferation, apoptosis, differentiation, ECM deposition and angiogenesis occur and involve multiple, highly regulated gene pathways.[11,12]

Interestingly, several events that occur in a coordinated manner in scar forming, healing skin wounds do not occur in non-healing chronic skin wounds. For instance in chronic wounds, inflammation appears to be prolonged for an extended time period leading to the production of pro- and anti-inflammatory cytokines resulting in impaired angiogenic response, decreased granulation tissue formation, and reduced keratinocyte and fibroblast migration, and proliferation.[19,20] In addition, in contrast to adult, scar forming and late fetal skin wound healing, early fetal skin wound repair results in regeneration without scar formation. There are several differences between early fetal and later stages of skin repair. Early fetal cutaneous wound healing is characterized by rapid re-epithelialization in which actin filaments are formed at the edge of the wound providing coordinated cell movement to close the defect.[21] In addition, since blood vessels are yet to form, there is no bleeding or fibrin clot formation to initiate inflammation, instead wound healing is further accomplished by interactions between two cell layers, the primordial epidermis and the underlying mesenchymal tissue.[22] Several studies demonstrated that the transition from scar less to scar forming repair coincides with the development and maturation of blood vessels, inflammatory system, and skin appendages such as the hair follicles.[23] Moreover, it was shown that fetal wounds contain Type III collagen not Type I, their ECM is rich in hyaluronic acid and glycosaminoglycans especially at the wound site into which rapidly proliferating mesenchymal cells migrate, differentiate, and mature.

Animal and human studies that employ the tools of contemporary molecular genetics have identified a number of genes that act at specific stages of skin wound repair and regulate its process leading either to scar less or scar forming wound (Figure 1, Table 1). Below we discuss some recent findings regarding genes and pathways that control, both, scar forming and scar-less cutaneous wound healing and specifically the role of some major growth factors and the transcription factors that mediate their signaling.

2. Signaling Factors and their Downstream Cascades Control Cutaneous Wound Healing

The Wnt pathway: WNTs are cysteine rich glycoproteins. They function as morphogens that are secreted from one cell to activate surface receptors in neighboring cells, thus regulating cell proliferation, survival, and differentiation and they initiate three different intracellular signaling cascades, (a) the canonical/β-catenin dependent Wnt pathway, (b) the planar cell polarity (PCP), and (c) the Wnt/Ca2+ pathway.[24] The key player of canonical Wnt pathway is β-catenin, which is stabilized in the cytoplasm in response to ligand-receptor binding and then translocated into nucleus where it binds to TCF-LEF transcription factors to regulate gene expression. In contrast, activation of non-canonical Wnt pathway leads to increase in intracellular calcium followed by stimulating the activities of protein kinase C (PKC) and

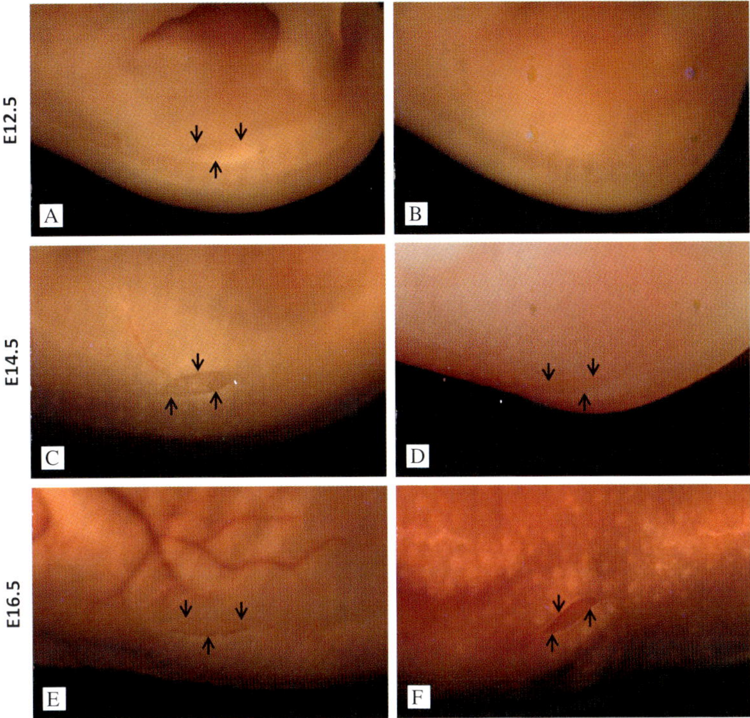

Figure 1. Wound healing after skin incision in mouse embryos. Black arrows depict the wound edges. (A, C and E) Immediately after skin incision were made at E12.5, E14.5, and E16.5 embryos respectively. (B) 12 h after wounding and subsequent culture at E12.5 embryos. The wound was completely closed. (D and F) E14.5 and E16.5 embryos were wounded and cultured for 24 h, respectively. The wounds were still open. Magnification in all pictures is 4-fold. Modified from Ref. 72.

phospholipase C (PLC) or activation of jun N-terminal kinase (JNK) that regulate PCP.

Cutaneous wounds express various WNTs during the early phases of healing. WNT 1, 3, 4, 5a and 10b are all present in murine cutaneous wounds up to 7 days after wounding.[25] In a study where a reporter BAT-gal mouse model was used to determine the activity of canonical WNT signaling in wound healing, it was shown that the activity of WNT signaling was increased in post-natal wounds but not in early fetal wounds.[26] The differential activity of WNT signaling in early fetal wounds versus post-natal wounds was further demonstrated by *in vitro* studies where WNTs had a greater effect on fibroblasts from post-natal mouse skin compared to fetal fibroblasts in promoting cell proliferation, hyaluronic acid, type I collagen

Table 1. Mouse models with kin wound healing abnormalities.

Gene	Mutation	Wound Phenotype	Reference
TiEG-1	Null	Impaired wound contraction	55
Klf4	Conditional KO	Delayed wound healing	56, 57
Nf-1C	Null	Wound closure accelerated	58
En-1	Null	Decrease in connective tissue deposition during healing	59
Foxo3	Null	Accelerated wound healing	60, 61
c-Myc	Conditional KO	Spontaneous wounds Delayed re-epithelialization	67
PU-1	Null	Scarless skin wound healing	64, 65, 66
HIF - 1α	Conditional KO	Impaired wound healing	70
GSK3b	Conditional KO	Increased scarring and myofibroblast formation	29
Notch-1	Hemizygous	Increased collagen deposition	39
Smad-3	Null	Small wound area	48
TGF-βRII	Conditional KO	Defects in dermal ECM increased scar tissue formation	50

synthesis and differential TGF-β isoform expression.[26] The role WNT signaling in wound healing was further highlighted by several studies showing the importance of β-catenin levels and activity during healing process and in particular during the proliferative phase of wound healing.[27] High levels of β-catenin lead to an enlarged, hypercellular dermal compartment, while low levels of β-catenin are associated with a smaller and less cellular dermal compartment.[28] Mice with a fibroblast-specific conditional deletion of GSK3b that result in elevated β-catenin levels had increased scarring and myofibroblast formation after wounding.[29] Interestingly, mice treated with adenovirus expressing the WNT signaling inhibitor Dick-kopf (DDK1, which binds LRP6/Arrow) did not show significant decrease in β-catenin protein levels implying that other factors play a role in regulating β-catenin levels, as well.[30,31] Indeed studies have shown that β-catenin levels may be stimulated by growth factors such as TGF-β1[28,32,33] and ECM components such as fibronectin.[30]

The Notch Pathway: the Notch signaling pathway is a well-known regulator of epidermal differentiation and maintenance of skin homeostasis during adult and embryonic development.[34–36] In mammals there are four Notch receptors and their ligands are members of Delta like and Jagged families. Ligand binding causes sequential proteolytic cleavage of Notch intracellular domain (NCID) from the cell membrane and its translocation to the nucleus where it associates with RBP-J and other cofactors to modulate the transcription of target genes such as Hes/Hey transcription repressors.[35–37] The role of Notch signaling in wound healing was recently explored by studies where the expression of Jagged 1 and Notch1 as well as that of signaling modulator, Lunatic Fringe reported in healing epidermis of wounded human skin grafted onto mice.[38] In addition, transgenic mice expressing Notch antisense sequence that causes a 50% reduction in Notch protein, led to delayed skin wound healing, while Notch 1 hemizygous mice exhibit increased collagen deposition and vascularity in healing wounds.[39] Furthermore, mice treated with the Notch ligand, Jagged, show accelerated wound closure and *in vitro* scratch wound assays with cultured cells, using a Notch inhibitor or activator, implied that Notch pathway mediates wound healing delay or acceleration respectively, through specific effects on fibroblast and vascular endothelial cells.[39]

The Hedgehog pathway: the role of SHH and its modulators in the development of hair follicle, sebaceous gland, skin development and epidermal stem cells is well documented.[40–45] Sonic Hedgehog signaling (Shh) pathways also have the ability to modulate several aspects of wound healing, dermal repair and wound vascularization.[40–45] Specifically, topical application of a human Shh-expressing plasmid to mouse wounds stimulated Shh expression in keratinocyte like and fibroblast like cells.[46] Diabetic mice treated with Shh demonstrated increased re-epithelialization and dermal healing including a larger and more collagen rich dermal compartment, compared to control mice.[46] The beneficial effects observed in Shh-treated wounds may be caused either by enhanced fibroblast activity, since Shh is known to stimulate proliferation of primary dermal fibroblasts *in vitro* or by its effect in recruiting bone marrow derived endothelial progenitor cells to wound vasculature, thus increasing vascularity.[46]

The TGF-βs: During wound healing TGF-βs are secreted by macrophages and fibroblasts and disrupted ECM. TGF-β1 is known as scar promoting factor, while TGF-β3 is associated with anti-scarring actions. Treatment of rat skin wounds with topical TGF-β1 improved dermal healing, angiogenesis, inflammatory cell infiltration, and epithelial closure of the wounds, while mouse models deficient in TGF-β pathway signaling exhibit defects in dermal healing.[47,48] Specifically, the dermis of Smad-3 knock out mice displays fewer fibroblasts, reduced ECM deposition, and smaller wound area compared to control mice, while inactivation of fibroblast-specific TGF-βRII causes defects in dermal ECM leading to increased scar tissue formation and diminished collagen deposition.[49,50] These results clearly demonstrate that TGF-β signaling is crucial in mediating cutaneous wound repair. In addition, the fact that topical application of TGF-β1 promotes epithelial wound closure on rats further highlights the important role of TGF-β1 as a positive regulator of adult wound healing.[47] Interestingly, one of the differences between adult and scar less fetal wound healing is the low TGF-β1 expression in early fetal wounds. Scarless embryonic wounds in rats have lower TGF-β1 and higher TGF-β3 levels compared to the scar-forming wounds that occur during late gestation or during adulthood.[51] In fact, TGF-β3 is known to reduce scarring in a rodent wound model,[52] while pharmaceutical application of TGF-β3 is improving the appearance of scars in phase II clinical trials, but not in phase III[53] and http://www.pharmatimes.com/ Article/11-02-15/Renovo_stock_demolished_by_Justiva_trial_failure.aspx. TGF-β3 is not the only growth factor used in human clinical trials showing some promise in the field of cutaneous wound healing. PDGF and rPDGF have shown to accelerate wound healing by stimulating the inflammatory response and for this reason PDGF approved by U.S. Food and Drug Administration to promote wound closure through topical application.[54] EGF, another growth factor also applied topically in chronic wounds resulted in complete healing in 8/9 wounds at a mean 34 days, thus shortened healing time.

It is obvious from the above that growth factors and their mediators act in an equilibrium and play a very significant role in skin wound healing process. For example, while PDGF was found in fetal, neonatal, and adult wounds, TGF and basic FGF were not detected in fetal wounds suggesting that it may be possible to manipulate adult wound to produce more fetal like, scar less wound healing by altering the level of growth factors involved.

3. Transcription Factors and their Role in Cutaneous Wound Healing

Interestingly, the transcription factors that mediate such signaling networks are also important for wound-healing management, although their role in wound healing is not yet studied extensively (Table 1). Below we briefly describe some interesting findings based on animal and human studies.

Kruppel-like family (kf) members Klf4 and Klf10 are transcription factors that are strongly induced in response to TGF-β/inducible early gene1 (Tieg-1) and are encoded by a member of the klf of transcription factors, the

Klf10.[56] Upon wound induction, mice lacking TIEG-1 showed impaired wound contraction, granulation tissue formation and re-epithelialization.[55] Another Klf family member has recently been shown to regulate wound inflammation. Mice lacking Klf4 in myeloid cells exhibited, delayed wound healing and decrease in the number of bulge stem cells, which normally contribute to wound healing.[56,57]

NF1-C is a member of the NF-1 transcription factor family and is predominantly expressed in dermal fibroblasts and represses gene expression in response to TGF-β.[58] When NF1-C knockout mice were subjected to full thickness excisional wounding, wound closure was strongly accelerated due to enhanced myofibroblast differentiation that occurred upon activation of TGF-β signaling.[58]

Other examples of transcription factors that control wound healing are the En1, FOXO3, c-Myc, HIF.1, and PU.1. En1 is a homeobox transcription factor expressed by cells of central dermomyotome and their progeny is known to contribute to dorsal dermis and in expressing extracellular collagens during normal skin development. When En1 is genetically removed there was significant decrease in connective tissue deposition during wound healing.[59] Deletion of FOXO3, on the other hand, led to accelerated wound healing in mice and reduced bacterial infections in the wound beds of these mice.[60,61] In support of the latter TNF-α signaling that is associated with decreased wound healing in diabetic mice is also accompanied by high nuclear levels of FOXO1 in fibroblasts.[62] PU.1 is a member of Ets family of transcription factors that controls the development/differentiation of inflammatory cells.[63] PU.1 knock out mice are completely deficient in macrophages and functional neutrophils[64] and when a small incision was induced in the skin of these mice complete scarless wound closure was observed.[65,66] c-Myc is a basic helix-loop-helix factor that promotes cell proliferation and programmed cell death.[67] Mice lacking c-Myc in keratinocytes had a fragile epidermis and developed spontaneous skin wounds.[67] Upon full thickness wound induction the mice showed delayed re-epithelialization as a result of impaired keratinocyte proliferation.[68] In early wounds there is hypoxia, due to vascular disruption and high oxygen consumption of the rapidly dividing cells. Hypoxia causes activation of HIF 1α, a basic helix-loop-helix transcription factor, which is expressed in different cell types of wound healing. This has been demonstrated using mice with a cell type specific loss of HIF-1α.[69] Loss of HIF-1α in keratinocytes caused severe defect in re-epithelialization in aged mice due to impaired keratinocyte migration.[70] Loss of either HIF-1α or HIF-1β in Tie 2-positive cells caused delayed closure of burn wounds that was associated with reduced vascularization.[71]

4. Epilogue

Wound healing is a complex process controlled by a sequence of cellular and molecular networks that act at particular places and times to guide skin cells to restricted fates. The mechanisms that control and determine the history of the cells so that the program is properly executed during healing towards regeneration instead of scar formation are not known. Despite the fact that recent studies led to the discovery of numerous genes and pathways involved in different stages of wound healing, the same genes are required for other processes, thus, further studies are required at the genetic and epigenetic level in order to understand how to manipulate cells towards skin regeneration.

5. Source of Funding

This project was supported by U.S. Army Medical Research and Material Command (USAMRMC; http://mrmc.amedd.army.mil/) award W81XWH-10-2-0058 and NIH RO1 DE 019226 to MB.

6. Conflict of Interest Disclosure Statement

The authors have declared that no competing interests exist.

References

1. Nichols, S. A. *et al.*, Early evolution of animal cell signaling and adhesion genes. *Proc Natl Acad Sci USA* **103**(33), 12451–12456 (2006).
2. Yokoyama, H. *et al.*, Prx-1 expression in Xenopus laevis scarless skin-wound healing and its resemblance to epimorphic regeneration. *J Invest Dermato* **131**(12), 2477–2485 (2011).
3. Seifert, A. W. *et al.*, Skin regeneration in adult axolotls: A blueprint for scar-free healing in vertebrates. *PloS one* **7**(4), e32875 (2012).
4. Richardson, R. *et al.*, Adult zebrafish as a model system for cutaneous wound-healing research. *J Invest Dermatol* **133**(6), 1655–1665 (2013).
5. Alman, B. A. Kelley, S. P. and Nam, D. Heal thyself: Using endogenous regeneration to repair bone. *Tissue Eng B: Rev* **17**(6), 431–436 (2011).
6. Carlson, B. M. Some principles of regeneration in mammalian systems. *Anat Rec B: New Anat* **287**(1), 4–13 (2005).
7. Han, M. *et al.*, Limb regeneration in higher vertebrates: Developing a roadmap. *Anat Rec B: New Anat* **287**(1), 14–24 (2005).
8. Mascré, G. *et al.*, Distinct contribution of stem and progenitor cells to epidermal maintenance. *Nature* **489**(7415), 257–262 (2012).

9. Taylor, G. *et al.*, Involvement of follicular stem cells in forming not only the follicle but also the epidermis. *Cell* **102**(4), 451–461 (2000).

10. Tumbar, T. *et al.*, Defining the epithelial stem cell niche in skin. *Science* **303**(5656), 359–363 (2004).

11. Gurtner, G. C., *et al.*, Wound repair and regeneration. *Nature* **453**(7193), 314–321 (2008).

12. Singer, A.J. and Clark, R.A.F., Cutaneous Wound Healing. *N Engl J Med* **341**(10), 738–746 (1999).

13. Baum, C. L. and Arpey, C. J. Normal cutaneous wound healing: Clinical correlation with cellular and molecular events. *Dermatologic Surg* **31**(6), 674–686 (2005).

14. Wilgus, T. A. Roy, S. and McDaniel, J. C. Neutrophils and wound repair: Positive actions and negative reactions. *Adv Wound Care* **2**(7), 379–388 (2013).

15. Sindrilaru, A. and Scharffetter-Kochanek, K. Disclosure of the culprits: Macrophages—Versatile regulators of wound healing. *Adv Wound Care* **2**(7), 357–368 (2013).

16. Ito, M. *et al.*, Stem cells in the hair follicle bulge contribute to wound repair but not to homeostasis of the epidermis. *Nat Med* **11**(12), 1351–1354 (2005).

17. Kalluri, R. and Weinberg, R. A. The basics of epithelial-mesenchymal transition. *J Clin Investig* **119**(6), 1420–1428 (2009).

18. Kong, W. *et al.*, Epithelial–mesenchymal transition occurs after epidermal development in mouse skin. *Exp Cell Res* **312**(19), 3959–3968 (2006).

19. Goova, M. T. *et al.*, Blockade of receptor for advanced glycation end-products restores effective wound healing in diabetic mice. *Am J Pathol* **159**(2), 513–525 (2001).

20. Blakytny, R. and Jude, E. The molecular biology of chronic wounds and delayed healing in diabetes. *Diabetic Med* **23**(6), 594–608 (2006).

21. McCluskey, J. and Martin, P. Analysis of the Tissue Movements of Embryonic Wound Healing—DiI Studies in the Limb Bud Stage Mouse Embryo. *Dev Biol* **170**(1), 102–114 (1995).

22. Gallicano, G. I. Bauer, C. and Fuchs, E. Rescuing desmoplakin function in extra-embryonic ectoderm reveals the importance of this protein in embryonic heart, neuroepithelium, skin and vasculature. *Development* **128**(6), 929–941 (2001).

23. Bucala, R., *et al.*, Circulating fibrocytes define a new leukocyte subpopulation that mediates tissue repair. *Molecular Med* **1**(1), 71–81 (1994).

24. Gordon, M. D. and Nusse, R. Wnt signaling: Multiple pathways, multiple receptors, and multiple transcription factors. *J Biol Chem* **281**(32), 22429–22433 (2006).

25. Okuse, T. *et al.*, Differential expression and localization of WNTs in an animal model of skin wound healing. *Wound Repair Regen* **13**(5), 491–497 (2005).

26. Carre, A. L. *et al.*, Interaction of Wingless Protein (Wnt), Transforming Growth Factor-β1, and Hyaluronan Production in Fetal and Postnatal Fibroblasts. *Plast Reconstr Surg* **125**(1), 74–88 (2010).

27. Amini-Nik, S., *et al.*, Pax7 Expressing cells contribute to dermal wound repair, regulating scar size through a β-catenin mediated process. *Stem Cells* **29**(9), 1371–1379 (2011).

28. Cheon, S. S. *et al.*, Beta-catenin regulates wound size and mediates the effect of TGF-beta in cutaneous healing. *Fed Am Soc Exp Biol J* **20**(6), 692–701 (2006).

29. Kapoor, M. *et al.*, GSK-3β in mouse fibroblasts controls wound healing and fibrosis through an endothelin-1–dependent mechanism. *J Clin Investig* **118**(10), 3279–3290 (2008).

30. Bielefeld, K. A. *et al.*, Fibronectin and β-catenin act in a regulatory loop in dermal fibroblasts to modulate cutaneous healing. *J Biol Chem* **286**(31), 27687–27697 (2011).

31. Bafico, A. *et al.*, Novel mechanism of Wnt signalling inhibition mediated by Dickkopf-1 interaction with LRP6/Arrow. *Nat Cell Biol* **3**(7), 683–686 (2001).

32. Cheon, S. S. *et al.*, Growth factors regulate β-catenin-mediated TCF-dependent transcriptional activation in fibroblasts during the proliferative phase of wound healing. *Exp Cell Res* **293**(2), 267–274 (2004).

33. Amini Nik, S. *et al.*, TGF-β modulates β-Catenin stability and signaling in mesenchymal proliferations. *Exp Cell Res* **313**(13), 2887–2895 (2007).

34. Blanpain, C. and Fuchs, E. Epidermal homeostasis: A balancing act of stem cells in the skin. *Nat Rev Mol Cell Biol* **10**(3), 207–217 (2009).

35. Okuyama, R., Tagami, H. and Aiba, S. Notch signaling: Its role in epidermal homeostasis and in the pathogenesis of skin diseases. *J Dermatol Sci* **49**(3), 187–194 (2008).

36. Watt, F. M., Estrach, S. and Ambler, C. A. Epidermal Notch signalling: Differentiation, cancer and adhesion. *Curr Opin Cell Biol* **20**(2), 171–179 (2008).

37. Gridley, T., Notch signaling in the vasculature. *Cur Top Dev Biol* **92**, 277–309 (2010).

38. Thélu, J., Rossio, P. and Favier, B. Notch signalling is linked to epidermal cell differentiation level in basal cell carcinoma, psoriasis and wound healing. *BMC Dermatol* **2**, 7–7 (2002).

39. Chigurupati, S. *et al.*, Involvement of Notch Signaling in Wound Healing. *PLoS One* **2**(11), e1167 (2007).

40. Athar, M. *et al.*, Hedgehog signalling in skin development and cancer. *Exp Dermatol* **15**(9), 667–677 (2006).

41. St-Jacques, B. *et al.*, Sonic hedgehog signaling is essential for hair development. *Curr Biol* **8**(19), 1058–1069 (1998).

42. Karlsson, L., Bondjers, C. and Betsholtz, C. Roles for PDGF-A and sonic hedgehog in development of mesenchymal components of the hair follicle. *Development* **126**(12), 2611–2621 (1999).

43. Niemann, C. *et al.*, Indian hedgehog and *β*-catenin signaling: Role in the seba-ceous lineage of normal and neoplastic mammalian epidermis. *Proc Nat Acad Sci USA* **100**(S1), 11873–11880 (2003).

44. Brownell, I., *et al.*, Nerve-derived Sonic hedgehog defines a niche for hair fol-licle stem cells capable of becoming epidermal stem cells. *Cell Stem Cell* **8**(5), 552–565 (2011).

45. Chiang, C. *et al.*, Essential role for Sonic hedgehog during hair follicle morpho-genesis. *Dev Biol* **205**(1), 1–9 (1999).

46. Asai, J. *et al.*, Topical sonic hedgehog gene therapy accelerates wound healing in diabetes by enhancing endothelial progenitor cell–mediated microvascular remodeling. *Circulation* **113**(20), 2413–2424 (2006).

47. Puolakkainen, P. A. *et al.*, Acceleration of wound healing in aged rats by topi-cal application of transforming growth factor-*β*1. *Wound Repair Regen* **3**(3), 330–339 (1995).

48. Ashcroft, G. S. *et al.*, Mice lacking Smad3 show accelerated wound healing and an impaired local inflammatory response. *Nat Cell Biol* **1**(5), 260–266 (1999).

49. Martinez-Ferrer, M., *et al.*, Dermal transforming growth factor-*β* responsive-ness mediates wound contraction and epithelial closure. *Am J Pathol* **176**(1), 98–107 (2010).

50. Denton, C. P. *et al.*, Inducible lineage-specific deletion of T[beta]RII in fibro-blasts defines a pivotal regulatory role during adult skin wound healing. *J Invest Dermatol* **129**(1), 194–204 (2008).

51. Soo, C. *et al.*, Ontogenetic transition in fetal wound transforming growth factor-*β* regulation correlates with collagen organization. *Am J Pathol* **163**(6), 2459–2476 (2003).

52. Shah, M. Foreman, D. M. and Ferguson, M. W. Neutralisation of TGF-beta 1 and TGF-beta 2 or exogenous addition of TGF-beta 3 to cutaneous rat wounds reduces scarring. *J Cell Sci* **108**(3), 985–1002 (1995).

53. McCollum, P. T. *et al.*, Randomized phase II clinical trial of avotermin versus placebo for scar improvement. *Br J Surg* **98**(7), 925–934 (2011).

54. Park, S. A. *et al.*, PDGF-BB does not accelerate healing in diabetic mice with splinted skin wounds. *PLoS One* **9**(8), e104447 (2014).

55. Hori, K. *et al.*, Impaired cutaneous wound healing in transforming growth factor-*β* inducible early gene1 knockout mice. *Wound Repair Regen* **20**(2), 166–177 (2012).

56. Liao, X. *et al.*, Krüppel-like factor 4 regulates macrophage polarization. *J Clin Investig* **121**(7), 2736–2749 (2011).

57. Li, J. *et al.*, Expression of Kruppel-like Factor KLF4 in mouse hair follicle stem cells contributes to cutaneous wound healing. *PLoS One* **7**(6), e39663 (2012).

58. Plasari, G. *et al.*, Nuclear factor I-C links platelet-derived growth factor and transforming growth factor *β*1 signaling to skin wound healing progression. *Mol Cell Biol* **29**(22), 6006–6017 (2009).

59. Sennett, R. and Rendl, M. A scar is born: Origins of fibrotic skin tissue. *Science* 348(6232), 284–285 (2015).

60. Lees, S. J. Childs, T. E. and Booth, F. W. Age-dependent FOXO regulation of p27(Kip1) expression via a conserved binding motif in rat muscle precursor cells. *Am J Physiol Cell Physiol* 295(5), C1238–C1246 (2008).

61. Furuyama, T. *et al.*, Effects of aging and caloric restriction on the gene expression of Foxo1, 3, and 4 (FKHR, FKHRL1, and AFX) in the rat skeletal muscles. *Microsc Res Tech* 59(4), 331–334 (2002).

62. Siqueira, M. F. *et al.*, Impaired wound healing in mouse models of diabetes is mediated by TNF-α dysregulation and associated with enhanced activation of forkhead box O1 (FOXO1). *Diabetologia* 53(2), 378–388 (2010).

63. Martin, P. and Leibovich, S. J. Inflammatory cells during wound repair: The good, the bad and the ugly. *Trends Cell Biol* 15(11), 599–607 (2005).

64. McKercher, S. R. *et al.*, Targeted disruption of the PU.1 gene results in multiple hematopoietic abnormalities. *Eur Mol Biol Organ J* 15(20), 5647–5658 (1996).

65. Leibovich, S. J. and Ross, R. The role of the macrophage in wound repair. A study with hydrocortisone and antimacrophage serum. *Am J Pathol* 78(1), 71–100 (1975).

66. Martin, P. *et al.*, Wound healing in the PU.1 null mouse — tissue repair is not dependent on inflammatory cells. *Cur Biol* 13(13), 1122–1128 (2003).

67. Zanet, J., *et al.*, Endogenous Myc controls mammalian epidermal cell size, hyperproliferation, endoreplication and stem cell amplification. *J Cell Sci* 118(8), 1693–1704 (2005).

68. Stojadinovic, O., *et al.*, Molecular pathogenesis of chronic wounds: The role of β-catenin and c-myc in the inhibition of epithelialization and wound healing. *Am J Pathol* 167(1), 59–69 (2005).

69. Tandara, A. A. and Mustoe, T. A. Oxygen in wound healing — more than a nutrient. *World J Surg* 28(3), 294–300 (2004).

70. Rezvani, H. R., *et al.*, Loss of epidermal hypoxia-inducible factor-1α accelerates epidermal aging and affects re-epithelialization in human and mouse. *J Cell Sci* 124(24), 4172–4183 (2011).

71. Sarkar, K., *et al.*, Tie2-dependent knockout of HIF-1 impairs burn wound vascularization and homing of bone marrow-derived angiogenic cells. *Cardiovasc Res* 93(1), 162–169 (2012).

72. Zhang, W. and Bei, M. 2015. Kcnh2 and Kcnj8 interactively regulate skin wound healing and regeneration. *Wound Repair Regen* 23(6), 797–806.

6. Nutrition, Metabolism, and Wound Healing Process

Yong-Ming Yu* and Alan J. Fischman[†]

*Burn Research Center,
Department of Surgery,
Massachusetts General Hospital,
Shriners Hospital for Children,
Harvard Medical School,
Boston Massachusetts 02114, USA
[†]Shriners Hospital for Children,
Boston Massachusetts 02114, USA

Abstract

A wound, by true definition, is a breakdown in the protective function of the skin; the loss of continuity of epithelium, with or without loss of underlying connective tissue (i.e. muscle, bone, nerves)[1] following injury to the skin or underlying tissues/organs caused by surgery, a blow, a cut, chemicals, heat/cold, friction/shear force, pressure or as a result of disease, such as leg ulcers or carcinomas.[2]

1. A Brief Review of the Physiology of Wound Healing Process

Wounds heal by primary intention or secondary intention depending upon whether the wound may be closed with sutures or left to repair, whereby damaged tissue is restored by the formation of connective tissue and re-growth of epithelium.[3,4] The wound healing process can be divided into five phases: hemostasis, inflammation, proliferation, contraction, and remodeling.[5–8] These phases could be overlapped in time, physiology, and cell type involved. The healing process is not uniform for all wounds depending on the etiology of the wound, presence of infection, and medical/surgical

141

interventions. As a consequence, the wounds may be closed immediately, or go through a granulation stage or even more delay in the primary closure. Medical and surgical interventions may also affect the time course of pathological changes in wounds.

1.1 *Hemostasis*

The responses to the insult of skin injury involved complicated mechanisms. The first natural reaction is vasoconstriction to immediately reduce blood loss. Hemorrhage after injury immediately stimulates the Hageman Factor (XII) to initiate the blood clotting cascade. Collagen is the major protein involved in wound healing. Its presence in the wound area stimulates the alternative complement pathway, platelet aggregation, and degranulation.[9,10] Multiple additional vasoactive and chemotactic components are released from and fibrinogen in the circulation is quickly coverts to fibrin, which, along with platelets, to form a scab[11,12] which provides a temporary protective barrier. The cut epidermal-edge cells migrates forward at the interface between scab and healthy granulation tissue through a complicated signaling axis which ultimately leads to wound healing.[13]

1.2 *Inflammation*

The inflammatory phase is also a natural response of the host to injury. Once hemostasis has been achieved, blood vessels dilate to allow essential cells, antibodies, white blood cells, growth factors, enzymes, and nutrients to reach the wounded area.[14,15] This causes increased exudates so the surrounding skin needs to be monitored for signs of maceration.[2] It is at this stage that the characteristic signs of topic inflammation appear: erythema, heat, oedema, pain, and functional disturbance. The predominant cells at work here are the phagocytic cells; 'neutrophils and macrophages'; mounting a host response and autolysing any devitalized tissue. Vasodilation is related to the release of prostaglandins, nitric oxide, and other inflammatory mediators. The release of bradykinin, histamines, and free radicals from leukocytes produces increased vascular permeability. As a consequence, plasma leaks into the interstitial space accompanied with increased margination of leukocytes (PMNs) and diapedesis.[15] At this stage, glucose and oxygen are crucial to the inflammatory process. More recently, it has been determined that even influx of a single amino acid such as arginine into the wound may serve as a precursor for important mediators such as nitric oxide.[16] Lipid metabolism in the wound is related to the formation of isoprostanes which serves as a biomarker of free radical formation from lipid metabolism. Peroxidation in wound area delays the healing process.[17,18] These inflammatory mediators may stimulate a cascade of events in the wound.[19] The influx of PMN is followed by that of monocytes. These cells transition to phagocytic macrophages that remove debris and bacteria from the wound. The macrophages secrete proteases, producing interferon, prostaglandins and cytokines. These cytokines, among other things, are chemoattractants for mesenchynal cells which differentiate into fibroblasts. The latter are the major players in the proliferative process of connective tissue formation.[20]

1.3 *Proliferation*

The proliferation stage is for "rebuilding" the injured area with the new granulation tissue which is comprised of collagen and extracellular matrix (ECM), into which a new network of blood vessels develop, known as "angiogenesis". The formation of healthy granulation tissue is dependent upon the fibroblast receiving sufficient levels of oxygen and nutrients supplied by the blood vessels. Healthy granulation tissue appears granular and uneven in texture; it is pink/red in color and does not bleed easily. The color and appearance of the granulation tissue often reflect how the wound is healing. Dark granulation tissue can be indicative of poor perfusion, ischaemia, and/or infection. Epithelial cells finally resurface the wound, a process known as "epithelialization". The secretion of proliferative cytokines (PDGF, interleukin-1 [IL-1], fibroblast growth factor [FGF]), and chemoattractants (transforming growth factor-[TGF-β], PDGF) all play important roles in this phase of wound healing.[21,22] This transition to the proliferative stage is not discrete. Macrophages may remain active at altered levels throughout the healing process, secreting collagenase and elastase into the wound to aid the influx of proliferative cells and remodeling of the wound. If the wound becomes infected, PMNs will return to control bacterial proliferation, and macrophage numbers will increase. This will stimulate a recurrent cascade of inflammatory process and the wound remains in even more inflammatory state to delay the wound healing process. Topical treatment by debriding the necrotic tissues in the wound and systemic treatment to overcome infection may help shorten the healing process by ameliorating the inflammation status of the wound. Obviously aggressive systemic support to improve tissue perfusion and facilitate the delivery of adequate nutrients to the wound area is necessary to meet the demand of energy and protein metabolism in promoting protein synthesis for the wound repair and healing process.

Collagen is one of the most complex proteins in the body, consisting of a triple-alpha helix of approximately 1000 amino acids per strand depending on the type.[10,23] Type I collagen dominates in the mature wound and accounts for 80% to 90% with the rest being primarily Type III.[24,25] Early in the healing process, the concentration of Type III collagen may be even higher. Smaller amounts of Type IV (basement membrane), Type V (blood vessels), and Type VII (epidermal basement membrane anchoring fibers) may be present as well.[24] Collagen has a unique structure of amino acid sequence. It consists of a repeating triplet structure of glycine $-X-Y$. Approximately 20% to 25% of the "Y" position is either proline or hydroxyproline. Hydroxyproline is found almost exclusively in collagen and is formed by post-translational hydroxylation of proline in the peptide chain.[26] The post-translational modification requires the mixed function of oxygenase prolyl hydroxylase as well as the cofactors of ferrous iron, α-ketoglutorate, free molecular oxygen, and vitamin C. Besides promoting collagen synthesis, more recent findings further suggest multiple mechanisms of Vitamin C in promoting wound healing process including promoting fibroblast proliferation, migration, and replication-associated base excision repair of potentially mutagenic DNA lesions involving the hypoxia-inducible transcription factor-1 and collagen receptor-related signaling pathways.[27] A similar process also creates the amino acid hydroxylsine from lysine in the peptide chain, an amino acid crucial to proper collagen cross-linking.[28] The cross-linking is essential to impart tensile strength and decreased turnover of collagen as seen in the mature collagen molecules. It is clear that these amino acid substrates or their precursors must be present in adequate amounts for a normal wound healing process. Without the availability of adequate amino acids and cofactors such as vitamin C, there would be inadequate collagen formation, leading to dehiscence of the wound.

1.4 *Contraction*

Contraction is a process whereby the wound essentially shrinks by recruiting adjacent tissue and pulling it inward. The key effect-cells are fibroblasts. This process is intimate with the phases of proliferation and remodeling. The specific cell involved is the myofibroblast.[29–31] The motor function of these cells is modified in a way that moves the edges of the wound toward the center, instead of being motile and moving themselves. This process is independent of collagen synthesis, but collagen in the ECM aids in locking the cells in place, facilitating the contraction process which is under the control of TGF-β and other cytokines.[14]

Contraction may be a major process in wound closure in some animal species and in certain locations of the human subjects. In most species of animals, such as rodents, the contraction of the panniculus carnosus, a unique component of the animal skin dominates the wound healing process, which complicates the extrapolation of the data from animals to humans wound healing. The dorsal side of human hands has loose mobile tissue, similar to the skin of a rat, which also allows contraction to be a major factor in healing. After wound healing, a tightened dorsal skin may affect the function of the hands. The contraction occurs less prominently at sites where the skin is less movable, such as the palm of the hand or the sole of the foot.

1.5 *Remodeling*

This phase involves remodeling of collagen from Type III to Type I. Before this phase, the wound has been characterized by high metabolic activity. The elevated metabolic activities support extensive proliferation of inflammatory cells, epithelial cells, endothelial cells for angiogenesis, and fibroblasts actively laying down collagen matrix of the wound. During the remodeling phase, the nutritional requirements for wound healing are diminished. At first the wound is a soft, cellular structure lacking strength. During the remodeling process, the scar is transformed into a mature healed wound with tensile strength approaching uninjured tissue.

Remodeling represents a new phase of wound healing. Although hemostasis may take minutes, inflammation days, and proliferation weeks, remodeling may continue for months to years. In contrast to progressively increased collagen synthesis, in a net sense, collagen accumulation decreases to a balance between synthesis and degradation, except in abnormal states such as keloid or hypertrophic scars. The initial mixture of Type I and Type III collagen is changed into a structure with primarily Type I collagen. The amount of ground substance is decreased, and collagen becomes more cross-linked.[31]

The remodeling process is under the control of multiple factors including interferon, matrix metallo-proteinases, TGF-β, PDGF, and IL-1, as well as other factors.[30] These factors control the collagen type and the synthesis and degradation rates, however, they may not play significant roles in collagen cross-linking which starts with hydroxylation of lysine to produce hydroxylysine. When the initial cross-links are formed, maturation of the cross-link to produce more mature cross-links is not under enzymatic control.[32] Therefore, the wound maturation involves direct change in chemical structure of collagen rather than an enzymatic controlled biochemical process.

Remodeling is a slower process as compared to those in the earlier phases, but it also plays an important role in the overall healing process. The imbalance between the synthesis and degradation processes in remodeling could occur in a healed wound for even years after the original injury, such as in scurvy wounds.

In summary, wound healing is a complex process which is orchestrated over time by an interaction of soluble mediators, blood cells, inflammatory cells and parenchymal cells.[33] Wound healing requires energy, protein, a variety of vitamins and trace elements, and certain amino acids (especially arginine, glutamine).[34,35]

2. Nutrition and Metabolism in Relation to Wound Healing

Creation of a wound is often associated with an injury to the host. Severe injury, such as burn and trauma, could trigger overwhelming metabolic response to the insults. Cuthbertson described the metabolic response as an "ebb" phase and a "flow" phase.[36,37] The "ebb" phase is the period of traumatic shock or hypometabolism during the first few hours or days after injury. This is soon replaced by the "flow" phase which is a period of hypermetabolism that may last for weeks, months or even years depending on the characteristics of the injury and obstacles to recovery.[38] In these patients, nutritional support becomes necessary to replenish the increased oxidation of substrates for energy production and net loss of proteins during the hypermetabolic and protein catabolic conditions.[39] In contrast, for patients with good pre-traumatic nutritional status undergoing routine elective surgery, it is not likely that perioperative nutritional support will have an important effect on outcome of the healing wound.[37,40,41] Similar to a human fetus, wounds can be considered as an effective parasite on the substrates available in the remainder of the body and thus wounds demand high priority of circulating nutrients. Therefore, whole body inflammatory status and nutritional condition has direct impact on wound healing.

The nutrients for wound healing may have two major sources. Endogenously, during metabolic stress, the body can effectively catabolize the carcass (mainly muscle, bone, and skin) to support the synthesis of acute phase proteins, immunoglobulins, inflammatory cells, and collagen, needed to fight infection and stimulate healing.[26,42,43,44] For proteins, intricate shuttling mechanisms have developed to allow redistribution of substrates from the periphery to internal organs. It is likely that similar mechanisms are applicable to other nutrients.[45] This may also apply to micronutrients. Since a large amount of zinc in the body is present in the skin, it is not surprising that if skin proteins are catabolized, this might allow mobilization of zinc to provide cofactors for enzymatic activity in other tissues. Another source of nutrients is from the exogenous nutritional support. Some nutrients may be specifically required; we will further discuss this point in the section below.

Since the endogenous nutrients are important for topical wound metabolism, the overall nutritional status of the patients would significantly impact the wound healing process. In a nutritionally depleted state, which could result from significant hypermetatabolism and accelerated protein catabolism in the "flow phase" of severe injury[38,39] or result from chronic malnutrition, the patients may eventually have inadequate reserves to respond appropriately to metabolic demands of the whole body as well as wound healing. This is similar to the described conversion of individuals with marasmus (protein-calorie malnutrition) to kwashiorkor (a more severe form of protein-calorie malnutrition characterized by protein deficiency greater than caloric deficiency) with the onset of infection.[46] These marginally compensated individuals lack the reserve to respond to infection and may succumb to an infection that might not be otherwise life threatening. Deficiency of even a single micronutrient such as vitamin C may produce disastrous results in the healing process.

A number of experiments indicated the relationship between wound healing process and whole body metabolic conditions in severe trauma and chronic malnutrition status.

Animal studies in rodents using positron emission tomography indicated that in acute stage of severe burn injury[47] glucose utilization of the unburned skin was not changed at 6 or 24 h after injury, but was increased at 3 weeks after injury. In contrast, the utilization of glucose by the wound was almost zero at 6 and 24 h after injury; however, at 3 weeks after injury, the glucose utilization at wound area were two to three times higher than that of the normal skin of sham animals. The increased glucose utilization was accompanied with significantly elevated hexokinase activity in wound tissue and histologically the formation of granulation tissue with chronic inflammation, neovascularization (arrows), and fibroblast proliferation. This implies the significantly increased metabolic activity and increased demand of nutrients for wound healing at the expense of nutrient utilization by other tissues. Obviously a depletion of body reserves with the prolonged catabolic and inflammatory states in critically ill patients would affect normal wound healing. Clinical evidence supported the notion that protein inadequacy impedes wound healing by prolonging the inflammatory phase, impairing fibroplasias, collagen, and proteoglycan syntheses, and by impairing wound remodeling.[48]

The effects of chronic malnutrition on wound healing has been convincingly proved by a series of classical experiments reported by Molnar *et al.*[49–51] The first series of investigations were focused on quantitative evaluation of collagen turnover rate *in vivo*. Since 70%–90% of the protein of skin is collagen, the rate of collagen synthesis reflects the process of wound healing. In order to accurately measure how fast the collagen was formed and degraded, the animals were exposed to pure $^{18}O_2$ inhalation in a chamber for 36 h. Based on the considerations (1) the major peptide structure of collagen is composed of the repeated units of glycine-X–Y; where X may be any amino acid but Y is approximately 25% proline or hydroxyproline[6,26]; (2) there is no transfer RNA for hydroxyproline, thus the only source of this amino acid in peptide chain is from the post-translational hydroxylation of the peptide-bound proline, (3) the inhaled $^{18}O_2$ labeled the OH-group of hydroxyproline, thus continuing monitoring the kinetics of $^{18}O_2$ hydroxyproline in collagen through sequential biopsy of the skin provided quantitative information on the turnover rate of collagen proteins in a living animal. In healthy rats, soluble collagen synthesis is about 95 mg per day, about 55% of the newly synthesized soluble collagen per day is subsequently matured into insoluble collagen, 45% of the soluble collagen is degraded into amino acids, which is equivalent to less than 30 mg per day. Using the established method, further studies revealed that in protein malnourished rats (where the animals received only 1/6 of the adequate amount of total protein intake for 180 days) the rates of collagen synthesis, degradation and maturation were reduced to only 11%, 16%, and 16% of those in well-nourished healthy controls.[51] These studies provided direct evidence that whole body nutritional status has significant impact on skin collagen metabolism and hence the healing process of the wound. The findings are in parallel with the clinical observations that improving whole body protein nutritional status accelerates the wound healing process in severely burned patients and other surgical patients. It should be noted that, because ^{18}OH-hydroxyproline does not re-incorporate into collagen after this amino acid is released from collagen degradation, scientifically, the $^{18}O_2$ inhalation method has provided the most accurate estimate of collagen metabolism. Other approaches to label collagen using either [2H]- or [13C] proline as probes had always been underestimating the rate of collagen metabolism due to the fact that proline can be released and re-incorporated into collagen proteins, which caused an underestimate of 44% in the measurements.[49] The method could potentially be applied to the investigations in human subjects and burn patients.

Further studies by Zhang *et al.* using differently labeled stable isotopes successfully[52–58] evaluated the *in vivo* synthesis rates of protein and DNA in the skin. The synthesis of DNA reflects the rate of cell proliferation in the wound area.

These *in vivo* findings of wound metabolism in relation to whole body metabolism can be summarized as follows:

(1) In healthy conditions, the biochemistry of cell proliferation (as indicated by DNA synthesis) and protein synthesis in the skin are differently regulated.

(2) Protein synthesis rates in the epidermis are about 10 times of those in the dermis, which suggests a very active metabolic process for wound repair occurring in the epidermis component of the skin. It also stresses the importance of an intact dermis in transporting nutrients and in supporting epidermis protein synthesis.

(3) In healthy animals, skin protein metabolism is maintained in a dynamically stable condition in response to short-term changes in nutritional status (starvation, amino acid, and insulin) as compared to the immediate responses seen in muscle protein. However, prolonged malnutrition results in a reduced rate of collagen protein synthesis accompanied by poor protein nutrition states.

(4) After thermal injury, the injured skin is in negative protein balance, in parallel with the net loss of protein from whole body and muscle tissues, and hence, a negative nitrogen balance is also found in skin tissue. However, in wound area, there is an increased demanding of nutrients to meet its regional repairing process.

Therefore, improvement of whole body protein nutritional status is important in the protein anabolism in the wound protein.

On the other hand, wounds itself, such as wounds involving extensive area and depth induced by burn injury could have significant impact on the whole body metabolism in the host and again, the disturbed whole body metabolic status can "feedback" to affect topical wound metabolism and healing process. Studies by Prelack *et al.* revealed that in severely burned children, muscle protein catabolism was significantly reduced when the open wound area was almost covered by grafted skin.[59] Therefore, accelerating the wound healing process, especially in burn patients, is critical to the success of the overall treatment and the survival of the patients. In this regard, the development of early incision and grafting proposed in the 1950s, has been one of the milestones of burn surgery because it significantly reduced the inflammation and accelerated wound healing.[60,61] It also emphasizes the merit of this book in summarizing our current understanding on the wound healing process; searching for strategies to promote the wound healing process and inventing multiple skin substitutes to cover the open wound. Rapid healing of the wound also reduces the whole body inflammation and improves metabolic statues, especially in patients with severe burn injury.

3. Special Nutrients Supporting Wound Healing

Nutrients may be arbitrarily divided into "macro-" and "micronutrients" However, the difference between these classifications is not well defined. Macronutrients include: carbohydrates, amino acids, and fat. Micronutrients include all the vitamins and trace elements.

Based on the forgoing discussion, the first goal of nutritional support is to improve the whole body nutritional states and avoid a nutritional depleted condition, a good whole body nutritional statues is the premise for promoting wound healing process. For critically ill patients, there have been numerous suggestions of energy and nitrogen nutrient requirements to "compensate" the accelerated protein loss and to meet the requirements of increased energy production in the hypermetabolic state. The improved whole body nutritional state has direct or indirect impact on topical wound healing process. For patients under chronic malnutrition condition, treatment of the underlying diseases, such as diabetes, plus providing adequate amounts of nutrients to improve the overall nutritional status is an important step in helping wound healing.[39] It remains to be determined how much the exogenous nutrients are used to directly support the anabolic process of wound healing, and how much were used to indirectly support the wound healing by first improving the overall nutritional status. Earlier studies in fetus — maternal nutrition revealed that in fed condition protein accumulation in fetal body is less active than that in fasting condition when the nutrient supply to fetus is totally dependent on the exchange between maternal body and fetus. Therefore, the maternal–fetal nutrient exchange may be more important than directly using the exogenous nutrients.[62,63] A similar situation might apply to wound healing process in critically ill patients who have increased demand of nutrients to support whole body metabolism and in chronic malnutrition state in which depletion of nutrient reserves in the host affects nutrient exchange between the wounds and other parts of the body, ultimately affecting wound healing. Therefore nutrition plays pivotal rules, either directly or indirectly in wound healing under diseased conditions.

Extensive basic science studies have reported that a number of special nutrients may be required for the wound healing process, through various mechanisms.

Amino acids are among the most important macronutrients to support the process of wound healing. They serve as building blocks for "rebuilding" the wound defects as well as exerting modulatory roles in promoting the repairing process. Among the amino acids, the effects of arginine and glutamine on wound healing have been extensively studied.

In healthy subjects, arginine is synthesized in the kidney and liver from gut-derived citrulline.[64] Therefore, arginine is not an essential part of our diets. However, in stressed conditions such as sepsis, trauma, and large wound such as burn injury, the significantly increased arginine degradation through the ornithine–glutamate–(TCA) cycle pathway and limited endogenous production led to higher demand of this amino acid.[65] Therefore, arginine is considered a conditionally indispensable (conditionally essential) amino acid in nutritional support to these patients.

There are a number of theoretical possibilities of how arginine may affect wound healing. Proline is necessary for the synthesis of collagen and arginine is a precursor via the pathway of Arginine/Ornithine/glutamine-semialdehyde/Proline.[66–70] It has also been suggested that when the production of collagen is high, arginine may be an essential precursor of collagen precursor proline.[71] Arginine is also a precursor for nitric oxide[72] an essential compound for proper wound healing as demonstrated with high activities of NO production and expression of arginanse I.[73,74] Expression of NO synthase creates a cytotoxic environment that may be important to the early phase of wound healing, while supply of arginine and increased arginase activity produces an environment favorable for fibroblast replication and collagen production. Zandifar reported that systemic L-arginine is more efficient than topical L-arginine in wound healing and the process was mediated through the increase of VEGF and NO in the wound fluid.[75,76] Some earlier studies had revealed that arginine also played roles in wound healing through a central nerve effect on stimulating growth hormone secretion, following an arginine-thymus-hypothalmus-pituitary axis.[77] Arginine also exerts its effect on wound healing through modulating immune functions.[78,79] Some clinical studies also demonstrated benefits of arginine supplementation in promoting the healing of chronic wounds.[79–83] However, in most of the clinical studies arginine was supplied in combination with other nutrients, therefore the observed benefits are likely a synergetic effects of all the nutrients and other factors given rather than a single amino acid.

The suggestion of supplementing glutamine on wound healing has been proposed. Glutamine is the most abundant amino acid in the body. It has unique roles of shuttling nitrogen in the body and serving as a precursor for protein synthesis. The fast turnover cells, including gut mucosa cells and neutrophils, preferentially use glutamine as an energy source at a faster rate than glucose. Glutamine also plays a role in leukocyte apoptosis, superoxide production, antigen processing, and phagocytosis.[84,85] Therefore, glutamine also has an immune modulating effect.[86] In rapidly proliferating cells, glutamine provides energy through the pathway of glutamine–glutamate–alpha-ketoglutarte–TCA cycle. The nitrogen can then be used for the production of other nitrogen-containing molecules, including the other needed amino acids, and purines and pyrimidine for DNA and RNA synthesis.[87,88] The metabolic functions of glutamine also include provision of NADPH and increasing insulin sensitivity.[87] The anti-inflammatory effect of glutamine includes activation of nuclear factor alpha, beta, and cytokine release.[89] Glutamine is a precursor of glutathione, an important anti-oxidant necessary for stabilization of cell membranes transport of amino acids, and as a co-factor for some enzymatic activities.[85] Patients after trauma and surgery are in a glutamine depleted state with reduced circulating glutamine levels and tissue availability,[90] therefore, glutamine supplementation is required to improve whole body nutritional state and theoretically will increase its tissue availability to exert the above-mentioned tropic functions for tissue regeneration for "amending" the injured skin. Certain animal studies have reported the efficacy of enhanced glutamine supplementation on wound healing,[91–93] but more clinical trials are required to prove its positive effects in wound healing process in patients with chronic and acute wounds.[94]

It might be worth considering that providing glutamine and arginine may increase the proline availability for collagen synthesis.[70] Since an important step in collagen synthesis is the post-translational modification of proline on collagen molecule, the interstitial proline content has been found to be related to wound healing. Metabolically, proline is a nutritionally dispensable amino acid. Its formation and disposal are closely related to the metabolism of ornithine and glutamate (Figure 1). Ornithine is one of the components of the urea cycle, and glutamate availability is closely related to the metabolism of α-glutarate, an intermediate of the TCA cycle (Figure 2). In patients with thermal injuries, hypermetabolism and increased nitrogen loss indicate an accelerated TCA cycle and accelerated urea cycle activities, which "drains" both ornithine and glutamate. Because proline occupies one corner of the metabolic "triangle" interacting with these three amino acids, the *de novo*

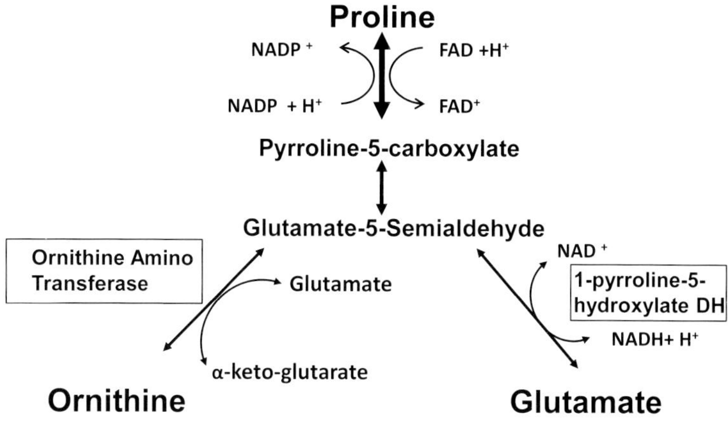

Figure 1. The "triangle" relationship among the metabolic pathways of proline, glutamate and ornithine.

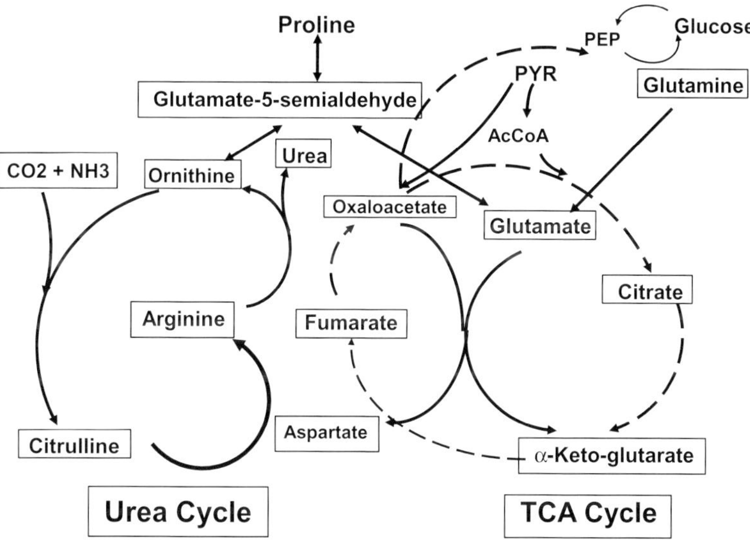

Figure 2. Metabolic pathways of glutamate, proline, and ornithine, with the coupling of urea and TCA cycles. In hypermatabolic and catabolic state after severe burn injury, both urea and TCA cycles are accelerated to meet the increased energy and urea productions. These processes may "drain" the proline pool and reduce its availability for collagen synthesis.

synthesis of tissue proline is likely to be compromised in the burn condition. This notion is supported by three facts: (1) Stable isotope tracer studies by Jaksic *et al.* revealed a negative proline balance in patients with burns, suggesting a status of proline depletion.[95] (2) Tracer studies by Yu *et al.* revealed an accelerated ornithine disposal via TCA cycle combined with limited arginine formation in severely burned patients.[65] (3) Studies also revealed a reduced tissue glutamine availability after thermal injury.[96] As a result, the availability of proline is likely to be limited for collagen synthesis in burn patients. Therefore combined arginine and glutamine supply may help replenish the proline pool for collagen synthesis in the wound healing process.

3.1 *Micronutrients*

Micronutrients are defined as "nutrients present and required in the body in minute quantities", which include vitamins and tracer minerals.[97] Micronutrients function primarily as cofactors in biochemical pathways and, as such, are critical to all of the activities of macronutrients, in relation to the metabolism in whole body and wound areas. For example, protein synthesis cannot continue without adequate quantities of vitamin B, zinc, and copper.[98,99] Micronutrients also play important roles in modulating pro/antioxidant status, inflammatory process, and angiogenesis[100–102] in trauma patients, these biological activities affect their wound healing process.[103] The effects of some important micronutrients in whole body nutrition and their relationships with wound healing are worth specific attention.

3.1.1 *Vitamins*

Vitamin C is involved in all phases of wound healing.[104,105] In the inflammatory phase, it is required for neutrophil apoptosis and clearance. During the proliferative phase, vitamin C contributes towards synthesis, maturation, secretion, and degradation of collagen. Its deficiencies affect the maturation phase by altering collagen production and scar formation. The body strives to maintain homeostasis of vitamin C, thereby ensuring its availability for collagen synthesis. After wounding, plasma and tissue levels of vitamin C diminish and, as a consequence, supplements may be useful for healing, although levels beyond saturation are excreted. Clinicians need to be aware of both the nutritional status of patients with either acute or chronic wounds and the possibility of any vitamin C deficiency which may hinder healing. In addition, collagen synthesis also requires other micronutrients, such as iron and copper. Tanaka and Molnar recommend that individuals with small wounds or pressure ulcers consider supplementation with 500 to 1,000 mg of ascorbic acid daily in divided doses for optimal utilization.[106] For patients with more severe wounds, such as burns over a large area, doses can be increased to 1–2 g/day.

Vitamin A is an essential fat-soluble vitamin that has been used topically in dermatology for many years for the management of pathological conditions as photodamage, psoriasis, and to aid healing after such treatments as dermabrasion. It has been found to stimulate epithelial growth, fibroblasts, and ground substance as well as having an anti-inflammatory effect in open wounds. The literature supports a positive effect of supplemental vitamin A in acute wounds and healing of fractures, burns, bowel, and radiation-induced injury.[107] Indeed, vitamin A seems to function as a hormone altering the activity of epithelial cells, melanocytes, fibroblasts, and endothelial cells through a family of retinoic acid receptors.[108] The effect of vitamin A on patients with chronic wounds is less clear except in the case of the patient being treated with corticosteroids for inflammatory diseases such as rheumatoid arthritis. For these patients, vitamin A has a unique role in nutritional management. Since the early studies of Ehrlich and Hunt,[109] investigators have noted that vitamin A supplementation can counteract the negative effects of glucocorticoids on wound healing. Wicke *et al.*[110] demonstrated in a rat model that administration of methylprednisone significantly reduced hydroxyproline synthesis by 50% on day 17 of treatment (methylprednisone) but supplementation of steroid-fed animals with either all trans-retinoic or 9-cis retinoic acids increased hydroxyproline content to near normal levels. Since certain steroids are considered as catabolic hormones and their levels increase in stressed condition,[111] vitamin A reversed the effect of steroids to decrease the levels of (TGF-b) and insulin-like growth factor-1 (IGF1) in the wound.[112] However, since TGF-b is one of the proposed mediators of chronic inflammation in the chronic wound, the net effect of Vitamin A in chronic wounds is worth further exploring. Stechmiller suggests oral administration of Vitamin A at 10,000–15,000 IU/day to enhance wound healing in patients receiving corticosteroids.[113] For patients with chronic wounds, daily supplementation of 20,000–25,000 IU of Vitamin A is suggested. Treatment must be provided as a short course of 10–14 days to be safe and used with caution in individuals who may have deficiency of the carrier protein, retinol binding protein, due to liver malfunction or severe protein-calorie malnutrition.[105]

3.1.2 *Trace minerals*

Zinc is an important agent in wound healing. In human body, approximately 300 enzymes require zinc for their activities. Zinc is an essential trace mineral for DNA synthesis, cell division, and protein synthesis, all necessary processes for tissue regeneration and repair.[114] Zinc deficiency has been associated with poor wound healing and decreased breaking strength of animal wounds[115] which can result from decreased protein and collagen synthesis during the healing process in zinc-deficient animals.[116] Studies have found that zinc levels were 50% higher in muscle and skin from abdominal wounds of rats during wound healing, hence mild zinc deficiency reduced its wound accumulation.[117] Sequential changes in zinc concentrations were studied in the incisional wound model in the rat. Zinc levels increased from wounding and peaked on the fifth day — at a time of high inflammation, granulation tissue formation, and epidermal cell proliferation.[118] Therefore zinc demands are the highest from time of wounding throughout the early inflammatory phase. Zinc concentrations returned to normal by the seventh day, when inflammation had regressed. It has been suggested that increased local demand for zinc resulting from surgery and wounding exposes otherwise marginal zinc deficiencies in humans Despite these findings, some clinicians still suggest that zinc supplementation is only recommended when a zinc deficiency is determined (serum and wound assay) as excess zinc supplementation can interfere with other cation absorption, specifically iron and copper.[113] Dorner *et al.* recommend zinc supplementation up to 40 mg elemental zinc/day (176 mg zinc sulfate) for up to 10 days to enhance wound healing if a zinc deficiency is indicated [(National Pressure Ulcer Advisory Panel White Paper. www.npuap.org (Assessed July 2013)]. Other recommendations are often higher: 220 mg of zinc sulfate (50 mg elemental zinc) twice a day for 2 weeks.[119] It is worth mentioning that many of the research trials that evaluate supplementation of zinc to assist in wound healing have zinc as a component of a multinutrient supplement. A number of these supplements contain arginine, glutamine, ascorbic acid, protein, and other micronutrients, in addition to zinc, and have had variable impacts on wound healing.[80,120–122] Research trials using these multinutrient supplements make it difficult to determine the individual nutrient contribution to wound healing. A review on the impact of zinc in wound healing, in both enteral and topical administration routes, has been written by Lansdown *et al.*, in which topical zinc application to surgical wounds seems to consistently augment wound healing.[123]

Copper is a component of a carbonyl cofactor that consists of an o-quinone in lysyl oxidase (LOX). The latter catalyzes the oxidation of peptidyl lysine to alpha-aminoadipic-delta-semialdehyde, the precursor to the covalent cross-linkages that stabilize fibers of elastin and collagen. The mechanism is that copper stabilizes a catalytically competent conformation and/or may align the carbonyl cofactor and the amine substrate for further processing. It is expected that enzyme-bound $Cu(II)$ might also function in the reoxidation of the substrate-reduced carbonyl cofactor, either by binding oxygen and/or by mediating the transfer of electrons from the reduced carbonyl cofactor to molecular oxygen.[124] This was supported by the finding using electron spin resonance (ESR) of the role of copper in other copper-dependent amine oxidases.[125] More recent findings further suggest that LOX like-2 (LOXL2) as part of the LOX family, which also comprises copper Cu(2+)- and lysine tyrosylquinone (LTQ)-dependent amine oxidases. LOXL2 functions similarly to LOX in the ECM by promoting cross-linking of collagen and elastin. LOXL2 has also been proposed to regulate extracellular and intracellular cell signaling pathways.[126] However, copper metabolism and its requirements for patients with acute wounds (such as burn injury) and chronic wounds need further investigation.

In addition, numeral trace mineral statuses are changed in diabetes condition, indicating its importance in glucose metabolism which may directly or indirectly impact local wound healing,[127] since glucose is a required energy source for this process. However, further investigation is required for assessing the necessity of its supplementation for diabetic patients and their effects on the healing of chronic wounds.

The general view is that micronutrients are ubiquitous in a normal diet, in many cases, they are only cofactors in chemical reactions, not altered or consumed and thus they may be reutilized, therefore, in well-nourished healthy subjects severe deficiencies of micronutrients are uncommon. However, many pathologic conditions may be accompanied with micronutrient deficiency, these include chronic fistula, severe burn injury with large volume of fluid replacement, prolonged period of artificial nutrition without proper supplementation of micronutrients. However, for each patient, a holistic assessment of micronutrient status is necessary. The general consensus if that there is no benefit from micronutrient supplementation if there is no deficient.[128–130] However, for some micronutrients, there may be of certain value for their administration at magnitudes greater than that needed for cofactors of biochemical reactions. For example, vitamin C in large quantities may be a useful free-radical scavenger, and vitamin A in large quantities may be useful to offset the adverse effects of corticosteroids on wound healing.

3.1.3 *Anabolic hormone and wound healing*

Patients with diseases associated with large area skin defects or chronic unhealed wounds are always in a catabolic state. Nutritional support had been the only approach to replenish the severe body waste. In the past decade, an approach of altering metabolism using hormonal and other agents has been developed, especially in treating the severely burned patients in accurate and later stages.[131] These anabolic agents played multiple roles in improving protein metabolism and reducing hypermetabolism of the burn patients which may last for one year after burn injury.[39] Concomitantly, it was found that these anabolic agents exerted positive effects in wound healing process.

The effects of some anabolic agents are discussed below.

Insulin

Insulin is a "traditional" anabolic hormone. In early 2000s, the use of intensive insulin therapy has been found to be beneficial to critically ill patients.[132] Insulin therapy has been used to burn treatment. Clinical evidences indicated that insulin improved whole body protein metabolism in a dose dependent fashion, meanwhile, it improved the wound healing in burn patients.[133] Animal studies revealed insulin administration resulted in almost 50-fold difference in factional synthesis rate in dermal and epidermal layers of skin.[56] Results of skin protein metabolism studies in burn patients were consistent with the findings in animals.[134] Since wound healing in the flow phase of burn injury was found to be mainly protein synthesis rather than proliferation, the positive effect of insulin on wound healing might mostly be in promoting wound protein anabolism. The improved wound metabolism at this stage was also correlated with the shortened hospital stay.[134]

Experimental topical application of insulin in animals showed some beneficial effects on wound healing, possibly by modulating the inflammatory status rather than its direct effect on promoting protein metabolism.[135,136] This effect is to be evaluated in human studies. The effects of insulin on chronic wounds have long been widely accepted because insulin therapy so far has been a major treatment for diabetes, correction of diabetes condition is associated with improved outcomes in many complications of diabetes conditions.

Growth hormone and insulin-like-growth factor

The idea of using growth hormone to improve the nutritional status of severely burnt patients dates back to early 1970s, as proposed by Wilmore.[137] In the early 1990s, following the availability of human recombinant growth hormone and IGF-1, these anabolic hormones have been applied in the treatment of burn patients. IGF-1 is a 7.7-KD single-chain polypeptide of 70 amino acids similar in sequence to proinsulin.

This hormone is synthesized by hepatocytes in response to growth hormones, it also acts to mediate growth hormone action during the hypermetabolic state by improving cell recovery and enhancing wound healing. It also improves the immune system and promotes protein anabolism. Furthermore, IGF-1 is bound to its binding protein IGFBP and has special receptors on fibroblasts and keratinocytes, which stimulate their proliferative activity and mitogenicity, as well as stimulate local collagen formation for wound healing.[138,139] In IGF-1 enriched cell cultures, the fibroblast and keratinocyte proliferation and fibroblast amino acid uptake, which usually occurs *in vivo* in wounds, were increased.[140,141] In addition, IGF-1 and other growth factors modulate epidermal and dermal cell survival and regeneration[33] as evidenced by the existence of high levels of IGF-1 below the wound and at wound edges.[142] IGF-1 cDNA transfer appears to facilitate this process by increasing local VEGF, FGF, and KGF concentrations, leading to the development of neo-angiogenesis and neo-vascularization in injured skin. Studies of exogenous liposomal IGF-1 cDNA gene transfer with growth factors does not appear to increase types I or III collagen, but does increase Type IV collagen, which is expressed in the early stages of re-epithelialization.[142] In addition, IGF-1 gene transfer has been found to increase the amount of anti-inflammatory cytokines in wounds, as well as decrease pro-inflammatory cytokine levels. This has led to the suggestion that this growth factor potentially facilitates the reduction of apoptotic outcomes to ultimately promote the reformation of skin epithelium.[143] It is yet to be further studied whether IGF-1 directly stimulates VEGF to increase neo-vascularization, or if increased VEGF expression causes an increase in granulation tissue formation.[144]

Human recombinant growth hormone has also been applied in the treatment of burn patients. Long term growth hormone treatment in burn patients was associated with increased muscle protein synthesis[145–147] and decreased donor site healing time by 1.5 days.[148] Patients receiving growth hormone treatment also demonstrated 100% increases in serum IGF-1 and IGFBP-3 as compared to healthy subjects, it is likely the beneficial effects of growth hormone are mediated by IGF-1.[149,150]

Oxadrolone and other androgen hormones

Anabolic steroids are synthetic derivatives of testosterone with anabolic and androgenic properties. They have been used to promote healing and weight gain in persons with burns, surgical wounds, and pressure ulcers.[151–153] Oxandrolone is approved by the U.S. Food and Drug Administration for treatment of involuntary weight loss or chronic infections. It is listed under the coverage authorization criteria for severe burns, and is a potent anabolic agent.[154] Based on clinical studies, oxandrolone 0.1 mg/kg twice daily is safe when given for up to 12 months.[155] A summary of eight studies in pediatric burn patients suggested that oral administration of oxandrolone 0.1 mg/kg twice daily increased protein synthesis, lean body mass accretion, and muscle strength; improved serum visceral protein concentrations; promoted weight gain; and increased bone mineral content. During the post-burn rehabilitation period, oxandrolone 0.1 mg/kg/day improved muscle strength, especially when combined with exercise. More recent studies further support the above positive effects of long term use for 5 years.[156] The above-mentioned anabolic effects of oxandrolone were also demonstrated in adult burn patients.[152,157] Accompanied with the protein anabolism, oxadrolone has been proved to promote wound healing in burn patients.[158,159] A double blind multicenter trial conducted by the American Burn Association confirmed the effect of oxandrolone on improving the healing of the burn wound.[160] For chronic wound healing, some clinical trials did not demonstrate positive results in pressure ulcers of patients with spinal cord injury.[161] but it may help wounds by increasing the appetite of the patients.[162] It is likely that the effect of oxandrolone on local wound healing is associated with promoting protein synthesis and improving whole body nutritional status. The latter has been fully demonstrated in the condition of severe burn injury, but the improvement of the nutritional status of the patients with chronic wounds after being treated with oxandrolone were not clearly demonstrated. Furthermore, clinical trials are also required to prove its effects of wound healing in pediatric burn patients.

Propranolol

It had long been recognized that catecholamines play a role as primary mediators of the hypermetabolic response in severe burns.[163] Soon after the burn injury, a 10-fold increase in plasma catecholamine concentration can be observed in patients, and this dramatic rise could explain the increased muscle protein catabolism and the elevated metabolic rate.[164] Adrenergic blockade, specifically the non-selective adrenergic antagonist propranolol has been used to minimize the effects of the elevated level of plasma catecholamine concentrations. This hypothetically results in an anticatabolic effect, thereby reducing the catecholamine-induced hypermetabolic response.[165] Since propranolol is a non-selective beta-antagonist, its non-selective activity on beta 1 and beta 2 adrenergic receptors theoretically counteracts the increased levels of catecholamines by blocking the positive inotropic and chronotropic effect of the sympathetic system.[166] Therefore, propranolol can safely be used in the treatment of burn patients with large wound areas. The most recent systematic review and metastasis showed the safety and positive effects of propranolol treatment, including the improved wound healing in adult burn patients.[167] Study in pediatric burn patients showed that propranolol use resulted in shortening the healing time, the time ready to graft, the area needed for the skin graft, and length of hospital stay.[168] Similar effects of propranolol were also seen in adult burn patients.[169] The mechanism of propranolol on wound healing may be related to the findings that β-adrenergic receptors widely exist in cutaneous fibroblasts, keratinocytes, and endothelial cells. Pullar and colleagues reviewed the effects of $\beta2$-adrenergic receptor modulation on wound repair. They reported that the effects included cell proliferation and migration via galvanotaxis, inflammation via neutrophil chemotaxis, wound contraction, and re-epithelialization via keratinocyte migration and fibroblast mediation, and angiogenesis via cyclic AMP-mediated vascular endothelial-derived growth factor expression.[170] In fact, $\beta2$-adrenergic receptor inhibition in a murine model resulted in increased dermal fibroblast function and re-epithelialization during the early stages of wound repair.[171] In addition, alterations in microvascular blood flow induced by β-blockade may also play a role in wound healing. In a rodent model, low (2 mg/kg) and high (20 mg/kg) doses of propranolol resulted in 35% lower cortical regional cerebral blood flow than the control treatment. Propranolol also significantly reduced mean oxygen consumption but did not affect oxygen saturation levels in wound tissues compared to placebo.[172] Propranolol abolished isoproterenol, induced increases in plasma von Willebrand factor antigen levels but had no effect on tissue factor or D-dimer expression in hypertensive patients. Because increased levels of von Willebrand factor are indicative of vascular injury, propranolol plays a role in protecting against endothelial cell damage.[173,174] Furthermore, animal experiments demonstrated that acute propranolol infusion directly stimulated protein synthesis in skin wound, but the molecular mechanism does not involve the signal transduction of downstream signals of beta-adrenergic receptors because the expressions of ERK1/2 and AKT were not affected by propranolol treatment,[175] other possible mechanisms need further exploration.

Overall, nutritional support plays important roles in promoting the wound healing process by improving the whole body nutritional status and providing special nutrients to the wounds serving as building blocks for tissue regeneration, maintaining enzyme functions for the biochemical process of skin repairing, modulating regional blood supply for substrate delivery, and ameliorating inflammation status of the open wound. The benefits of nutritional support may result from the synergistic effects of multiple nutrients instead of a single individual component.[174] Furthermore, the local wound healing process is mainly dictated by the underlying disease which leads to skin injury. Aggressive treatment of the underlying diseases, such as early excision and grafting of the burn wounds, control of the diabetes status, improvement of gut function for digestion and nutrient absorption might exert more important role than providing nutrients itself in the treatment of wound. A better nutritional status of the host provides a platform for the local wound healing process but nutrition treatment *per se* is an adjunct measure which cannot replace the aggressive surgical procedures and other direct treatments to the local wounds. On the other hand, healing of the local wound may have "feedback" effects to improve the overall health condition of the patients. Therefore, proper nutritional support combined with adequate wound treatment can result in a "win–win" condition to benefit the patients; but improper whole body nutritional support and inadequate topical wound management could cause a vicious cycle of malnutrition combined with wound deterioration.

References

1. Leaper, D. J. *Wounds: Biology and Management.* Oxford University Press, (1998).
2. Hutchinson, J. The Wound Programme. Dundee, Center for Medical Eduction (1992).
3. Gray, D. and Cooper, P. "Nutrition and wound healing: What is the link?" *J Wound Care* 10(3), 86–89 (2001).
4. Cooper, P. A review of different wound types and their principles of management. *Wound Healing: A Systemic Approach to Advanced Wound Healing and Management.* UK: Cromwell Press (2005).
5. Gottschlich, M. M. Nutrition in the burned pediatric patient. In: Queen, P. M. a. L. and Gaithersburg, C. E. (eds.) *Handbook of Pediatric Nutrition.* Aspen, pp. 536–559 (1993).
6. Michael, G. *Biochemical Pathways: N tals of Biochemistry and Molecular Biology.* New York, John Wiley &. Sons, Inc. NY: (1999).
7. Wilson, J. A. and J. J. Clark. "Obesity: Impediment to postsurgical wound healing." *Adv Skin Wound Care* 17(8), 426–435 (2004).
8. Gupta, A. and R. Raghubir. "Energy metabolism in the granulation tissue of diabetic rats during cutaneous wound healing." *Mol Cell Biochem* 270(1–2), 71–77 (2005).

9. Wakefiled, T. W. Homeostasis. In: Greenfield, L. J. Mulholland, M. W. Oldham, K. T. Zelenock, G. B. and Lillemoe, K. D. (eds.), *Surgery: Scientific Principals and Practice.* Philadelphia: Lippincott Williams & Wilkins, 86 (2001).

10. Stephens, D. J. "Cell biology: Collagen secretion explained." *Nature* **482**(7386), 474–475 (2012).

11. Clark, R. A. Lanigan, J. M. *et al.* "Fibronectin and fibrin provide a provisional matrix for epidermal cell migration during wound reepithelialization." *J Invest Dermatol* **79**(5), 264–269 (1982).

12. Herndon, D. N. Nguyen, T. T. *et al.* "Growth factors. Local and systemic." *Arch Surg* **128**(11), 1227–1233 (1993).

13. Nunan, R. J. Campbell, *et al.* "Ephrin-Bs Drive Junctional Downregulation and Actin Stress Fiber Disassembly to Enable Wound Re-epithelialization." *Cell Rep* **13**(7), 1380–1395 (2015).

14. Nurden, A. T. "Platelets, inflammation and tissue regeneration." *Thromb Haemost* **105** Suppl 1, S13–S33 (2011).

15. Matsumoto, S. R. Tanaka, R. *et al.* "The effect of control-released basic fibroblast growth factor in wound healing: Histological analyses and clinical application." *Plast Reconstr Surg Glob Open* **1**(6), e44 (2013).

16. Neilly, P. J. Kirk, J. S. *et al.* "The L-arginine/nitric oxide pathway — biological properties and therapeutic applications." *Ulster Med J* **63**(2), 193–200 (1994).

17. Loo, A. E. Wong, Y. T. *et al.* "Effects of hydrogen peroxide on wound healing in mice in relation to oxidative damage." *PLoS One* **7**(11), e49215 (2012).

18. Milne, G. L. Dai, Q. *et al.* "The isoprostanes — 25 years later." *Biochim Biophys Acta* **1851**(4), 433–445 (2015).

19. Piraino, F. and Selimovic, S. "A current view of functional biomaterials for wound care, molecular and cellular therapies." *Biomed Res Int* **2015**, 403801 (2015).

20. Rosique, R. G. Rosique, M. J. *et al.* "Curbing Inflammation in Skin Wound Healing: A Review." *Int J Inflam* **2015**, 316235 (2015).

21. Postlethwaite, A. E. Keski-Oja, J. *et al.* "Stimulation of the chemotactic migration of human fibroblasts by transforming growth factor beta." *J Exp Med* **165**(1), 251–256 (1987).

22. Bodnar, R. J. "Chemokine Regulation of Angiogenesis During Wound Healing." *Adv Wound Care* **4**(11), 641–650 (2015).

23. Wu, G. Bazer, F. W. *et al.* "Proline and hydroxyproline metabolism: Implications for animal and human nutrition." *Amino Acids* **40**(4), 1053–1063 (2011).

24. Miller, E. J. and Rhodes, R. K. "Preparation and characterization of the different types of collagen." *Methods Enzymol* **82** Pt A, 33–64 (1982).

25. Gay, S. and Miller, E. J. "What is collagen, what is not." *Ultrastruct Pathol* **4**(4), 365–377 (1983).

26. Munro, H. N. Allison, J. B. Protein turnover in mammalian tissues. In: Munro, H. N. and Allison, J. B., (eds.), *Mammalian Protein Metabolism*. Academic Press, pp. 264–271 (1964).

27. Duarte, T. L. Cooke, M. S. *et al.* "Gene expression profiling reveals new protective roles for vitamin C in human skin cells." *Free Radic Biol Med* **46**(1), 78–87 (2009).

28. Perdivara, I. Yamauchi, M. *et al.* "Molecular characterization of collagen hydroxylysine–glycosylation by mass spectrometry: Current status." *Aust J Chem* **66**(7), 760–769 (2013).

29. Gabbiani, G. Ryan, G. B. *et al.* "Presence of modified fibroblasts in granulation tissue and their possible role in wound contraction." *Experientia* **27**(5), 549–550 (1971).

30. Gospodarowicz, D. Neufeld, G. *et al.* "Fibroblast growth factor: Structural and biological properties." *J Cell Physiol Suppl* **5**, 15–26 (1987).

31. Hinz, B. Phan, S. H. *et al.* "Recent developments in myofibroblast biology: Paradigms for connective tissue remodeling." *Am J Pathol* **180**(4), 1340–1355 (2012).

32. Molnar, J. A. Skin collagen turnoever in healthy and protein calorie malnourished rats. *Applied Biology*, Cambridge, Massachusetts Institute of Technology. Ph.D. Thesis (1985).

33. Singer, A. J. and Clark, R. A. "Cutaneous wound healing." *N Engl J Med* **341**(10), 738–746 (1999).

34. Biesalski, H. (2010). "Micronutrients, wound healing, and prevention of pressure ulcers." *Nutrition* **26**(9), 858.

35. Guo, S. and Dipietro, L. A. "Factors affecting wound healing." *J Dent Res* **89**(3), 219–229 (2010).

36. Cuthbertson, D. P. "Observations on the disturbances of metabolism produced by injury to the limbs." *Quat J Med* (1), 233 (1932).

37. Bessey, P. Q. Metabolic respnses to trauma and infection In: Fisher, J. E. (ed.), Nutrition and Metabolism in the Surgical Patients. Boston: Little, Brown and Company, 577–600 (1996).

38. Yu, Y. M. Tompkins, R. G. *et al.* "The metabolic basis of the increase of the increase in energy expenditure in severely burned patients." *JPEN J Parenter Enteral Nutr* **23**(3), 160–168 (1999).

39. Herndon, D. N. and Tompkins, R. G. (2004). "Support of the metabolic response to burn injury." *Lancet* **363**(9424), 1895–1902.

40. Albina, J. E. "Nutrition and wound healing." *JPEN J Parenter Enteral Nutr* **18**(4), 367–376 (1994).

41. Clark, M. A. Plank, L. D. *et al.* "Wound healing associated with severe surgical illness." *World J Surg* **24**(6), 648–654 (2000).

42. Cuthbertson, D. P. and Tilstone, W. J. "Nutrition of the injured." *Am J Clin Nutr* **21**(9), 911–922 (1968).

43. Kinney, J. M. and Dudrick, S. J. "Trauma workshop report: Metabolic response to trauma, and nutrition." *J Trauma* **10**(11), 1065–1068 (1970).

44. Moore, F. D. "Endocrine-metabolic response to surgery and injury: Summary remarks." *JPEN J Parenter Enteral Nutr* **4**(2), 173–174 (1980).

45. Cuthbertson, D. P. Fell, G. S. *et al.* "Metabolism after injury. I. Effects of severity, nutrition, and environmental temperatue on protein potassium, zinc, and creatine." *Br J Surg* **59**(12), 926–931 (1972).

46. Bhattacharyya, A. K. "Protein-energy malnutrition (Kwashiorkor-Marasmus syndrome): Terminology, classification and evolution." *World Rev Nutr Diet* **47**, 80–133 (1986).

47. Carter, E. A. Tompkins, R. G. *et al.* "Thermal injury in rats alters glucose utilization by skin, wound, and small intestine, but not by skeletal muscle." *Metabolism* **45**(9), 1161–1167 (1996).

48. Ruberg, R. L. "Role of nutrition in wound healing." *Surg Clin North Am* **64**(4), 705–714 (1984).

49. Molnar, J. A. Alpert, N. *et al.* "Synthesis and degradation rates of collagens in vivo in whole skin of rats, studied with 1802 labelling." *Biochem J* **240**(2), 431–435 (1986).

50. Molnar, J. A. Alpert, N. M. *et al.* "Relative and absolute changes in soluble and insoluble collagen pool size in skin during normal growth and with dietary protein restriction in rats." *Growth* **51**(1), 132–145 (1987).

51. Molnar, J. A. Alpert, N. M. *et al.* "Synthesis and degradation of collagens in skin of healthy and protein-malnourished rats in vivo, studied by 18O2 labelling." *Biochem J* **250**(1), 71–76 (1988).

52. Macallan, D. C. Fullerton, C. A. *et al.* "Measurement of cell proliferation by labeling of DNA with stable isotope-labeled glucose: Studies *in vitro*, in animals, and in humans." *Proc Natl Acad Sci USA* **95**(2), 708–713 (1998).

53. Zhang, X. J. Chinkes, D. L. *et al.* "Metabolism of skin and muscle protein is regulated differently in response to nutrition." *Am J Physiol* **274**(3 Pt 1), E484–492 (1998).

54. Zhang, X. J. Chinkes, D. L. *et al.* "Insulin but not growth hormone stimulates protein anabolism in skin wound and muscle." *Am J Physiol* **276**(4 Pt 1), E712–E720 (1999).

55. Zhang, X. J. Chinkes, D. L. *et al.* "Anabolic action of insulin on skin wound protein is augmented by exogenous amino acids." *Am J Physiol Endocrinol Metab* **282**(6), E1308–E1315 (2002).

56. Zhang, X. J. Chinkes, D. L. *et al.* "Measurement of protein metabolism in epidermis and dermis." *Am J Physiol Endocrinol Metab* **284**(6), E1191–E1201 (2003).

57. Zhang, X. J. Chinkes, D. L. *et al.* "Fractional synthesis rates of DNA and protein in rabbit skin are not correlated." *J Nutr* **134**(9), 2401–2406 (2004).

58. Zhang, X. J. Chinkes, D. L. *et al.* "Leucine supplementation has an anabolic effect on proteins in rabbit skin wound and muscle." *J Nutr* **134**(12), 3313–3318 (2004).

59. Prelack, K. Yu, Y. M. *et al.* "The contribution of muscle to whole-body protein turnover throughout the course of burn injury in children." *J Burn Care Res* **31**(6), 942–948 (2010).

60. Herndon, D. N. Barrow, R. E. *et al.* "A comparison of conservative versus early excision. Therapies in severely burned patients." *Ann Surg* **209**(5), 547–552; discussion 552–543 (1989).

61. Burke, J. F. "Burn treatment's evolution in the 20th century." *J Am Coll Surg* **200**(2), 151–153 (2005).

62. Ling, P. R. Bistrian, B. R. *et al.* "Effect of fetal growth on maternal protein metabolism in postabsorptive rat." *Am J Physiol* **252**(3 Pt 1): E380–E390.

63. Ling, P. R. Bistrian, B. R. *et al.* "Effect of continuous feeding on maternal protein metabolism and fetal growth in the rat". *Am J Physiol* **256**(6 Pt 1): E852–E862 (1989).

64. Beaumier, L. Castillo, L. *et al.* "Arginine: New and exciting developments for an "old" amino acid." *Biomed Environ Sci* **9**(2–3), 296–315 (1996).

65. Yu, Y. M. Ryan, C. M. *et al.* "Arginine and ornithine kinetics in severely burned patients: Increased rate of arginine disposal." *Am J Physiol Endocrinol Metab* **280**(3), E509–E517 (2001).

66. Zieve, L. "Conditional deficiencies of ornithine or arginine." *J Am Coll Nutr* **5**(2), 167–176 (1986).

67. Albina, J. E. Mills, C. D. *et al.* "Temporal expression of different pathways of 1-arginine metabolism in healing wounds." *J Immunol* **144**(10), 3877–3880 (1990).

68. Albina, J. E. Abate, J. A. *et al.* "Role of ornithine as a proline precursor in healing wounds." *J Surg Res* **55**(1), 97–102 (1993).

69. Maggie L. Dylewski, Y. Protein and wound healing. In: Molnar, J. A. *Nutrition and Wound Healing*. Boca Raton: CRC Press, pp. 49–64 (2006).

70. Dylewski, M. L. Y. Y. M. Protein and wound healing. In: Molnar, J. A. *Nutrion and Wound Healing*. Boca Raton: CRC Press, pp. 49–64 (2007).

71. Barbul, A. "Proline precursors to sustain Mammalian collagen synthesis." *J Nutr* **138**(10), 2021S–2024S (2008).

72. Argaman, Z. Young, V. R. *et al.* "Arginine and nitric oxide metabolism in critically ill septic pediatric patients." *Crit Care Med* **31**(2), 591–597 (2003).

73. Debats, I. B. Wolfs, T. G. *et al.* "Role of arginine in superficial wound healing in man." *Nitric Oxide* **21**(3–4), 175–183 (2009).

74. Zunic, G. Supic, G. *et al.* "Increased nitric oxide formation followed by increased arginase activity induces relative lack of arginine at the wound site and alters whole nutritional status in rats almost within the early healing period." *Nitric Oxide* **20**(4), 253–258 (2009).

75. Shearer, J. D. Richards, J. R. *et al.* "Differential regulation of macrophage arginine metabolism: A proposed role in wound healing." *Am J Physiol* **272**(2 Pt 1), E181–E190 (1997).
76. Zandifar, A. Seifabadi, S. *et al.* "Comparison of the effect of topical versus systemic L-arginine on wound healing in acute incisional diabetic rat model." *J Res Med Sci* **20**(3), 233–238 (2015).
77. Barbul, A. Rettura, G. *et al.* "Wound healing and thymotropic effects of arginine: A pituitary mechanism of action." *Am J Clin Nutr* **37**(5), 786–794 (1983).
78. Barbul, A. Lazarou, S. A. *et al.* "Arginine enhances wound healing and lymphocyte immune responses in humans." *Surgery* **108**(2), 331–336; discussion 336–337 (1990).
79. Kirk, S. J. Hurson, M. *et al.* "Arginine stimulates wound healing and immune function in elderly human beings." *Surgery* **114**(2), 155–159; discussion 160 (1993).
80. van Anholt, R. D. Sobotka, L. *et al.* "Specific nutritional support accelerates pressure ulcer healing and reduces wound care intensity in non-malnourished patients." *Nutrition* **26**(9), 867–872 (2010).
81. Yatabe, J. Saito, F. *et al.* "Lower plasma arginine in enteral tube-fed patients with pressure ulcer and improved pressure ulcer healing after arginine supplementation by Arginaid Water." *J Nutr Health Aging* **15**(4), 282–286 (2011).
82. Leigh, B. Desneves, K. *et al.* (2012). "The effect of different doses of an arginine-containing supplement on the healing of pressure ulcers." *J Wound Care* **21**(3), 150–156 (2012).
83. Sipahi, S. Gungor, O. *et al.* "The effect of oral supplementation with a combination of beta-hydroxy-beta-methylbutyrate, arginine and glutamine on wound healing: A retrospective analysis of diabetic haemodialysis patients." *BMC Nephrol* **14**, 8 (2013).
84. Ardawi, M. S. "Glutamine and glucose metabolism in human peripheral lymphocytes." *Metabolism* **37**(1), 99–103 (1988).
85. Newsholme, P. "Why is L-glutamine metabolism important to cells of the immune system in health, postinjury, surgery or infection?" *J Nutr* **131**(9 Suppl), 2515S–2522S; discussion 2523S–2514S (2001).
86. Lorenz, K. J. Schallert, R. *et al.* "Immunonutrition — the influence of early postoperative glutamine supplementation in enteral/parenteral nutrition on immune response, wound healing and length of hospital stay in multiple trauma patients and patients after extensive surgery." *GMS Interdiscip Plast Reconstr Surg DGPW* **4**, Doc15 (2015).
87. Caldwell, M. D. "Local glutamine metabolism in wounds and inflammation." *Metabolism* **38**(8 Suppl 1), 34–39 (1989).
88. Soeters, P. B. and Grecu, I. "Have we enough glutamine and how does it work? A clinician's view." *Ann Nutr Metab* **60**(1), 17–26 (2012).

89. Wischmeyer, P. E. "Glutamine: Mode of action in critical illness." *Crit Care Med* **35**(9 Suppl), S541–S544 (2007).

90. Lacey, J. M. and Wilmore, D. W. "Is glutamine a conditionally essential amino acid?" *Nutr Rev* **48**(8), 297–309 (1990).

91. Peng, X. Yan, H. *et al.* "Clinical and protein metabolic efficacy of glutamine granules-supplemented enteral nutrition in severely burned patients." *Burns* **31**(3), 342–346 (2005).

92. Pattanshetti, V. M. Powar, S. R. *et al.* "Enteral glutamine supplementation reducing infectious morbidity in burns patients: A randomised controlled trial." *Indian J Surg* **71**(4), 193–197 (2009).

93. Goswami, S. Kandhare, A. *et al.* "Oral l-glutamine administration attenuated cutaneous wound healing in Wistar rats." *Int Wound J* **13**(1), 116–124 (2016).

94. Armstrong, D. G. Hanft, J. R. *et al.* "Effect of oral nutritional supplementation on wound healing in diabetic foot ulcers: A prospective randomized controlled trial." *Diabet Med* **31**(9), 1069–1077 (2014).

95. Jaksic, T. Wagner, D. A. *et al.* "Proline metabolism in adult male burned patients and healthy control subjects." *Am J Clin Nutr* **54**(2), 408–413 (1991).

96. Fong, Y. M. Minei, J. P. *et al.* "Skeletal muscle amino acid and myofibrillar protein mRNA response to thermal injury and infection." *Am J Physiol* **261**(3 Pt 2), R536–R542 (1991).

97. Teitelbaum, D. Guenter, P. *et al.* "Definition of terms, style, and conventions used in A.S.P.E.N. guidelines and standards." *Nutr Clin Pract* **20**(2), 281–285 (2005).

98. Harris, E. D. Rayton, J. K. *et al.* "Copper and the synthesis of elastin and collagen." *Ciba Found Symp* **79**, 163–182 (1980).

99. Mochizuki, S. Takano, M. *et al.* "The effect of B vitamin supplementation on wound healing in type 2 diabetic mice." *J Clin Biochem Nutr* **58**(1), 64–68 (2016).

100. Saghiri, M. A. Asatourian, A. *et al.* "Functional role of inorganic trace elements in angiogenesis — Part I: N, Fe, Se, P, Au, and Ca." *Crit Rev Oncol Hematol* **96**(1), 129–142 (2015).

101. Saghiri, M. A. Asatourian, A. *et al.* "Functional role of inorganic trace elements in angiogenesis-Part II: Cr, Si, Zn, Cu, and S." *Crit Rev Oncol Hematol* **96**(1), 143–155 (2015).

102. Saghiri, M. A. Orangi, J. *et al.* "Functional role of inorganic trace elements in angiogenesis part III: (Ti, Li, Ce, As, Hg, Va, Nb and Pb)." *Crit Rev Oncol Hematol* **98**, 290–301 (2016).

103. Blass, S. C. Goost, H. *et al.* "Extracellular micronutrient levels and pro-/anti-oxidant status in trauma patients with wound healing disorders: Results of a cross-sectional study." *Nutr J* **12**(1), 157 (2013).

104. MacKay, D. and Miller, A. L. "Nutritional support for wound healing." *Altern Med Rev* **8**(4), 359–377 (2003).

105. Molnar, J. A. Underdown, M. J. *et al.* "Nutrition and Chronic Wounds." *Adv Wound Care* **3**(11), 663–681 (2014).

106. Tanaka, H. Molnar, J. A. Vitamin C and Wound Healing. In: Molnar, J. A. (eds.), *Nutrition and Wound Healing* Boca Raton: CRC Press, 121–148 (2007).

107. Gottschlich, M. M. Fat-soluble vitamins and wound healing. In: Molnar, J. A. *Nutrition and Wound Healing*. Boca Raton: CRC Press, pp. 149–172 (2007).

108. Reichrath, J. Lehmann, B. *et al.* "Vitamins as hormones." *Horm Metab Res* **39**(2), 71–84 (2007).

109. Ehrlich, H. P. and Hunt, T. K. "Effects of cortisone and vitamin A on wound healing." *Ann Surg* **167**(3), 324–328 (1968).

110. Wicke, C. Halliday, B. *et al.* "Effects of steroids and retinoids on wound healing." *Arch Surg* **135**(11), 1265–1270 (2000).

111. Boonen, E. and Van den Berghe, G. "Endocrine responses to critical illness: Novel insights and therapeutic implications." *J Clin Endocrinol Metab* **99**(5), 1569–1582 (2014).

112. Wicke, C. Huners, M. *et al.* "Production and structure elucidation of glyco-glycerolipids from a marine sponge-associated microbacterium species." *J Nat Prod* **63**(5), 621–626 (2000).

113. Stechmiller, J. K. "Understanding the role of nutrition and wound healing." *Nutr Clin Pract* **25**(1), 61–68 (2010).

114. Prasad, A. S. "Zinc: An overview." *Nutrition* **11**(1 Suppl), 93–99 (1995).

115. Agren, M. S. and Franzen, L. "Influence of zinc deficiency on breaking strength of 3-week-old skin incisions in the rat." *Acta Chir Scand* **156**(10), 667–670 (1990).

116. Fernandez-Madrid, F. Prasad, A. S. *et al.* "Effect of zinc deficiency on nucleic acids, collagen, and noncollagenous protein of the connective tissue." *J Lab Clin Med* **82**(6), 951–961 (1973).

117. Senapati, A. and Thompson, R. P. "Zinc deficiency and the prolonged accumulation of zinc in wounds." *Br J Surg* **72**(7), 583–584 (1985).

118. Lansdown, A. B. Sampson, B. *et al.* "Sequential changes in trace metal, metallothionein and calmodulin concentrations in healing skin wounds." *J Anat* **195**(Pt 3), 375–386 (1999).

119. Posthauer, M. E. "Do patients with pressure ulcers benefit from oral zinc supplementation?" *Adv Skin Wound Care* **18**(9), 471–472 (2005).

120. Collins, C. E. Kershaw, J. *et al.* "Effect of nutritional supplements on wound healing in home-nursed elderly: A randomized trial." *Nutrition* **21**(2), 147–155 (2005).

121. Cereda, E. Gini, A. *et al.* "Disease-specific, versus standard, nutritional support for the treatment of pressure ulcers in institutionalized older adults: A randomized controlled trial." *J Am Geriatr Soc* **57**(8), 1395–1402 (2009).

122. Blass, S. C. Goost, H. *et al.* "Time to wound closure in trauma patients with disorders in wound healing is shortened by supplements containing antioxidant micronutrients and glutamine: A PRCT." *Clin Nutr* **31**(4), 469–475 (2012).

123. Lansdown, A. B. Mirastschijski, U. *et al.* "Zinc in wound healing: Theoretical, experimental, and clinical aspects." *Wound Repair Regen* **15**(1), 2–16 (2007).

124. Kagan, H. M. and Trackman, P. C. "Properties and function of lysyl oxidase." *Am J Respir Cell Mol Biol* **5**(3), 206–210 (1991).

125. Williamson, P. R. and Kagan, H. M. "Reaction pathway of bovine aortic lysyl oxidase." *J Biol Chem* **261**(20), 9477–9482 (1986).

126. Moon, H. J. Finney, J. *et al.* "Human lysyl oxidase-like 2." *Bioorg Chem* **57**, 231–241 (2014).

127. Kaur, B. and Henry, J. "Micronutrient status in type 2 diabetes: A review." *Adv Food Nutr Res* **71**, 55–100 (2014).

128. Wilkinson, E. A. and Hawke, C. I. "Oral zinc for arterial and venous leg ulcers." *Cochrane Database Syst Rev*(2), CD001273 (2000).

129. Fuhrman, P. Parker, M. "Micronutrient assessment." *Support Line* **26**(1), 17–24 (2004).

130. Shenkin, A. "The key role of micronutrients." *Clin Nutr* **25**(1), 1–13 (2006).

131. Pereira, C. T. Murphy, K. D. *et al.* "Altering metabolism." *J Burn Care Rehabil* **26**(3), 194–199 (2005).

132. van den Berghe, G. Wouters, P. *et al.* "Intensive insulin therapy in critically ill patients." *N Engl J Med* **345**(19), 1359–1367 (2001).

133. Tuvdendorj, D. Zhang, X. J. *et al.* "Intensive insulin treatment increases donor site wound protein synthesis in burn patients." *Surgery* **149**(4), 512–518 (2011).

134. Tuvdendorj, D. Chinkes, D. L. *et al.* "Donor site wound protein synthesis correlates with length of acute hospitalization in severely burned children." *Wound Repair Regen* **18**(3), 277–283 (2010).

135. Chen, X. Liu, Y. *et al.* "Topical insulin application improves healing by regulating the wound inflammatory response." *Wound Repair Regen* **20**(3), 425–434 (2012).

136. Chen, X. Zhang, X. *et al.* "Effect of topical insulin application on wound neutrophil function." *Wounds* **24**(7), 178–184 (2012).

137. Wilmore, D. W. Moylan, J. A. Jr. *et al.* "Anabolic effects of human growth hormone and high caloric feedings following thermal injury." *Surg Gynecol Obstet* **138**(6), 875–884 (1974).

138. Steenfos, H. H. "Growth factors and wound healing." *Scand J Plast Reconstr Surg Hand Surg* **28**(2), 95–105 (1994).

139. Pierre, E. J. Perez-Polo, J. R. *et al.* "Insulin-like growth factor-I liposomal gene transfer and systemic growth hormone stimulate wound healing." *J Burn Care Rehabil* **18**(4), 287–291 (1997).

140. Misra, P. Nickoloff, B. J. *et al.* "Characterization of insulin-like growth factor-I/somatomedin-C receptors on human keratinocyte monolayers." *J Invest Dermatol* 87(2), 264–267 (1986).

141. Tavakkol, A. Elder, J. T. *et al.* "Expression of growth hormone receptor, insulin-like growth factor 1 (IGF-1) and IGF-1 receptor mRNA and proteins in human skin." *J Invest Dermatol* **99**(3), 343–349 (1992).

142. Jeschke, M. G. Schubert, T. *et al.* "Interaction of exogenous liposomal insulin-like growth factor-I cDNA gene transfer with growth factors on collagen expression in acute wounds." *Wound Repair Regen* **13**(3), 269–277 (2005).

143. Spies, M. Nesic, O. *et al.* "Liposomal IGF-1 gene transfer modulates pro- and anti-inflammatory cytokine mRNA expression in the burn wound." *Gene Ther* **8**(18), 1409–1415 (2001).

144. Cioffi, W. G. Gore, D. C. *et al.* "Insulin-like growth factor-1 lowers protein oxidation in patients with thermal injury." *Ann Surg* **220**(3), 310–316; discussion 316–319 (1994).

145. Aili Low, J. F. Barrow, R. E. *et al.* "The effect of short-term growth hormone treatment on growth and energy expenditure in burned children." *Burns* **27**(5), 447–452 (2001).

146. Hart, D. W. Herndon, D. N. *et al.* "Attenuation of posttraumatic muscle catabolism and osteopenia by long-term growth hormone therapy." *Ann Surg* **233**(6), 827–834 (2001).

147. Branski, L. K. Herndon, D. N. *et al.* "Randomized controlled trial to determine the efficacy of long-term growth hormone treatment in severely burned children." *Ann Surg* **250**(4), 514–523 (2009).

148. Herndon, D. N. Barrow, R. E. *et al.* "Effects of recombinant human growth hormone on donor-site healing in severely burned children." *Ann Surg* **212**(4), 424–429; discussion 430–421 (1990).

149. Takala, J. Ruokonen, E. *et al.* "Increased mortality associated with growth hormone treatment in critically ill adults." *N Engl J Med* **341**(11), 785–792 (1999).

150. Elijah, I. E. Branski, L. K. *et al.* "The GH/IGF-1 system in critical illness." *Best Pract Res Clin Endocrinol Metab* **25**(5), 759–767 (2011).

151. Demling, R. H. and DeSanti, L. "The rate of restoration of body weight after burn injury, using the anabolic agent oxandrolone, is not age dependent." *Burns* **27**(1), 46–51 (2001).

152. Ferrando, A. A. Sheffield-Moore, M. *et al.* "Testosterone administration in severe burns ameliorates muscle catabolism." *Crit Care Med* **29**(10), 1936–1942 (2001).

153. Hart, D. W. Wolf, S. E. *et al.* "Anabolic effects of oxandrolone after severe burn." *Ann Surg* **233**(4), 556–564 (2001).

154. Fox, M. Minot, A. S. *et al.* "Oxandrolone: A potent anabolic steroid of novel chemical configuration." *J Clin Endocrinol Metab* **22**, 921–924 (1962).

155. Miller, J. T. and Btaiche, I. F. "Oxandrolone in pediatric patients with severe thermal burn injury." *Ann Pharmacother* **42**(9), 1310–1315 (2008).

156. Porro, L. J. Herndon, D. N. *et al.* "Five-year outcomes after oxandrolone administration in severely burned children: A randomized clinical trial of safety and efficacy." *J Am Coll Surg* **214**(4), 489–502; discussion 502–484 (2012).

157. Herndon, D. N. Ramzy, P. I. *et al.* "Muscle protein catabolism after severe burn: Effects of IGF-1/IGFBP-3 treatment." *Ann Surg* **229**(5), 713–720; discussion 720–712 (1999).

158. Demling, R. and De Santi, L. "Closure of the "non-healing wound" corresponds with correction of weight loss using the anabolic agent oxandrolone." *Ostomy Wound Manage* **44**(10), 58–62, 64, 66 passim (1998).

159. Demling, R. H. and Orgill, D. P. "The anticatabolic and wound healing effects of the testosterone analog oxandrolone after severe burn injury." *J Crit Care* **15**(1), 12–17 (2000).

160. Wolf, S. E. Edelman, L. S. *et al.* "Effects of oxandrolone on outcome measures in the severely burned: A multicenter prospective randomized double-blind trial." *J Burn Care Res* **27**(2), 131–139; discussion 140–131 (2006).

161. Bauman, W. A. Spungen, A. M. *et al.* "The effect of oxandrolone on the healing of chronic pressure ulcers in persons with spinal cord injury: A randomized trial." *Ann Intern Med* **158**(10), 718–726 (2013).

162. Krasner, D. L. and Belcher, A. E. "Oxandrolone restores appetite. An increase in weight helps heal wounds." *Am J Nurs* **100**(11), 53 (2000).

163. Wilmore, D. W. Long, J. M. *et al.* "Catecholamines: Mediator of the hypermetabolic response to thermal injury." *Ann Surg* **180**(4), 653–669 (1974).

164. Pereira, C. Murphy, K. *et al.* "Post burn muscle wasting and the effects of treatments." *Int J Biochem Cell Biol* **37**(10), 1948–1961 (2005).

165. Rojas, Y. Finnerty, C. C. *et al.* "Burns: An update on current pharmacotherapy." *Expert Opin Pharmacother* **13**(17), 2485–2494 (2012).

166. Satwani, S. Dec, G. W. *et al.* "Beta-adrenergic blockers in heart failure: Review of mechanisms of action and clinical outcomes." *J Cardiovasc Pharmacol Ther* **9**(4), 243–255 (2004).

167. Flores, O. Stockton, K. *et al.* "The efficacy and safety of adrenergic blockade after burn injury: A systematic review and meta-analysis." *J Trauma Acute Care Surg* **80**(1), 146–155 (2016).

168. Williams, F. N. Herndon, D. N. *et al.* "Propranolol decreases cardiac work in a dose-dependent manner in severely burned children." *Surgery* **149**(2), 231–239 (2011).

169. Ali, A. Herndon, D. N. *et al.* "Propranolol attenuates hemorrhage and accelerates wound healing in severely burned adults." *Crit Care* **19**: 217 (2015).

170. Pullar, C. E. Manabat-Hidalgo, C. G. *et al.* "Beta-Adrenergic receptor modulation of wound repair." *Pharmacol Res* **58**(2), 158–164 (2008).

171. Pullar, C. E. Le Provost, G. S. *et al.* "Beta2AR antagonists and beta2AR gene deletion both promote skin wound repair processes." *J Invest Dermatol* **132**(8), 2076–2084 (2012).

172. Chi, O. Z. Liu, X. *et al.* "The effects of propranolol on heterogeneity of rat cerebral small vein oxygen saturation." *Anesth Analg* **89**(3), 690–695.

173. von Kanel, R. Dimsdale, J. E. *et al.* "Effects of nonspecific beta-adrenergic stimulation and blockade on blood coagulation in hypertension." *J Appl Physiol* (1985) **94**(4), 1455–1459 (2003).

174. Ellinger, S. "Micronutrients, Arginine, and Glutamine: Does Supplementation Provide an Efficient Tool for Prevention and Treatment of Different Kinds of Wounds?" *Adv Wound Care* **3**(11), 691–707 (2014).

175. Zhang, X. J. Meng, C. *et al.* "Acute propranolol infusion stimulates protein synthesis in rabbit skin wound." *Surgery* **145**(5), 558–567 (2009).

7. Polarization Sensitive Optical Coherence Tomography for Imaging of Wound Repair

Martin Villiger* and Brett E. Bouma

Wellman Center for Photomedicine
Massachusetts General Hospital and Harvard Medical School
50 Blossom Street, 02114 Boston, MA, USA
**mvilliger@mgh.harvard.edu*

Abstract

To improve our understanding of wound repair and develop novel therapeutic approaches, advanced tools to monitor and assess the healing process are needed. Optical coherence tomography is an optical imaging modality that provides images of the subsurface microstructure of biological tissue and has shown great potential for imaging the healing of skin-wounds. Polarization sensitive optical coherence tomography further analyzes the polarization states of the scattered light, and can provide additional insight into the arrangement and content of collagen in the tissue, providing important hallmarks of wound healing and tissue remodeling. This chapter explains the working principle of this imaging modality and discusses its prospects to help in research on wound healing.

1. Introduction

This chapter explains the working principle of optical coherence tomography (OCT) and polarization sensitive OCT (PS-OCT), and discusses its merits as an imaging tool to assist research in wound healing. OCT provides cross-sectional views of the subsurface microstructure of biological tissue at high spatial resolution of ~10–20 μm.[1] It employs light in the near infrared spectral

"

region (700–2500 nm). The propagation of light in tissue is defined by the sample's refractive index. Rapid spatial variations of the refractive index scatter the incident light, part of which reaches the detector and generates intrinsic contrast without the need for any labeling. Scattering, however, also scrambles the forward propagating beam and limits the imaging depth to 1–2 mm. This is sufficient to image the entire thickness of the skin in small animal models, and even in many sites on the human body, making OCT very appropriate to investigating and monitoring wound healing in skin. The high speed of OCT enables comprehensive imaging of large fields of view of several square centimeters in living subjects. The fast imaging speed and the absence of any labeling make it well suited for preclinical and clinical imaging in small animals and human patients, respectively. Imaging with OCT is non-invasive and facilitates longitudinal studies of wound repair.

PS-OCT in addition measures the polarization state of the light scattered by the tissue.[2] In birefringent tissue, the refractive index depends on the polarization state of the light. Collagen and similar tissue components with fibrillar architecture exhibit birefringence. This creates a relative delay between light with different polarization states. PS-OCT detects this delay and offers a measure of tissue birefringence, providing valuable insight into the tissue's collagen architecture. The primary role of collagen in the formation of granulation tissue and wound remodeling make this contrast a very attractive feature for the investigation of wound healing.

OCT continues to progress and is actively researched by many academic groups. It is a mature technology and conventional OCT systems are commercially available from several suppliers. PS-OCT is less well established, yet there exists a commercially available system from Thorlabs. There are only a few additional components needed to upgrade a conventional OCT system into a PS-OCT system, although it is outside the scope of this chapter to provide detailed instructions on how to realize this upgrade.

The goal of this chapter is first to convey a basic understanding of the working principle of OCT and the required instrumentation, and is covered in Section 2. We then continue to discuss PS-OCT and the reconstruction of tissue birefringence and depolarization, as well as their interpretation and physical meaning in Section 3. After briefly reviewing applications of (PS-) OCT to imaging of wound healing, Section 4 illustrates the merits of PS-OCT with practical examples of animal studies on wound healing and scar formation in burn and excisional wounds, respectively.

2. The Principle of OCT Imaging

2.1 *Coherence gating*

OCT provides cross-sectional views of the subsurface microstructure of biological tissue. It employs interferometric detection and broad bandwidth laser light to discriminate the signal from different depths through an effect called "coherence gating." Similar to ultrasound imaging, OCT measures the delay of the signal reflected at different depths to recover the axial reflectivity profile of the sample. The speed of sound is sufficiently low to directly record the delay between the various echoes of the sample. In contrast, OCT employs interferometric detection and compares the signal reflected from the sample with a static reference signal, because the speed of light precludes direct detection of the delay. Figure 1 displays the basic configuration of an OCT instrument. Light from a source covering a wide wavelength spectrum is separated by a beam splitter into sample and reference arm. In the sample arm, the light propagates through the sample tissue and is scattered and reflected at different depths. Part of the backscattered light reaches the

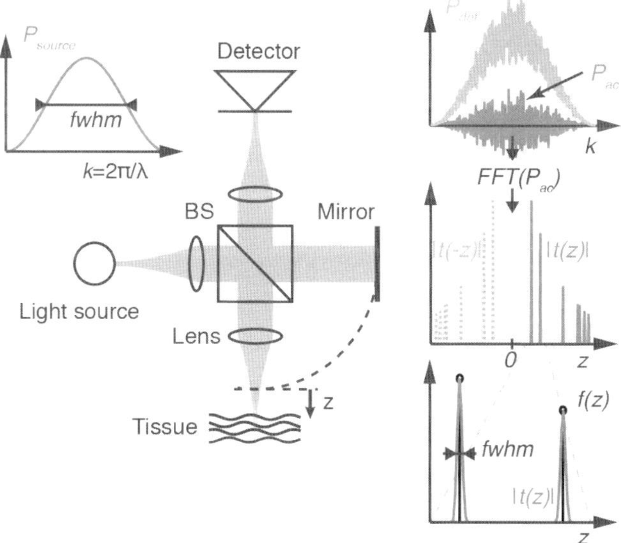

Figure 1.   Basic OCT setup with a beam splitter separating the light into the sample arm illuminating the tissue, and the reference arm reflecting off a mirror, respectively. The plots display the source spectrum (top left) and the resulting interference pattern (top right). The reconstructed tomogram $t(z)$ is an image of the sample reflectivity profile $f(z)$ (middle plot on right-hand side). Each discrete reflection in the sample is broadened in the tomogram to a full width at half maximum (fwhm) characteristic of the OCT system.

detector, where it interferes with the light from the reference arm that was simply reflected off a mirror. The interference signal is converted by the detector into an electrical signal and is of the form

$$P_{\text{det}} = \underbrace{|U_r|^2 + |U_s|^2}_{P_{\text{dc}}} + \underbrace{U_s^* U_r + U_s U_r^*}_{P_{\text{ac}}} = P_{\text{dc}} + P_{\text{ac}}, \tag{1}$$

where U_r and U_s are proportional to the electromagnetic field returned from the reference and the sample, respectively, and the star indicates the complex conjugate. The first term (P_{dc}) comprises the intensities of both arms. P_{ac} is the interference term of interest. The first of the two complex conjugate interference terms can be expressed as

$$U_s^* U_r(k) \propto \alpha(k) \int f(z) e^{-i2kz} dz. \tag{2}$$

Here, $\alpha(k)$ is the spectral shape of the source spectrum, and $f(z)$ is the reflectivity profile of the sample, defined by variations of the sample's refractive index. It specifies for each depth z, how much light is reflected. $i = \text{sqrt}(-1)$, and $k = 2\pi/\lambda$ is the wavenumber, with λ the wavelength. Light reflected at a larger depth accumulates additional phase, which is expressed through the exponential term. z is the sample depth, measured relative to the reference arm length. I.e. the sample signal from a single reflector positioned at $z = 0$ travelled the exact same distance when arriving at the detector as the reference arm light. Increasing z by placing this single reflector further away from the interferometer will result in a periodic modulation of the detected signal as a function of the wavenumber. The rate of the modulation increases linearly with the offset z.

The integral of Eq. (3) can be identified as the Fourier transformation of the sample reflectivity profile. Recording the signal as a function of the wavenumber is equivalent to taking the Fourier transform of the axial reflectivity profile. Accordingly, inverse Fourier transformation of the recorded signal will reconstruct the tomogram $t(z)$ of the sample reflectivity profile:

$$t(z) = \int U_s^* U_r(k) \, e^{i2kz} dk = FT^{-1} \left\{ U_s^* U_r(k) \right\}. \tag{3}$$

The reconstructed amplitude $|t(z)|$ is proportional to the amplitude of the electromagnetic field scattered from the sample at the same depth. However, the depth-profile after reconstruction is recovered as a function of the optical path length, which is the physical offset times the refractive index of the sample medium. The true physical offset can be recovered by scaling with the sample refractive index. Correspondingly, the squared norm $T(z) = |t(z)|^2$ is proportional to the intensity scattered from the sample as a function of depth. The interferometric detection conveys excellent sensitivity to OCT.[3,4] Even a very faint sample signal is amplified by the mixing process with the strong reference signal and measurements close to the shot-noise limit are feasible. This translates to high imaging speeds and a large dynamic range. To accommodate this large dynamic range, tomograms are displayed in logarithmic scale: $10 \times \log_{10}(T(z)) = 20 \times \log_{10}(|t(z)|)$.

There are two strategies to record the spectral modulation of Eq. (1) as a function of the wavenumber k. Optical Frequency Domain Imaging (OFDI)[5] and other swept source implementations of OCT use a wavelength-swept laser source that sweeps a narrow line-width laser line over a broad wavelength range in a short time. Accordingly, the interference signal recorded as a function of time corresponds to the signal as a function of k. Alternatively, spectrometer-based OCT systems use a diffraction grating to split the broad bandwidth signal into individual spectral channels that are recorded in parallel with a line-camera.

The fringe signal P_{ac} contains two complex conjugate interference terms. If the first term results in a tomogram $t(z)$, then the second term produces a tomogram $t^*(-z)$, mirrored about the zero differential path length, as illustrated in Figure 1. Generally, this results in a disturbing overlap that can be avoided by adjusting sufficient path length difference between the reference and the sample arm, or by employing additional strategies to separate the two terms.[6,7]

Although the reconstructed tomogram $t(z)$ is a close image of the original reflectivity profile $f(z)$, the finite spectral width of the light source results in some loss of the finest details. $t(z)$ can indeed be expressed as the convolution of $f(z)$ with a point spread function (PSF), defined as the inverse Fourier transformation of the source spectrum. As illustrated in Figure 1, this PSF defines the axial width of the peaks in the tomogram and this width is used to define the axial resolution of OCT. Assuming a spectrum with a Gaussian shape the full width at half maximum (fwhm) of the resulting Gaussian peak in $t(z)$ can be characterized by:

$$\Delta z_{\text{fwhm}} = \frac{4\ln(2)}{\Delta k_{\text{fwhm}}} = \frac{2\ln(2)\lambda_c^2}{\pi \Delta\lambda_{\text{fwhm}}}. \tag{4}$$

Here, Δk_{fwhm} and $\Delta\lambda_{\text{fwhm}}$ are the widths of the Gaussian spectrum as a function of the wavenumber and wavelength, respectively, and λ_c is the mean wavelength. The wider the spectrum covered by the light source, the better the structure of the axial tomogram is resolved. Most OCT sources have spectral widths of ~100 nm fwhm, providing an axial resolution below 10 μm.

2.2 *Imaging lateral structure*

In order to resolve structure in the lateral direction, the probe beam is focused into a small spot on the sample using an objective lens. The spot is scanned in the lateral direction using a galvanometric mirror, while constantly acquiring depth-profiles. Using a pair of mirrors, located in the front focal plane of the objective lens, enables to laterally displace the focused beam across the sample along both lateral directions, as illustrated in Figure 2. At each scan location, an axial depth profile or A-line is acquired. Assembling all A-lines of a single lateral scan results in a cross-sectional image or B-scan. The collection of multiple B-scans while scanning along the remaining lateral direction represents a three-dimensional tomogram and is sometimes referred to as C-scan. Figure 2 displays the resulting volume tomogram of a burn scar in a rat. For visualization, the resulting volume is re-sliced along various spatial directions to define "en face" and "longitudinal" views, in addition to the more common "B-scan" view, which is usually previewed in real-time, while the acquisition is performed.

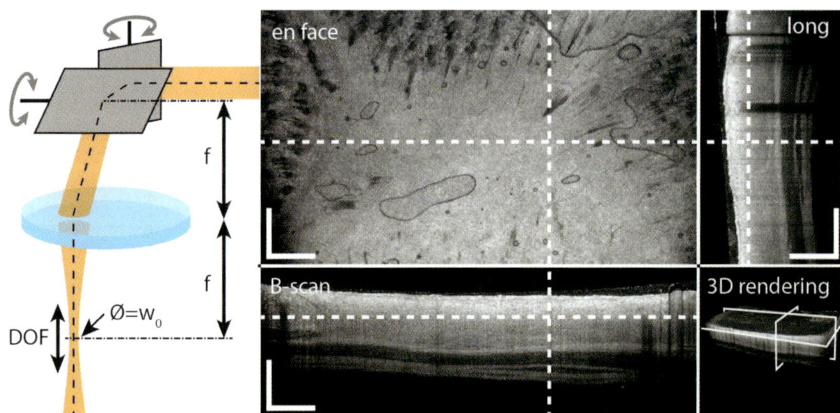

Figure 2. Volumetric imaging with OCT. (Left) Schematic of galvonometric scanner, deflecting the incident beam at the focal plane in front of the objective lens, achieving a lateral displacement of the focal volume in the sample. The depth of field (DOF) indicates the axial range within which the focus diameter w_0 extends by only a factor of square root of two. (Right) Volume scan of a burn scar in a rat, showing en-face, B-scan, and longitudinal cross sections, together with a 3D rendering of the imaged volume. Scale bars: 1 mm.

The extent of the laser spot on the tissue determines the lateral resolution. A beam of light emitted from a laser source features well-defined wave fronts, and its propagation can be accurately described with the Gaussian beam formalism.[8] Importantly, focusing a Gaussian beam results in a minimum radius or waist at the focus, designated w_0, related to both f, the focal length of the objective lens and w_{coll}, the waist of the collimated Gaussian beam incident on the lens, by $w_0 = \lambda f/(\pi w_{coll})$. The collapse of a beam approaching a focus is mirrored symmetrically by expansion of the diameter (diffraction) as the beam propagates beyond the focus. The smaller the focal spot diameter, the more rapid the collapse and expansion. At a distance $z_0 = (1/2)kw_0^2$, the waist has increased by a factor sqrt(2). The Rayleigh-range of length $2z_0$, centered on the focal plane is used to define the DOF. Structure within the DOF is imaged with a lateral definition of within a factor of sqrt(2) of that at the focal plane. Sample regions outside the DOF are considered out of focus, and suffer from dramatically reduced lateral resolution and signal amplitude. If the DOF is matched to the typical image penetration depth of 1–2 mm, then the lateral resolution is limited to approximately 20 μm.

A key feature of OCT is the decoupling of axial and lateral resolution. Whereas the source characteristics determine the axial resolution, the lateral resolution is defined through the focusing optics. The DOF and the lateral focus are directly coupled, however, and impose a limit on the lateral focusing.

2.3 *Polarization of light*

It is important to consider that electromagnetic waves are vector waves and their oscillations are beating in a spatial direction. In the realm of paraxial optics, which is well suited to describe OCT and only excludes very tightly focused beams, the field oscillations are orthogonal to the direction of the beam propagation. Within that orthogonal plane, they can vibrate along different directions, or follow a specific elliptical pattern as the light propagates. Most incoherent light sources such as incandescent light bulbs or sunlight are unpolarized, which means that their polarization states vary rapidly. Reflected and scattered light can become polarized, and polarization filters are frequently used in photography or in sunglasses to block this polarized light. If the vibrations are confined to a single direction along the propagating beam, the light is said to be linearly polarized. The Jones formalism more generally describes the polarization of a coherent wave as the superposition of two beams with linear polarizations, but polarized orthogonally to each other, along x and y, respectively:

$$\mathbf{u}(z) = \begin{bmatrix} U_x(z) \\ U_y(z) \end{bmatrix} = \begin{bmatrix} U_{x0}e^{i\phi_x}e^{-ikz} \\ U_{y0}e^{i\phi_y}e^{-ikz} \end{bmatrix} = \begin{bmatrix} \cos\gamma \\ \sin\gamma\, e^{i\delta} \end{bmatrix} U_0 e^{i\phi_x}e^{-ikz} \tag{5}$$

γ determines the ratio between the two orthogonal polarization states, and $\delta = \phi_y - \phi_x$ defines a phase offset. Lower case bold letters designate vectors. Linearly polarized light has a zero phase offset, and γ defines the direction of the linear polarization along the x and y coordinates. For a non-zero δ, the light is elliptically polarized, where the position of the maximum instantaneous field amplitude describes an elliptical movement around the axis of propagation. For $\gamma = \pi/4$ and $\delta = \pm\pi/2$, the light is right and left circularly polarized, respectively.

The propagation of the light through an optical element or a sample can alter the polarization state and is described by a 2×2 complex valued Jones matrix:

$$\mathbf{u}_{\text{out}} = \mathbf{J} \cdot \mathbf{u}_{\text{in}}. \tag{6}$$

$\mathbf{J}$ can be retarding, diattenuating, or a combination of both (matrices are designated by upper case bold letters). Retardation describes a differential phase or delay, and diattenuation a relative attenuation between two eigenstates. The eigenstates describe the two polarization states that pass through the system described by $\mathbf{J}$ without changing their polarization state, and are orthogonal to each other for homogeneous Jones matrices.[9]

The intensity or power of the field described by a Jones vector is the sum of the squared amplitudes of both field components. Accordingly, only light fields polarized parallel to each other can interfere:

$$\begin{aligned} P = \left|\mathbf{u}_{tot}\right|^2 &= \left|\mathbf{u}_1 + \mathbf{u}_2\right|^2 \\ &= \left|\mathbf{u}_1\right|^2 + \left|\mathbf{u}_2\right|^2 + U_1^{x*}U_2^{x} + U_1^{x}U_2^{x*} + U_1^{y*}U_2^{y} + U_1^{y}U_2^{y*}. \end{aligned} \tag{7}$$

In order to efficiently interfere the reference and sample field of Eq. (1), the two fields should be polarized parallel to each other.

In addition to the Jones formalism, the Stokes formalism is frequently employed to describe the polarization states of light. A Stokes vector can be constructed from its corresponding Jones vector as:

$$
\mathbf{s} = \begin{bmatrix} I \\ Q \\ U \\ V \end{bmatrix} = \begin{bmatrix} |U^x|^2 + |U^y|^2 \\ |U^x|^2 - |U^y|^2 \\ 2\mathrm{Re}\left\{U^{x*}U^y\right\} \\ -2\mathrm{Im}\left\{U^{x*}U^y\right\} \end{bmatrix}.
\tag{8}
$$

The first I-component describes the intensity of the beam, and the three last components define its polarization state. Normalizing these last three components enables the powerful visualization of the polarization states as a 3D vector on the Poincare sphere.

In analogy with the Jones matrix, the Mueller matrix $\mathbf{M}$ describes the relation between input and output Stokes vectors, propagating through an optical element: $\mathbf{s}_{\mathrm{out}} = \mathbf{M} \cdot \mathbf{s}_{\mathrm{in}}$. A retarding element rotates the Q, U, V-vector by the retardation angle around an optic axis on the Poincare sphere.

When constructing a Stokes vector directly from a Jones vector, the intensity is equivalent to the norm of the Q, U, V-vector, which means that the light field is fully polarized. However, unlike the Jones formalism, which is restricted to the description of fully coherent and polarized fields, the Stokes formalism extends to partially coherent and partially polarized light. In this case, it is insightful to describe the degree of polarization:

$$
\mathrm{DOP} = \frac{\sqrt{Q^2 + U^2 + V^2}}{I}.
\tag{9}
$$

The DOP is unity for a fully polarized beam, but decreases towards zero as the light becomes depolarized. It is a measure of the variation of the polarization states.

2.4 Instrumentation

2.4.1 Interferometer and receiver

The basic building block of an OCT instrument is an interferometer. To render it rugged and robust, most OCT systems are built with fiber optics

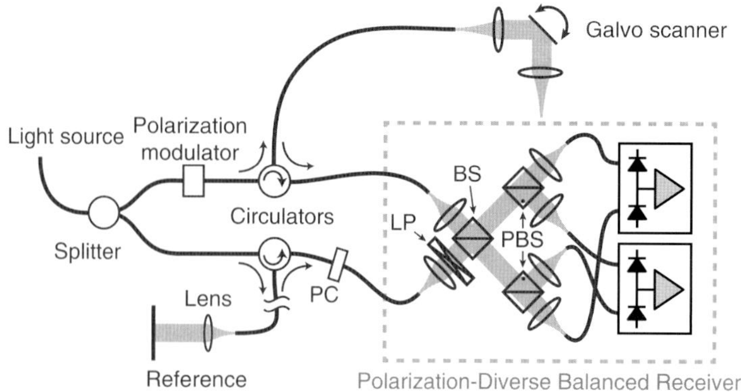

Figure 3. Schematic layout of a fiber-based interferometer for OCT imaging. BS: Beam splitter; PBS: Polarizing beam splitter; PC: Polarization controller; LP: Linear polarizer.

components. Single mode fibers guide coherent light in a narrow diameter inner core and can be interfaced with a range of fiber-based optical components. These do not have to be aligned as the light is inherently guided within the fiber and make the system resilient to thermal drift and mechanical variations. Figure 3 displays the schematic of a fiber-based OCT system for use with a wavelength-swept laser source. The fiber carrying the source light is divided into two separate fibers by a fiber coupler, creating the sample and reference paths. Both signals are fed through a fiber circulator. These non-reciprocal elements enable to efficiently relay the signals to the sample and the reference mirror, and the signals reflected from there to the detector. Unlike the double pass through the beam splitter of the free-space configuration, which directs half of the reflected signal back towards the light source, the circulators prevent this signal loss and provide an inherent signal advantage. The sample arm is interfaced with the sample scanner, whereas in the reference arm, the light is reflected off a mirror. The reference mirror is mounted on a translation stage to enable rapid and easy adjustment of the precise path length.

At the detector, the sample field is combined with the reference field to interfere, by means of a fiber coupler or a free-space beam splitter. Both of the two outputs of the splitter feature an interference signal, however, with opposite signs. If the interference is constructive in one output, it is destructive at the other. This opens the possibility of balanced detection: Detecting independently both interference signals, but amplifying and digitizing only their difference effectively eliminates the P_{dc} of Eq. (1) and conserves only the interference terms of interest. This approach also removes amplitude noise that is common to both channels and preserves dynamic range. The electronic signal from the fast dual-balanced detectors is digitized with a fast acquisition card and moved to the host computer for reconstruction of the tomogram.

Conventional optical fibers do not conserve the polarization state of the guided light. Indeed, mechanical stress and even slight variations in the fiber curvature will change its polarization state. Whereas the reference arm can be well protected and adjusted to output a known polarization state, the sample arm frequently is subject to movement when adjusting the sample scanner, and further the sample itself can alter the polarization state of the light. As a result, the polarization of the sample signal may not be aligned with that of the reference signal. Using a polarization diverse receiver avoids this shortcoming by projecting the sample signal on two polarization states orthogonal to each other, by using a polarizing beam splitter. The reference arm is polarized with a linear polarizer oriented at 45° with respect to the linear polarizations of the polarizing beam splitter, to provide an equal reference signal in both polarization channels. This polarization splitting is performed identical on both outputs of the first mixing beam splitter to achieve polarization diverse balanced detection. This way, the sample signal is efficiently detected, independent of its polarization state. In addition, from the detected amplitudes and phases of the two channels, the Jones vector of the sample light can be reconstructed.

For PS-OCT, an additional polarization modulator may be placed in the sample arm before the circulator, alternating the polarization state incident on the sample in between A-lines, for reasons discussed in the following sections.

2.4.2 *Light source*

The spectral width and central wavelength of the light source define the axial resolution of OCT, as defined by Eq. (4). OCT is primarily performed around 800 nm and 1300 nm. Imaging at 800 nm provides better axial resolution, with reports of sub-micron resolutions.[10,11] On the other hand, imaging around 1300 nm offers increased imaging depth. Scattering is stronger at shorter wavelengths, and the wave front of the incident sample beam degrades more rapidly at 800 nm than at 1300 nm. Although the absorption of water increases with longer wavelengths, 1300 nm avoids the stronger absorption of blood at shorter wavelengths, and overall results in large imaging depths of up to 2 mm, which is favorable for imaging of the skin and the endothelium. The axial resolution at 1300 nm is typically 5–10 μm.

For OCT with a wavelength-swept laser source, such as OFDI, the A-line rate is defined by the wavelength tuning speed of the light source. A-lines rate in the MHz range have been demonstrated, but the digitization of the resulting interference signal becomes challenging.[12,13] Also, at constant source power, an increase of sweep speed translates directly to a reduced number of photons per wavenumber, and results in a decreased signal strength and depth of imaging. Current commercial instruments operate at A-line rates around 100 kHz. The imaging speed for spectrometer-based systems is defined by the acquisition rate of the line-camera in the spectrometer and is generally lower than that of swept source systems.

Most sources produce an output power between 20–60 mW, resulting typically in ~8–20 mW power incident on the sample. The absorption at the OCT wavelengths is sufficiently low that the delivered power does not induce any notable tissue heating.

To reliably reconstruct the tomogram of the sample, the spectrum as a function of the wavenumber has to be recovered. The ideal wavelength tuning of a wavelength-swept laser source varies the wavenumber linearly with time. Any deviation has to be compensated with either a non-uniform sampling clock or interpolation of the digitized signal. Both can become challenging for strong non-linearity. Similar limitations apply to spectrometer-based systems.

The spectral resolution of the recorded spectrum defines the imaging range, i.e. the axial range within which a sample can be imaged. It is limited by the instantaneous line-width for a wavelength-swept laser source, and the spectrometer resolution for camera-based systems. The strongest signal with the highest signal-to-noise ratio (SNR) is achieved at a zero path-length difference between the sample and the reference signal. With increasing offset, the interference fringes feature an increased modulation frequency, and are eventually washed out when approaching the spectral resolution limit of the system. Imaging ranges of 6–10 mm are frequently achieved in state of the art OCT systems at 1300 nm.

2.4.3 *Sample scanning*

As previously discussed and illustrated in Figure 2, the probing beam is scanned across the sample with a galvanometric mirror system. In a fiber-based OCT system, the light from the sample fiber is collimated and directed onto the scanning mirrors, and then focused onto the sample with an objective lens. The focal length and numerical aperture of the objective lens define the lateral resolution as well as the field of view. It is possible to achieve lateral scans of 3 cm at a lateral resolution of ~30 μm. If the lateral resolution is improved by shortening the focal length, while maintaining the same numerical aperture, then the field of view is reduced by the same factor as the resolution is improved.

Because the scanner is interfaced with the imaging interferometer through an optical fiber, it can be made mobile and integrated into a handheld scanner for imaging of different sites in patients or large animals. Alternatively, the scan unit can also be static and the sample placed below the objective lens.

Although optical imaging can be performed without any sample contact, it is generally preferable, if possible, to immerse the sample in a transparent medium (water or ultrasound gel) and image it through a glass plate. This reduces the scattering of the probe beam at the sample surface and decreases sample motion, which can be critical when imaging live samples such as small animals or human patients.

3. Polarization Sensitive Optical Coherence Tomography

3.1 *Polarization properties of tissue*

From the two channels of the polarization diverse detection two depth-profiles can be reconstructed for each A-line. At each depth-location of these profiles, the two complex values of the reconstructed tomograms correspond indeed to the two elements of the Jones vector of the light reflected at this depth. Following Eq. (8), the corresponding Stokes vector can be readily retrieved. Observing how these Stokes (or Jones) vectors vary as a function of depth and lateral position provides insight into the polarization properties of the imaged tissue. In a birefringent sample, light polarized along the fast axis of the medium travels at a larger speed than light polarized along the slow axis, defining effectively a differential refractive index Δn. A birefringent sample acts as a retarding element and causes the polarization states to change with depth. Tissue with a regular fibrillar architecture such as collagen or muscle exhibits birefringence. Light polarized along or orthogonal to these fibers travels at a different speed within the sample. The important role of collagen in wound healing and scarring make this contrast mechanism very interesting for the study of these mechanisms. If collagen is present, but in a less ordered, intercalated arrangement, such as normal healthy skin, then the resulting variation of the polarization states along depth is less pronounced. Because within a single resolution volume of OCT various collagen bundles may have different orientations, the resulting polarization state is a superposition of the effect of the individual bundles and becomes more arbitrary. Also, between neighboring A-lines, the detected polarization states may vary significantly. Averaging the Stokes vectors of such neighboring A-lines results in a reduced DOP, revealing the depolarizing property of such a tissue. In comparison, adjacent Stokes vectors of tissue with regularly arranged

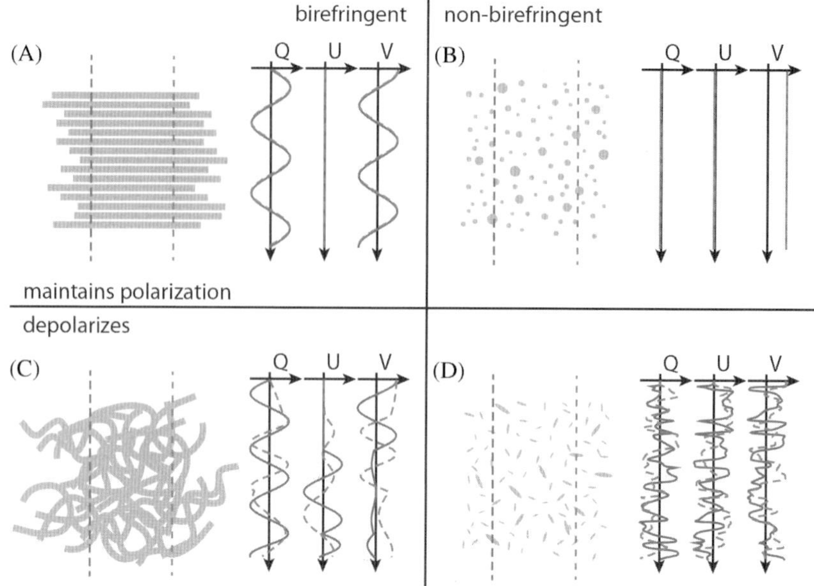

Figure 4. Tissue polarization properties and impact on the measured polarization states. (A) Regularly arranged fibrillar collagen results in a continuous variation of the polarization states. The normalized Q, U, and V components are visualized for two distinct lateral positions on the sample schematic. (B) Irregularly arranged collagen bundles result in irregular variations of the polarization states along depth, and distinct between different lateral locations. (C) Non-birefringent tissue with isotropic scatterers generates constant polarization states. (D) Anisotropic scatterers induce a random variation of the polarization states, both along depth and in the lateral direction.

collagen bundles are almost identical and yield a very high DOP. These mechanisms of light-tissue interactions are illustrated in Figure 4. In comparison, scattering in non-birefringent tissue with isotropic scatterers results in constant Stokes vectors along depth, indicating that the tissue does not alter the polarization state of the probing light. However, if the microscopic scatterers that compose the tissue are anisotropic and scatter light with a certain polarization states more strongly than that with the orthogonal polarization, a similar depolarization effect is achieved. Although such a tissue is not birefringent, it randomly alters the polarization states along depth, and does so differently for adjacent A-lines.

For the present purpose of imaging wound healing and scarring, the most relevant mechanism is birefringence, induced by the presence of collagen. It is primarily fibrillar collagen (types I and III) that generates tissue birefringence.[14] Depending on its spatial ordering, in addition to birefringence, it can also induce depolarization.[15]

3.2 *Reconstructing tissue birefringence*

The variation of the detected polarization state along depth not only depends on the sample properties, but also the polarization state incident on the sample. Although the tissue may be birefringent, if the incident state is linearly polarized and accurately aligned with the collagen fibers, then it propagates along the fast or slow axis and does experience any alteration. In contrast, linearly polarized light oriented at 45° to the fibrils would exhibit the most significant variation. To recover an unambiguous picture of the sample's polarization properties, two input polarization states are used in most PS-OCT systems.[16] Various modalities and implementations have been developed for PS-OCT. Here, we focus on PS-OCT using an inter-A-line modulation scheme of the incident polarization states and reconstruction using the Stokes formalism.[17] As indicated in Figure 3, a polarization modulator is used in the sample arm to vary the polarization state incident on the sample in between A-lines by 90° on the Poincare sphere. This would correspond, for instance, to linearly and circularly polarized light. Although the precise polarization state at the sample remains unknown because of the propagation through the sample fiber, the angle between the two modulated states is preserved (if the fiber is free of diattenuation, which is a safe assumption).

Figure 5 displays a 6-month old burn scar on the back of a rat. The Q component of one input polarization state is displayed and reveals a rapid variation along depth in the lower left corner. In order to obtain a clear

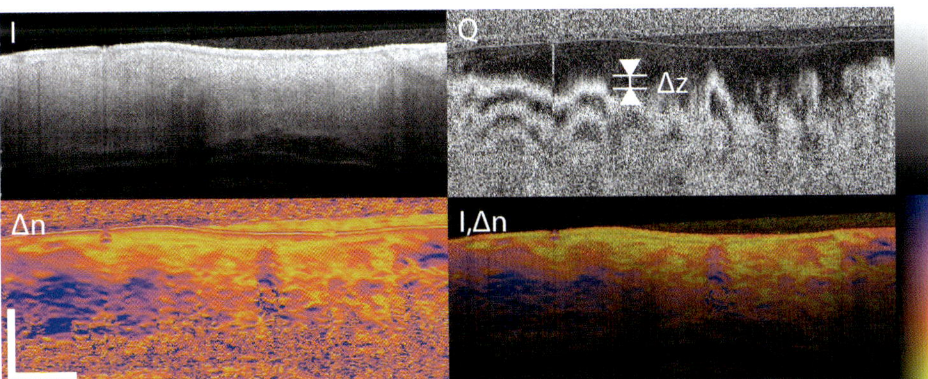

Figure 5. Principle of PS-OCT, illustrated on a 6-months old burn scar in a rat. I displays a cross-sectional structural intensity tomogram. Q shows the normalized Q-component of the Stokes vector of one input polarization state. Δn is the tissue birefringence. The colormap displays a range of Δn from 0–2.2×10^{-3}. I, Δn is the overlay of the structural intensity signal in brightness with the birefringence colormap. Scale bars: 1 mm.

picture of the depth-resolved birefringence of the tissue, we compare the Stokes vectors for the two measured input polarization states across a differential depth Δz of ~30 μm at each position in the tomogram. Assuming that the tissue acts as a pure retarder, the pairs of Stokes vectors at the differential depths define a rotation axis and a rotation angle on the Poincare sphere that maps them onto each other. Using trigonometric operations, this rotation angle φ can be recovered and related to the actual tissue birefringence $\Delta n = \varphi/(2k_c\Delta z)$, where $k_c = 2\pi/\lambda_c$ is the mean wavenumber (λ_c is the mean wavelength) of the source spectrum. Δn indicates the refractive index difference between the slow and fast axis of the birefringent medium. Δn is displayed in a (near) isoluminant colormap[18] in Figure 5, and reveals high birefringence in the lower left corner, as expected from the variation of the Q-component. Close to the surface, there is only little birefringence, but at larger depth there is increased birefringence throughout the entire imaged region. The isoluminant colormap enables to overlay the birefringence map with the structural intensity signal as a brightness channel, as shown in the last panel of Figure 5. This combined view offers an intuitive view of the sample, revealing both structural features and tissue birefringence in a single image. The intensity overlay masks regions of little or no signal, where the recovered birefringence is noisy and random.

Like other coherent imaging methods, the tomograms of OCT are dominated by speckle. Speckle refers to the granular or grainy appearance present in the images of many tissues imaged with OCT. The signal of homogenous tissue regions is strongly textured, exhibiting a grainy pattern with spots or "speckle" of bright, high intensity signal, alternating with regions of very low signal. The size of speckle corresponds to the spatial resolution of OCT along the axial direction and transverse direction, respectively. The measured polarization states in dark speckle are dominated by noise. Because of the prevalence of image pixels with low intensity, averaging is needed to recover meaningful measurements of the depth-dependent polarization states and reconstruct the depth-resolved tissue birefringence. This is most conveniently achieved by incoherently averaging the reconstructed Stokes vectors over a small neighborhood of pixels, covering a surface equivalent to that of about five speckle. In addition to this intrinsic noise, system induced polarization distortions can further limit the recovery of meaningful tissue birefringence and even lead to artificial birefringence features.[19] We have developed a strategy to overcome these limitations and recover reliable and robust imaged of tissue birefringence by using an approach termed "spectral binning".[20] The reconstruction process is slightly more involved in this case, but the result is conceptually equivalent to the discussed approach. Indeed, all presented PS-OCT results in this chapter have been processed with spectral binning.

3.3 *Measuring depolarization properties*

In addition to improving the accuracy of the measured polarization states, the averaging of the Stokes vectors also provides insight into depolarization effects. As previously discussed, the DOP resulting from the averaged Stokes vectors is reduced in regions of randomized polarization states. This DOP or DOP-uniformity can be used as additional contrast.[21] Together with the tissue birefringence and intensity overlay, Figure 6 also displays the DOP of the rat burn, of an extended cross-section, using the same colormap and overlay with the structural intensity signal. Unlike the birefringence, which is depth-resolved, the DOP is a cumulative measure, indicating at each sample depth how much the detected polarization states have been depolarized by traveling through all preceding tissue layers. With increasing imaging depth the SNR is reduced, and despite averaging it eventually leads to random polarization states and reduced DOP. This can be observed on the left portion of the imaged burn scar in Figure 6, where the DOP remains close to unity until deep inside the sample, and then decreases together with the vanishing intensity signal. On the right side, however, the DOP decreases more rapidly with depth, and already features low values at depths where the intensity signal is still appreciable, indicating true sample-induced depolarization.

The lower row of Figure 6 shows histological sections matching the cross-section imaged with PS-OCT. Specific regions of interest are highlighted and magnified in Figure 7. The histological sections were either stained with

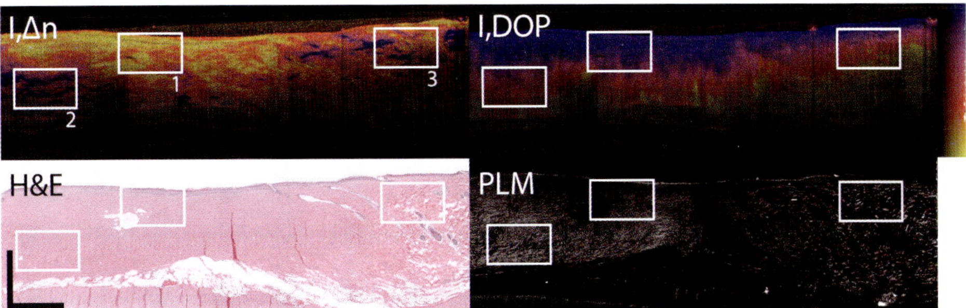

Figure 6. Birefringence and intensity overlay (I, Δ*n*) and DOP and intensity overlay (I, DOP) of a 6 months old burn scar in a rat. Matching histology (H&E) and unstained section, imaged with PLM. Regions of interest (ROI) are magnified in Figure 7. Color range indicates birefringence from $0\text{--}2.2 \times 10^{-3}$ and DOP from 0.5–1. Scale bars: 1 mm.

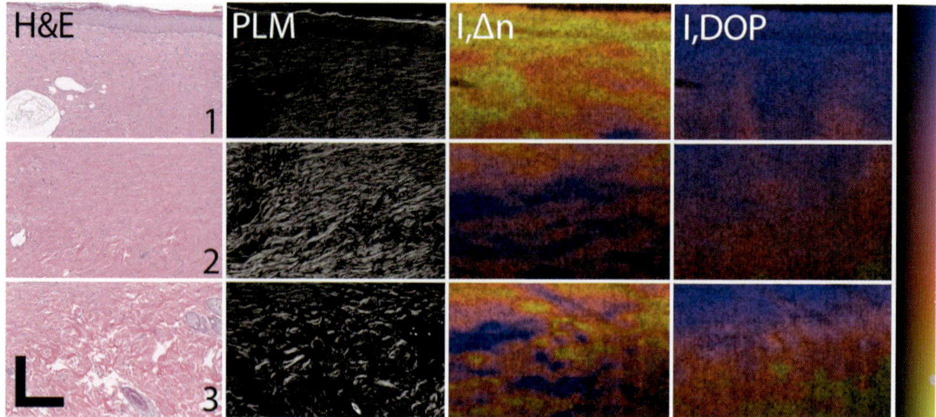

Figure 7. Magnified ROIs indicated in Figure 6, comparing histology (H&E, PLM) with PS-OCT (I, Δn, and I, DOP). Color range indicates birefringence from 0–2.2 × 10^{-3} and DOP from 0.5–1. Scale bars: 250 μm.

hematoxylin and eosin (H&E) and imaged with a bright light microscope, or remained unstained and were imaged with polarized light microscopy (PLM), using crossed circular polarizers. In PLM, regions of the sample that are birefringent produce a bright signal, whereas sample structure without birefringence remains dark. PLM thus measures the same tissue property as PS-OCT, albeit in a different sample orientation. PS-OCT measures the birefringence for light traveling orthogonal to the tissue surface into and out of the tissue, whereas PLM images cross-sections of tissue, corresponding to light traveling in parallel to the tissue surface.

Comparison with histology reveals that the imaged region corresponds to the edge of the scar, with normal-looking skin on the right-hand side. The zone with high birefringence in PS-OCT also results in a bright PLM signal, and matches the center of the scar, with dense maturing collagen fibers, oriented primarily parallel to the tissue surface. Closer to the surface, the collagen bundles appear smaller and less mature, resulting in less birefringence, both in PS-OCT and PLM. Both these regions are characterized by a high DOP, decreasing only very slightly along depth. In contrast, the region of normal skin generates a heterogeneous PS-OCT signal, with spots of very high birefringence, alternating with less birefringent zones. In a similar way, the PLM reveals a few very bright bundles, but organized in a loose mesh, surrounding regions without any signal. The DOP decreases rapidly in this area, consistent with the perturbing effect of these mature collagen bundles, oriented at varying angles, on the polarization states detected with PS-OCT.

4. Imaging Wound Healing

As discussed in the previous sections, OCT offers fast and non-invasive optical imaging of subsurface microstructures to a depth of 1–2 mm. Whereas this imaging depth is modest compared to medical imaging methods such as ultrasound, computed tomography, and magnetic resonance imaging, it largely outperforms other optical imaging methods, such as confocal microscopy, which is limited to imaging in the first few 100 μm of tissue. The spatial resolution of OCT is around 10 μm, again in between the superior resolution of high-resolution optical microscopy, and the much poorer resolution of conventional medical imaging techniques. The excellent sensitivity of OCT enables high volume imaging rates, and lateral fields of views of several square centimeters can be scanned in a matter of seconds. It is the ability to comprehensively image such large (in optical standards) fields of view that distinguishes OCT from optical microscopy.

OCT lacks the molecular specificity and contrast available with nonlinear high-resolution microscopy. These microscopy methods offer very detailed views of the structural and molecular mechanisms involved in wound healing, however, they are generally limited to fields of views of square millimeters, and suffer from long acquisition times. Whereas this may be insightful in preclinical research, it is generally incompatible with a clinical setting and imaging in human patients.

OCT measures intrinsic sample contrast, which makes it amenable to clinical imaging. The structural signal provides valuable insight into wound healing and scar formation. With PS-OCT, this structural signal is further complemented with additional contrast related to the collagen content and structural arrangement. Although this structural arrangement is beyond the spatial resolution of PS-OCT, the measures of birefringence and depolarization give access to these features, all while preserving a large field of view.

The original and driving application for OCT is ophthalmology, where it has rapidly become the gold standard for assessing retinal morphology in clinical practice.[22] Other emerging applications are imaging of the coronary arteries and the esophagus with catheter-based OCT.[23] In the coronary arteries, OCT reveals the morphology of atherosclerotic plaques and can improve the guidance of percutaneous coronary interventions. In the esophagus, it serves to monitor and identify dysplastic regions of the esophageal epithelium in Barrett's esophagus, a precursor lesion of esophageal cancer.

Skin, scars, and wound healing have been imaged since early on in the development of OCT technology. Yeh *et al.* imaged wound healing after thermal injuries in an *in-vitro* skin-equivalent tissue model with OCT and multiphoton microscopy.[24] OCT clearly identified thermally injured regions with a hypoechoic signal at early time-points. However, during the ongoing wound healing, the structural intensity signal did not exhibit much alteration, while the deposited collagen continuously matured, whereas this could be clearly visualized using nonlinear microscopy.

Cobb *et al.* used an early high-resolution OCT system to image cutaneous wound healing after punch wounds in the ears of mice and compared the OCT imaging with matching histological sections.[25] The authors developed an algorithm to automatically detect the epidermal–dermal junction. The authors also identified granulation tissue and remodeling in the dermis, although this requires a high degree of skill in reading OCT tomograms to identify the subtle differences in the structural intensity signal.

The development of therapeutic methods to treat scars or rejuvenate skin would benefit from monitoring tools. Tsai *et al.* have used OCT to monitor the healing process in human skin after fractional laser treatments.[26] The authors investigated ablative and non-ablative treatment modes and compared the time-course of the micro-structural changes of the miniature ablation sites.

Recently, Greaves *et al.* performed a rigorous study assessing the viability of OCT as an alternative to invasive histological assessment of acute wound healing in human skin.[27] Comparison with histology confirmed that the features observed in histology could likewise be visualized in OCT, and could offer a non-invasive, less painful, and faster alternative to punch-biopsies. However, detailed comparison of the geometry of the observed features did not yield exact statistical equivalence.

PS-OCT has been investigated as a tool to assess burn depth in thermal injuries in skin.[17] Thermal denaturation of collagen destroys the natural birefringence of skin, which can be used to identify the extent of the burn. Recent efforts investigated this signature in pediatric burns.[28] PS-OCT has also been used for quantitative assessment of wound healing of punch-wounds in ears of rabbits.[29] Quantitative measures of the tissue birefringence were used to compare the influence of different topically applied drugs on the time-course of the wound healing.

The described PS-OCT studies used previous versions of PS instrumentation and reconstruction methods. Rather than visualizing the local tissue birefringence, they extracted the cumulative retardation from the tissue surface to a given sample depth. This view offers only a very convolved picture of the sample birefringence and is difficult to interpret. We think that our recent efforts in suppressing polarization distortion artifacts and reconstructing improved views of depth-resolved tissue birefringence, combined with the assessment of tissue depolarization, renews the interest in PS-OCT as a powerful instrument for monitoring of wound healing and warrants additional studies of its use in preclinical and clinical research. In the following sections, we will briefly highlight two experiments conducted with our PS-OCT platform for the assessment of wound healing and scar formation of burn injuries and excisional wounds in rats, respectively.

4.1 *Imaging burn wounds*

Figure 8 displays cross-sectional views, measured with PS-OCT at various time-points after inducing third degree burns on the back of a rat. These results were adapted from Golberg *et al.*[30] At each time point, an entire volume was imaged, and the identical matching longitudinal cross-section could be selected for all time-points. Immediately after applying the burn, the structural signal exhibited a homogeneous and diffuse texture and less scattering than control normal skin. The birefringence was uniformly low and the DOP remained high until deep into the tissue, in very stark contrast to control normal skin. At 24 h, the scattering had somewhat recovered, but otherwise both the structural and polarization features remained similar to those at 0 h. After 1 week, a scab had formed, that was clearly visualized with OCT, and which fell off between 2 and 3 weeks after creating the injuries. At 5 weeks, the forming scar exhibited uniform and homogenous high scattering. The birefringence was mostly low, with a few locations that showed a slight increase in birefringence. The DOP remained very high until deep into the tissue. This is consistent with a healing scar that primarily consists of newly secreted collagen bundles that are still very thin and immature, leading to no noticeable birefringence. After 6 months, the structural

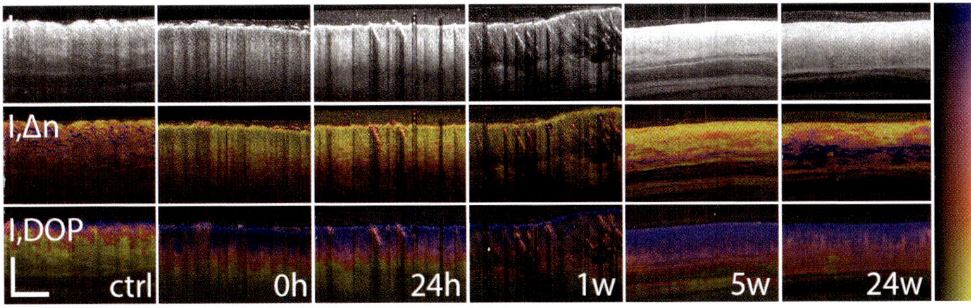

Figure 8. PS-OCT time series of burn wound in rat, compared to another location of normal control skin (ctrl), showing intensity (I), birefringence and intensity overlay (I, Δn), and DOP and intensity overlay (I, DOP). Color range indicates birefringence from 0–2.2×10^{-3} and DOP from 0.5–1. Scale bars 1 mm. Figure adapted from Ref. 30.

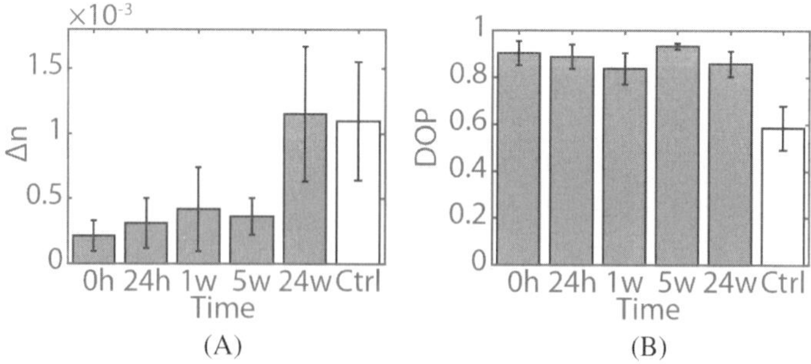

Figure 9. Quantitative analysis of volume data of burn lesions displayed in Figure 8, expressing the mean birefringence and DOP of a cylindrical volume, aligned on the lesion center, extending from 250 μm to 750 μm with a diameter of 1.4 mm. Figure adapted from Ref. 30.

intensity signal still looked very similar to that at 5 weeks. The birefringence, however, revealed that the collagen bundles had matured, and became thicker, and now generated a significant amount of birefringence. These findings were confirmed with histology.[30]

To highlight the quantitative nature of the polarization measurements, we defined cylindrical regions of interest with a diameter of 1.4 mm on the volume tomograms, extending from the surface into the scar tissue, aligned on the lesion center. We then averaged the birefringence and DOP within the depth region spanning 500 μm from 250 μm below the surface to 750 μm. The results and standard deviations of the values in the analyzed tissue volumes are displayed in Figure 9. As expected, the birefringence exhibits a gradual increase, almost exceeding the birefringence of control normal skin at the last time-point. The DOP remains very high throughout all time-points, and significantly above the values of normal skin.

4.2 *Imaging excisional wounds*

In a similar study, we investigated excisional wounds. The results in this section are adapted from Lo *et al.*[15] To induce a scar that bears more similarity with hypertrophic scars (HTS) in humans, a tension device was installed on day 4 after inducing a linear full-thickness incision on the back of a rat. This device was daily adjusted to maintain constant tension applied to the wound, for up to 10 days. Figure 10 shows the birefringence and DOP images of the wound right after the incision, after 10 days of applied tension, and 1 month after removing the tension device. Similar to the burn wounds, the freshly created lesions are characterized by low birefringence and high DOP, which enable an accurate quantitative assessment of the cross-sectional lesion size, as we have confirmed by comparison with

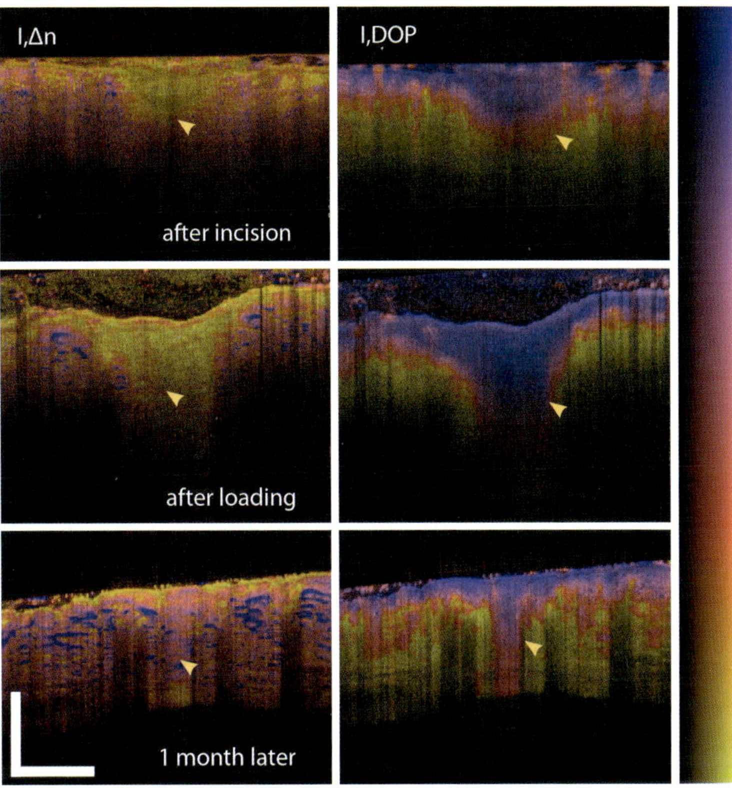

Figure 10. PS-OCT of an excisional wound, right after incision (top row), after installing a tension device for 10 days (middle row), and 1 month after removing this device (bottom row). The columns show the overlay of birefringence with intensity (I, Δn) and the overlay of DOP and intensity (I, DOP), respectively. Color range indicates birefringence from $0-2.2\times10^{-3}$ and DOP from 0.5–1. Scale bars 1 mm. Figure adapted from Ref. 15.

histology.[15] After 1 month of healing, the birefringence had increased, consistent with the maturation of the collagen. The healing also led to a slight decrease in the DOP signal, partly due to the thinning of the remaining scar.

To better visualize this ongoing contraction of the scar, we generated en face views of the healing scar at weekly time points, up to 1 month after removing the tension device. Figure 11 displays the average projections along depth from 150 μm to 250 μm of the birefringence and intensity overlay, as well as the DOP and intensity overlay. The initially low birefringence is approaching the birefringence of the surrounding normal skin at 4 weeks. However, the scar remains characterized by a more homogenous birefringence signal as compared to the very heterogeneous appearance of the birefringence in normal skin. The DOP still offers very pronounced contrast to normal skin even at 4 weeks, but nicely illustrates the continuous thinning of the scar with progress in healing.

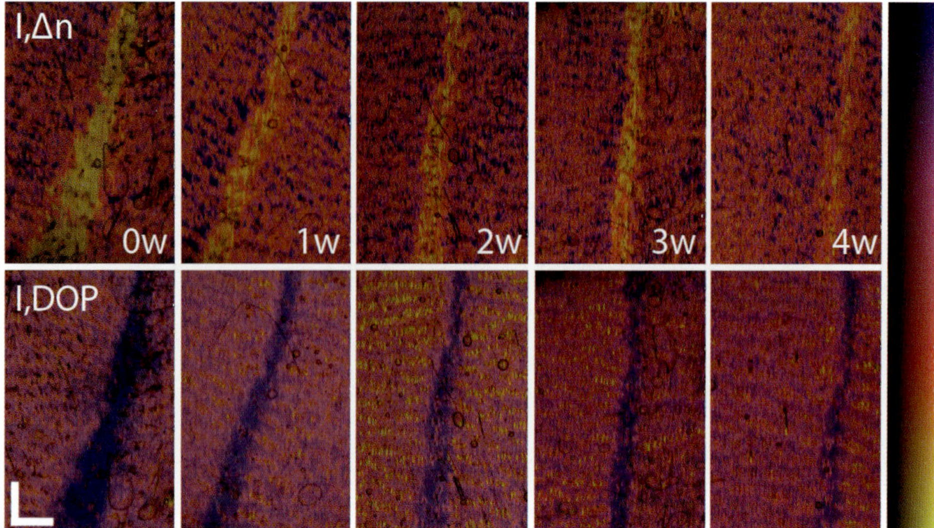

Figure 11. PS-OCT en face views of an excisional wound, imaged every week up to 1 month after removal of the tension device. The views show the overlay of birefringence with intensity (I, Δn) and the overlay of DOP and intensity (I, DOP), respectively, after averaging the intensity, the birefringence, and the DOP from 150 μm to 250 μm along depth. Color range indicates birefringence from 0–2.2×10^{-3} and DOP from 0.5–1. Scale bars 1 mm. Figure adapted from Ref. 15.

5. Conclusions

The spatial resolution, field of view, and imaging depth of OCT are midway between high-resolution optical microscopy techniques and ultrasound imaging. OCT is well suited for imaging *in vivo* in small animals, but also in a clinical setting in human patients, due to its high imaging speed and intrinsic contrast. PS-OCT offers additional metrics related to the content and conformation of collagen, which are of particular interest in the study of wound healing and scarring.

Many features that can be appreciated in histological sections of healing wounds imaged with bright field microscopy can also be assessed non-invasively, *in vivo*, and *in situ* with PS-OCT, offering a wide range of imaging metrics for the analysis of longitudinal observations of wound repair.

References

1. Huang, D. *et al.* Optical coherence tomography. *Science* **254**, 1178–1181 (1991).
2. de Boer, J. F. and Milner, T. E. Review of polarization sensitive optical coherence tomography and Stokes vector determination. *J Biomed Opt* **7**, 359–371 (2002).

3. Leitgeb, R. Hitzenberger, C. K. and Fercher, A. F. Performance of Fourier domain vs. time domain optical coherence tomography. *Opt Express* **11**, 889–894 (2003).

4. de Boer, J. F. *et al.* Improved signal-to-noise ratio in spectral-domain compared with time-domain optical coherence tomography. *Opt Letters* **28**, 2067–2069 (2003).

5. Yun, S. Tearney, G. de Boer, J. Iftimia, N. and Bouma, B. High-speed optical frequency-domain imaging. *Opt Express* **11**, 2953–2963 (2003).

6. Yun, S. Tearney, G. de Boer, J. and Bouma, B. Removing the depth-degeneracy in optical frequency domain imaging with frequency shifting. *Opt Express* **12**, 4822–4828 (2004).

7. Vakoc, B. J. Yun, S. H. Tearney, G. J. and Bouma, B. E. Elimination of depth degeneracy in optical frequency-domain imaging through polarization-based optical demodulation. *Opt Lett* **31**, 362–364 (2006).

8. Saleh, B. E. A. and Teich, M. C. *Fundamentals of Photonics*. John Wiley and Sons, (2013).

9. Lu, S.-Y. and Chipman, R. A. Homogeneous and inhomogeneous Jones matrices. *J Opt Soc Am A* **11**, 766–773 (1994).

10. Drexler, W. *et al. In vivo* ultrahigh-resolution optical coherence tomography. *Opt Lett* **24**, 1221–1223 (1999).

11. Liu, L. *et al.* Imaging the subcellular structure of human coronary atherosclerosis using micro-optical coherence tomography. *Nat Med* **17**, 1010–1014 (2011).

12. Wojtkowski, M. High-speed optical coherence tomography: basics and applications. *Appl Opt* **49**, D30–61 (2010).

13. Klein, T. *et al.* Multi-MHz retinal OCT. *Biomed. Opt Express* **4**, 1890 (2013).

14. Nadkarni, S. K. *et al.* Measurement of collagen and smooth muscle cell content in atherosclerotic plaques using polarization-sensitive optical coherence tomography. *J Am Coll Cardiol* **49**, 1474–1481 (2007).

15. Lo, W. C. Y. *et al.* Longitudinal, 3D imaging of collagen remodeling in murine hypertrophic scars *in vivo* using polarization-sensitive optical frequency domain imaging. *J Invest Dermatol* **136**, 84–92 (2016).

16. Saxer, C. E. *et al.* High-speed fiber-based polarization-sensitive optical coherence tomography of *in vivo* human skin. *Opt Lett* **25**, 1355–1357 (2000).

17. Park, B. H. Saxer, C. Srinivas, S. M. Nelson, J. S. and de Boer, J. F. *In vivo* burn depth determination by high-speed fiber-based polarization sensitive optical coherence tomography. *J Biomed Opt* **6**, 474–479 (2001).

18. Geissbuehler, M. and Lasser, T. How to display data by color schemes compatible with red-green color perception deficiencies. *Opt Express* **21**, 9862–9874 (2013).

19. Villiger, M. *et al.* Artifacts in polarization-sensitive optical coherence tomography caused by polarization mode dispersion. *Opt Lett* **38**, 923–925 (2013).

20. Villiger, M. *et al.* Spectral binning for mitigation of polarization mode dispersion artifacts in catheter-based optical frequency domain imaging. *Opt Express* **21**, 16353–16369 (2013).

21. Gotzinger, E. *et al.* Retinal pigment epithelium segmentation by polarization sensitive optical coherence tomography. *Opt Express* **16**, 16410–16422 (2008).

22. Drexler, W. and Fujimoto, J. G. State-of-the-art retinal optical coherence tomography. *Prog Retin Eye Res* **27**, 45–88 (2008).

23. Yun, S. H. *et al.* Comprehensive volumetric optical microscopy *in vivo*. *Nat. Med* **12**, 1429–1433 (2006).

24. Yeh, A. T. *et al.* Imaging wound healing using optical coherence tomography and multiphoton microscopy in an *in vitro* skin-equivalent tissue model. *J Biomed Opt* **9**, 248 (2004).

25. Cobb, M. J. *et al.* Noninvasive assessment of cutaneous wound healing using ultrahigh-resolution optical coherence tomography. *J Biomed Opt* **11**, 064002 (2006).

26. Tsai, M.-T. *et al.* Monitoring of wound healing process of human skin after fractional laser treatments with optical coherence tomography. *Biomed Opt Express* **4**, 2362–2375 (2013).

27. Greaves, N. S. *et al.* Optical coherence tomography: A reliable alternative to invasive histological assessment of acute wound healing in human skin? *Br J Dermatol* **170**, 840–850 (2014).

28. Kim, K. H. *et al. In vivo* imaging of human burn injuries with polarization-sensitive optical coherence tomography. *J Biomed Opt* **17**, 066012 (2012).

29. Oh, J.-T. Lee, S.-W. Kim, Y.-S. Suhr, K.-B. and Kim, B.-M. Quantification of the wound healing using polarization-sensitive optical coherence tomography. *J Biomed Opt* **11**, 041124 (2006).

30. Golberg, A. *et al.* Skin regeneration with all accessory organs following ablation with irreversible electroporation. *J Tissue Eng Regen Med* (2016).

8. Functional Imaging of Wound Metabolism

Jake Jones[*], Vasily Belov[†] and Kyle P. Quinn[*]

*Department of Biomedical Engineering, University of Arkansas, 120 John A. White, Jr. Engineering Hall, Fayetteville, AR, USA 72701
†Department of Radiology, Massachusetts General Hospital, Harvard Medical School, and Shriners Hospitals for Children — Boston, 51 Blossom Street, Boston, MA, USA 02114

Abstract

Many of the etiologies that contribute to delayed or impaired wound healing result in metabolic challenges to the wound site. The effect of underlying disease states as well as the natural changes in metabolic requirements during the different phases of wound healing have been characterized through a variety of imaging techniques. Positron emission tomography (PET), hyperspectral imaging, and two-photon excited autofluorescence imaging are three emerging modalities with applications in the field of wound healing. Each of these technologies is capable of providing real-time, quantitative metabolic information through non-invasive imaging that is sorely lacking in the wound healing field. In this chapter, we highlight recent advances in these imaging modalities and their applications in monitoring inflammatory and proliferative phases of wound healing. The continued development and validation of metabolic imaging biomarkers based on these modalities can serve an important role in guiding wound care in the clinic and product development in the laboratory.

1. Introduction

Wound healing is a complex and metabolically demanding process involving the interactions and coordination of a variety of cell types following injury to cutaneous tissue in an effort to re-establish equilibrium and the protective function of the damaged tissue. Improper or delayed healing can result from a number of etiologies that present metabolic challenges at the wound site and expose the patient to greater risk of infection, disability, and/or death if not treated in time. In this chapter, we provide an overview of imaging approaches that can be utilized to characterize tissue metabolism and function during different aspects of the wound healing process. Following an overview of wound metabolism, we highlight recent progress in positron emission tomography (PET), hyperspectral imaging (HSI), and fluorescence microscopy in mapping key aspects of the wound healing process. Finally, we provide an outlook on the future of these technologies for preclinical research and as a potential for guiding care in the clinic.

1.1 *Overview of wound metabolism*

The normal wound healing process can be broken down into four overlapping phases: hemostasis, inflammation, proliferation, and tissue remodeling.[1] Hemostasis occurs immediately as blood spills into the injury and is achieved with the formation of a platelet plug that prevents further fluid loss and reestablishes the protective barrier function of the skin. At this time, any neutrophils that were within the local area begin their task of cleaning the wound site by phagocytosis of both invading organisms and toxins. During the onset of this inflammatory phase, additional neutrophils are recruited to the wound and there is an influx of fluid to the wound area. Monocytes cross into the wound site and differentiate into macrophages that both debride and clean the wound. This inflammatory cell activity comes at a high metabolic cost. Phagocytosis and cellular migration require a steady input of adenosine triphosphate (ATP) to power the structural changes associated with ingestion and motility.[2,3] Earlier research has shown that macrophages in particular are known to increase cellular respiration and glucose catabolism significantly during this process.[4] Oxygen uptake is greatly increased in phagocytes as well during the inflammatory phase of wound healing.[5] Nicotinamide adenine dinucleotide phosphate (NADPH) oxidase associated with the phagosome membrane utilizes some of this oxygen in the creation of superoxide which gives rise to radical oxygen species (ROS) with antimicrobial and inflammatory-stimulating functions.[6]

As debris and microbes are cleared from the wound site, macrophages and platelets release platelet-derived growth factor (PDGF) and epidermal growth factor (EGF) to stimulate the migration of fibroblasts into the wound heralding the proliferative phase of wound healing.[7] This proliferative phase initiates at approximately 4 days post injury[8] and includes the formation of the granulation tissue and reepithelization. Construction of the granulation tissue begins as the fibroblasts that entered the wound begin to synthesize collagen and proliferate.[7] In addition to the ATP and amino acids required for proliferation and collagen synthesis, oxygen is needed for hydroxylation of proline and lysine in collagen maturation.[5] The large demand for oxygen throughout the phases of healing makes the wound area noticeably hypoxic compared to the uninjured tissue.[3] Interestingly enough, brief periods of hypoxia at the onset of wounding have been shown to increase proliferation of dermal fibroblasts.[9] Around this time, endothelial cells begin the process of angiogenesis activated by vascular endothelial growth factor (VEGF) from keratinocytes, macrophages, fibroblasts, and other endothelial cells.[7] Transforming growth factor beta 1 (TGF-β1) and PDGF released by macrophages stimulate some of the fibroblasts within the wound site to differentiate into myofibroblasts.[7] Myofibroblast contraction is an ATP dependent process triggered by additional signaling of TGF-β1 and PDGF that begins the process of wound closure.[8] Reepithelialization of the wound occurs as keratinocytes from around the wound site undergo significant changes to morphology. These cells form peripheral cytoplasmic actin filaments as well as deconstructing cellular desmosomes allowing the cells to migrate across the eschar left from hemostasis.[8] As the early cells migrate out, expression of keratinocyte growth factor 2 (KGF-2) and inter-leukin 6 (IL-6) from fibroblasts stimulates proliferation of additional keratinocytes.[7] Once this initial activation occurs, keratinocytes are able to proliferate through self-expression of IL-6 and NO and continue migration until the wound site is covered. Remodeling is the longest phase of wound healing and can take anywhere from days to months depending on the size of the wound. Working to reestablish the structural integrity of the area, fibroblasts continue to replace and reinforce the provisional extracellular matrix (ECM) composed of mostly fibronectin, collagen Type III, and hya-luronic acid with a denser matrix containing thicker, tougher collagen Type I.[10] The end result of the normal wound healing response is usually a largely acellular, fibrotic scar that patches the wound site indefinitely.

1.2 *Abnormal wound metabolism*

A number of pathogenic conditions can lead to failures in the healing process resulting in a chronic wound. Chronic healing environments such as those seen in venous stasis ulcers, diabetic foot ulcers, and pressure ulcers are characterized by prolonged inflammation and prevention of granulation tissue formation.[1,11] Prolonged hypoxia stemming from excessive inflammation is common in these sites, amplifying the metabolic demand of the region and detrimentally affecting neutrophil and macrophage function.[12] Patients afflicted by venous stasis specifically display disproportionate leukocyte accumulation in their lower extremities combined with increased leukocyte activation.[13] There are significant oxygen and nutrient demands as leukocytes migrate and create ROS[14] which may contribute to sustained inflammation. Another disease state in which ROS burden is common is diabetes mellitus. Hyperglycemia causes oxidative stress by the overproduction of ROS that cannot be controlled by the normal anti-oxidant capacity.[15] Oxidized or nitrosylated tissues suffer from decreased biological activity causing severe functional damage and contributing to losses in energy metabolism.[15] Matrix metalloprotease (MMP) production may also contribute to prolonged inflammation if overproduced since MMPs are normally responsible for removing granulation tissue. Evidence exists that unusually high, and likely pathogenic, levels of MMPs are found in diabetic foot ulcers[16,17] and venous stasis ulcers[13] contributing to continued inflammation and draining metabolic resources. Improper healing can also be brought about in the form of fibrosis, or the formation of disorganized and excess scar tissue such as that found in keloids and hypertrophic scars. Metabolic differences exist between fibroblasts isolated from keloids and fibroblasts isolated from healthy human skin in that the keloid fibroblasts produce at least twice as much collagen as the normal types.[18] Hypertrophic scars likewise exhibit increased levels of collagen, but differ from keloids in that keloids lack normal quantities of myofibroblasts, while hypertrophic scars contain them in abundance.[1,19]

Given the many differences in cell function and metabolism during impaired wound healing, a variety of non-invasive imaging modalities have been employed to understand and detect wound pathophysiology through quantitative measurements of metabolism. PET, commonly used to identify cancer metastases, is capable of detecting differences in blood flow and glucose uptake within wounds. HSI has been utilized to assess tissue oxygenation based on oxy- and deoxy-hemoglobin absorption. Additionally, autofluorescence imaging is capable of identifying alterations in wound metabolism at the cellular level based on the autofluorescence of intrinsic metabolic cofactors. These imaging techniques hold promise to assist with the early detection of abnormal healing at different stages of the process and may help in guiding wound care.

2. PET Imaging of Wound Metabolism

PET can be very instrumental in evaluating the following characteristics of wound healing metabolism: blood flow to the wound site, cellular energetics, proliferation, tissue oxygenation, and inflammation. PET instrumentality is based on the fact that it is fully quantitative, and is the most sensitive and specific *in vivo* imaging modality that has been used extensively in nuclear medicine since PET inception in 1970s[20] for functional evaluation of molecular pathways of the imaged target.[21] The fundamental principle underlying PET imaging is the use of radiolabeled biological agents that can localize in specific tissues on the basis of the agent's biochemical and physiological properties, and whose quantities in these organs can be tracked over time as long as the decay half-life of the positron-emitting radioisotope permits. The ring-shaped array of PET detectors, with each detector consisting of another array of photoelectric crystals, allows registration of the coincidences of 511 keV photons emitted in the opposite directions after the interaction of a positron with a negatively charged electron. The raw PET data represents the set of such coincidence events, logged for time and location. Using reconstruction algorithms, the position of all the decayed radioisotope molecules can be determined, and tomographic images can then be created. Isotope quantities in the target tissues can be detected down to 100 picomoles. The highest anatomical resolution that currently can be achieved with clinical PET scanners is 4–8 mm^3, whereas pre-clinical small animal imagers can provide 1–2 mm^3. Thus, PET is a method of non-invasive physiological monitoring enabling accurate quantitation of the radiotracer uptake with high sensitivity and spatiotemporal resolution. The accuracy of the analysis by PET can be further improved by combining PET with anatomical imaging modalities such as magnetic resonance imaging (MRI) and computed tomography (CT). Below we will demonstrate how these unique characteristics of PET are used for evaluation of wound metabolism parameters, such as blood perfusion, angiogenesis, and hypoxia; cellular proliferation and overall wound metabolism; as well as infection and inflammation.

2.1 *Imaging of blood perfusion, angiogenesis, and hypoxia*

Measurement of blood flow and perfusion characteristics in a non-invasive, quantitative, and serial manner with PET is advantageous in assessing revascularization after trauma as well as the possible risk of avascular necrosis.[22] One of the applications where PET can be especially instrumental is imaging bone healing with [^{18}F]Fluoride ($t_{1/2}$ = 109.8 min). This realization came with the first reports tracking back to the end of the 1990s and the early 2000s showing ability of PET to provide pathophysiologically relevant information on healing bone grafts and fractures in terms of bone metabolism (see below) and bone blood flow.[23–26] Bone neovascularization plays a significant role in the healing process due to the strong relationship between bone blood flow and bone remodeling.[27] Reduced perfusion can be related to an increased risk of fracture[28] and osteoporosis[29] and to delayed or impaired healing.[22]

PET imaging with [^{18}F]Fluoride is based on a 3-compartment, 4-parameter model proposed by Hawkins *et al.*[30] These compartments are: plasma (C_p), extravascular space (C_e), and bone (C_b). Following a rapid clearance of [^{18}F]Fluoride from the plasma, fluoride undergoes ionic exchange with hydroxyl groups of hydroxyapatite $Ca_{10}(PO_4)_6(OH)_{10}$ to form fluorapatite $Ca_{10}(PO_4)_6F_2$ (incorporated fraction). It has been shown in the dual-tracer studies with [^{15}O]H$_2$O and [^{18}F]Fluoride that the forward transfer of [^{18}F] Fluoride to the bone (K_1, mL/(min*cm^3)) can be related to bone blood flow f (mL/(min*cm^3)) by:

$$K_1 = f \times E = f \times \left(1 - e\left(-\frac{PS}{f} \right) \right),$$

where E is the so-called unidirectional extraction fraction of the tracer, PS is the permeability-surface area product of the capillary surface, and f is the arterial flow to the tissue.[31,32] The rate constant K_1 can be estimated with dynamic PET imaging and by fitting the time-activity curves of [^{18}F]Fluoride concentration in the arterial blood and the bone tissue to the compartment model using a nonlinear regression algorithm. It is of note that K_1 can adequately represent bone blood flow under low and normal flow conditions when extraction of [^{18}F]Fluoride ion approaches 100% in a single capillary passage.[32] However, at high blood flow rates, due to diffusional limitations, correction in the form of extraction fraction E should be applied.

A number of groups have demonstrated the utility of [^{18}F]Fluoride PET imaging of blood perfusion for the assessment of bone viability after trauma or reconstructive surgery. Brenner *et al.*[33] assessed the time course of allogenic bone graft healing in 11 patients imaged repeatedly at either 6 and 12 months or 6, 12, and 24 months after surgery. The mean K_1 value (0.0804 mL/min/mL) was significantly lower the critical flow rates (~0.2 mL/min/mL), above which diffusional limitations should be considered. However, increased graft-to-normal bone ratios were observed for the whole duration of observation in all patients. Importantly, in two cases of non-union K_1 values almost doubled at 12 and 24 months, which was explained by increased blood flow, vascular permeability, and bone turnover in the presence of inflammation. Bone blood flow measurement by dynamic [18F]Fluoride PET for the assessment of the extent of viable bone in the femoral head was used in 5 patients with hip fracture and clinical concern about osteonecrosis.[34] As a result, impaired blood flow predicted an eventual need for joint replacement surgery.

A number of methods of functional imaging of the vasculature by PET have been developed originally for tumor imaging. However, most of these methods are also applicable for wound healing evaluation. Tissue blood perfusion is a function of angiogenesis and can be measured using radiolabeled water ([15O]H$_2$O). [15O]H$_2$O ($t_{1/2}$ = 122.24 s) is biologically and metabolically inert, and can also freely diffuse into and out of tissue water. Thus, "tissue water" can be modeled as a single compartment including both tissue and its draining fluids (blood, lymphatics)[35]:

$$C_t = P \bullet \int \left(C_i - C_e \right) dt,$$

where C_t is a cumulative tissue clearance, mol/mL$_{tissue}$; C_i is an influx concentration, mol/mL$_{tissue}$; C_e is an efflux concentration, mol/mL$_{tissue}$; and P is a perfusion (carrier flow adjusted for the volume of tissue), mL$_{carrier}$*min^{-1}*mL$_{tissue}$$^{-1}$. Tissue perfusion can be assessed by either static or dynamic PET imaging. Both methods have been described in detail by Laking and Price.[36] Among practical examples of the [15O]H$_2$O imaging is the study by Blokhuis *et al.*[26] where the authors employed this technique for quantification of bone healing. They showed a more rapid increase of blood flow in the damaged area of normally healing tibia as compared to the fractured tibia resulting in a non-union. The difference was especially marked at an early stage of healing. Although not yet routinely performed, it has been suggested that [15O]H$_2$O PET can also be instrumental for imaging blood flow after spinal trauma.[37]

Oxygen-15–carbon monoxide ([15O]CO) can be employed for PET imaging of blood volume.[36] [15O]CO ($t_{1/2}$ = 122.24 s) binds irreversibly with hemoglobin to form [15O]CO–Hb carboxyhemoglobin. The latter remains exclusively within the vasculature and therefore can be used as a tracer of vascular volume. After a short inhalation of a fixed dose of $[^{15}\text{O}]\text{CO}$, the tissue of interest is imaged by PET over a further 5–6 min. Arterial $[^{15}\text{O}]$ CO–Hb concentration time activity curve can be derived via blood sampling or image-based arterial input function (measured in a large vessel or the left ventricle). Tissue vascular volume can then be derived as:

$$V_v = \frac{V_t \cdot \int C_t}{R \cdot \int C_a}$$

where V_v: volume of vessels, $\text{mL}_{\text{vessels}}$, V: volume of tissue within region of interest (ROI) on PET scan, $\text{mL}_{\text{tissue}}$, R: ratio of small vessel to large vessel hematocrit, $\int C_t(t)$: integrated tissue activity during period of scan, $\text{Bq*mL}^{-1}\text{*min}$, $\int C_a(t)$: integrated arterial activity during period of scan, $\text{Bq*mL}^{-1}\text{*min}$. Alternatively, blood volume imaging can also be performed using radiolabeled red blood cells or albumin because both are too big to leave normal blood vessels.[35]

Blood perfusion allows for functional assessment of angiogenesis; however, usage of specific markers of angiogenesis as molecular targets for radiotracers can make imaging molecular pathways of angiogenesis possible. The markers playing the most pivotal role in regulating angiogenesis, for which promising PET tracers have been developed, are VEGF and integrin $\alpha_V\beta_3$. The angiogenic action of VEGF is mediated via binding with VEGF receptor (VEGFR), which is overexpressed in ischemic or healing tissue. Thus, VEGF-based molecular imaging probes can be used for assessment of VEGFR expression and thereby serve as a valuable indicator of repair mechanisms. One of the promising probes is $^{64}\text{Cu–DOTA–VEGF}_{121}$ ($t_{1/2}$ = 12.7 h),[38] whose ability to visualize VEGFR expression in tissues besides cancer has been validated in myocardial infarction,[39] stroke,[40] and hindlimb ischemia[41] models. In all cases, PET imaging showed significantly higher $^{64}\text{Cu-DOTA–VEGF}_{121}$ uptake in ischemic tissues compared to normal tissue during the first 2 weeks, 16 days, and 3 weeks, respectively, indicating active neovascularization confirmed by other methods. Another approach for assessing VEGF/VEGFR interactions is to image VEGF itself by developing the probes able to bind to VEGF. Examples include radiolabeled specific monoclonal antibodies $[^{124}\text{I}]\text{VG76e}$[42] ($t_{1/2}$ = 4.2 d), $[^{124}\text{I}]$ HuMV833[43] ($t_{1/2}$ = 4.2 d), and $^{89}\text{Zr-bevacizumab}$[44] ($t_{1/2}$ = 3.3 d), the first two of which showing only a limited promise in imaging VEGF expression during tumor neovascularization. Several probes targeted to $\alpha_V\beta_3$, another important biological marker of angiogenesis, have also been developed for

PET imaging of angiogenesis.[35,45–47] Almutairi *et al.*,[47] for example, developed biodegradable probe consisting of dendritic core chemoselectively functionalized with heterobifunctional polyethylene oxide (PEO) chains forming a protective shell. Each of the eight branches of the dendritic core was functionalized with radiohalogen [76]Br, and the targeting peptides of cyclic arginine–glycine–aspartic acid (RGD) motifs were installed at the terminal ends of PEO chains. As a result, such sophisticated probe showed high specific uptake in angiogenic muscles in an *in vivo* murine hindlimb ischemia model.

Hypoxia is another important parameter of wound metabolism and the repair process. Many cell types have adaptive mechanisms (switching to anaerobic metabolism, reduced growth, neovascularization) to survive moderate hypoxia often present in wounds; however, severe, acute hypoxia (stroke, myocardial infarction) often results in cellular death. Severe hypoxia can significantly delay wound healing thus creating high demand for reliable, quantitative methods of hypoxia imaging. The best probe for measuring hypoxia would be the one competing directly with intracellular O_2.[48] In other words, the probe should not be trapped when the oxygen supply is adequate but would be retained when O_2 level is low. Furthermore, because of the direct relationship between blood perfusion and hypoxia, the probe delivery to the cells should be independent of blood flow. Nitroimidazoles[49] match the above criteria and ^{18}F-fluoromisonidazole ([^{18}F]FMISO) ($t_{1/2}$ = 109.8 min), in particular, showed a great promise in PET studies becoming the most widely used hypoxia-imaging agent.[48,50]

2.2 *Imaging of infection and inflammation*

Fluorine-18-fluorodeoxyglucose ([^{18}F]FDG) PET imaging originated about three decades ago as a powerful imaging tool in cancer. With time, it evolved to a much broader range of applications including imaging of septic and aseptic inflammatory processes. Despite the sparse data, [^{18}F]FDG ($t_{1/2}$ = 109.8 min) is increasingly being used as an excellent tracer for the assessment of infection and inflammation in wounds due to the ability of inflammatory cells to preferentially metabolize glucose and a widespread distribution of [^{18}F]FDG.[51]

Biological phenomenon that underlies PET imaging of infection and inflammation is the increased glucose uptake in the activated macrophages, neutrophils, and lymphocytes.[52] Immune cells utilize glucose for ATP production and macromolecule biosynthesis needed for immune responses. The increased glucose consumption is facilitated by overexpression of glucose transporters (GLUT 1, GLUT 3, GLUT 5), with the upregulation pattern being different in different types of immune cells.[53] [^{18}F]FDG is taken up by cells where it is phosphorylated to [^{18}F]FDG-6-phosphate, a molecule not able to enter further glycolytic pathways. In the absence of glucose-6-phosphatase, which catalyzes the reverse reaction to [^{18}F]FDG, the formation of [^{18}F]FDG-6-phosphate is the final step. This results in [^{18}F]FDG-6-phosphate becoming trapped intracellularly due to its highly negative charge and thereby low membrane permeability.[54] Within 1 h after injection, most of the radiation emitted from cells and tissues with low concentrations of glucose-6-phosphatase, such as heart, brain, many tumors, inflammatory cells, comes from intracellular [^{18}F]FDG-6-phosphate. This enables quantification of the overall glucose consumption by the cell and relative rates of glycolysis. Tissues with high levels of this enzyme, such as liver, kidney, intestine, and muscle, accumulate small amounts of [^{18}F]FDG-6-phosphate and contribute a low level of background radioactivity to PET images.[55]

Among the advantages of [^{18}F]FDG-PET compared to other nuclear medicine and anatomic imaging techniques in the assessment of inflammation are: (1) high resolution images, (2) high target-to-background contrast ratios, (3) high sensitivity, and (4) quantitation.[51] The first three aspects make the method very accurate in detecting inflammation, while the latter aspect is pivotal for assessing condition progress and response to the therapy. From a clinical perspective, the primary shortcoming of [^{18}F]FDG-PET imaging of wounds is difficulty in differentiating infection from non-infectious inflammation. Imaging at later time points (several weeks post operation), when [^{18}F]FDG uptake due to the post-operative non-infectious inflammation is considerably smaller than uptake in the infection site, may help to minimize this issue.[56] The location where activity measurements are performed can provide another context regarding the presence of infection versus aseptic inflammation. For example, in orthopedic wounds after hip arthroplasty surgery, evidence of infection is usually noted at the bone-prosthesis interface, whereas most non-infectious inflammatory reactions are found outside the bone-prosthesis junction.[51,57–59] This important knowledge resulted in relatively high levels of sensitivity, specificity, and accuracy of PET imaging of aseptic inflammation in different prostheses. For example, Zhuang *et al.*[59] reported 90.0%, 89.3%, and 89.5%, respectively, in hip prostheses and 90.0%, 72.0%, and 77.8%, respectively, in knee prostheses. Chacko *et al.*'s[57] numbers for sensitivity and specificity of detecting hip infection by PET were 92% and 97%, whereas, in Pill and coworkers[60] numbers were 95.2% and 93%, respectively. Thus, [^{18}F]FDG-PET is a promising diagnostic tool for differentiating septic from aseptic painful prostheses.

The usefulness of [^{18}F]FDG-PET for detecting and monitoring chronic inflammatory processes was investigated by Yamada *et al.*[61] The authors subcutaneously inoculated the rats with turpentine and observed a gradual increase of FDG uptake in the inflammatory tissue until 60 min, followed by a decrease. Postoperatively, surgical wound healing causes moderate [^{18}F]FDG accumulation, generally lower than that seen with malignancy. For example, Ozawa *et al.* employed [^{18}F]FDG-PET to study glucose metabolism in keloids.[62] Keloids are skin abnormalities that are unique to humans and are characterized by excessive deposition of collagen in the dermis and subcutaneous tissues secondary to trauma, inflammation, surgery, or burns. The authors found that [^{18}F]FDG uptake in the keloids of five patients was moderately elevated (mean SUV of 1.79), which they hypothesized was driven by the presence of macrophages in the lesions.

There are several aspects of [^{18}F]FDG-PET imaging that can facilitate wider application of the method for preclinical and clinical assessment of inflammation. In contrast to malignancy imaging, the blood glucose levels such as those observed during hyperglycemia or hypoglycemia do not appear to adversely affect [^{18}F]FDG uptake by inflammatory lesions.[63,64] The likely explanation for this is that inflammatory cells have the ability to store and mobilize their intracellular glycogen storage to produce glucose during periods of low plasma glucose.[65] This aspect might be particularly advantageous for inflammation assessment by PET when the healing process is complicated by diseases or conditions affecting blood glucose levels. Additionally, the combination of [^{18}F]FDG-PET with anatomic imaging modalities such as CT and MRI can significantly improve the speed and accuracy of wound localization, making PET imaging of inflammation an advantageous method for the clinical imaging of wound healing.

2.3 *Imaging of tissue proliferation and overall wound metabolism*

As mentioned in Section 1, cellular proliferation requires substantial glucose metabolism. This feature underlies PET imaging of proliferation using [^{18}F]FDG. Skin lesions developing hypermetabolism may be detected by [^{18}F]FDG-PET imaging.[66] Examples include but not limited to: cutaneous and subcutaneous thermal injury,[67] acne vulgaris,[68] mosquito bite,[69] fat lobules and lobular panniculitis,[70] and herpes zoster.[71] In the recent study, Golberg *et al.*[72] employed [^{18}F]FDG-PET to assess the impact of pulsed electric fields

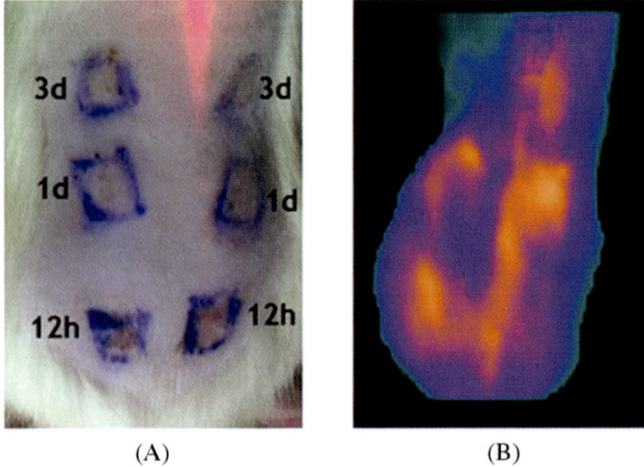

Figure 1. Skin metabolism after the treatment with PEFs. (A) Skin areas of rat at different time points after the PEF treatment. (B) Respective [^{18}F]FDG uptake measured by PET. Reproduced from Ref. 72.

(PEF) on total skin metabolism in rats at 12 h, 1 day, 3 days, and 7 days after the treatment (Figure 1). Although elevated glucose uptake in the PEF-treated areas of the skin was observed for all studied time points, maximum uptake was characteristic for the first 3 days after the treatment followed by diminished levels 1 week after the treatment. These observations implied active proliferation processes characteristic of skin repair and rejuvenation.

One of the challenges in [^{18}F]FDG imaging of tissue proliferation is differentiating it from the additional [^{18}F]FDG uptake by cells involved in inflammation as described in the previous section. This problem can be addressed by employing as the marker of tissue proliferation of the uptake of molecules involved in synthetic pathways. These include labeled amino acids for measuring transport and protein synthesis as well as labeled nucleosides needed for DNA synthesis. Tracing the uptake of these molecules *in vivo* has become the basis of PET imaging of proliferation.[73] Despite extensive use of these agents for evaluation of malignant growth, literature lacks reports on their utilization for monitoring proliferation and metabolism in wounds. Nevertheless, this approach might prove promising due to the improved specificity to proliferation and availability of suitable agents.

In addition to monitoring infection, the functional imaging of bone metabolic activity by PET may have benefits in monitoring the proliferative phase of bone healing after fracture or assessing graft viability. The major disadvantage of conventional imaging methods such as plain radiography,

CT, MRI, or bone scintigraphy is that these modalities are based on examination of morphological signs, which occur late in the normal course of healing and therefore are unreliable for the early prediction of impaired healing. Quantitative assessment of bone metabolism by PET can be carried out by using the bone-seeking $[^{18}F]$Fluoride ion. The method is based on the strong relationship between the amount of $[^{18}F]$Fluoride chemisorption and incorporation into the bone and the activity of osteoblasts and osteoclasts. In accordance with the 3-compartment, 4-parameter model that has been previously applied to blood perfusion measurements,[30] rate constants describing the transport and binding of the $[^{18}F]$Fluoride ion are as follows: k_1 and k_2 for the forward and reverse capillary transport, k_3 for the binding to the bone matrix, and k_4 for the release reaction. These microparameters can be estimated with dynamic PET imaging and by fitting the time-activity curves of $[^{18}F]$Fluoride concentration in the arterial blood and the bone tissue to the compartment model using a nonlinear regression (NLR) algorithm. The net transport of $[^{18}F]$Fluoride ion into the bone is indicative of the level of bone remodeling and mineralization and therefore can serve as a measure of bone metabolism. It can be described by fluoride bone influx rate (macroparameter) calculated by Patlak graphical analysis (K_{pat})[30,74,75] or based on microparameters (K_i): $K_i = K_1 k_3/(k_2 + k_3)$.

Piert *et al.*[32] undertook a special study in the normal bone aimed at investigating the relevance of the above quantitative estimates of bone metabolic activity provided by $[^{18}F]$Fluoride PET. They demonstrated significant correlation between K_i and K_{pat} and invasive histomorphometric indices of bone formation regarded as a gold standard for bone metabolic analysis. In a study of healing cancellous and full bone grafts in 34 patients, Brenner *et al.*[33] found significant correlation between K_i and K_{pat} and demonstrated time dependences of bone metabolism changes consistent with the healing biology specific for each graft type. At 24 months after surgery, bone metabolism levels in normally healing bones were significantly lower than in the beginning. A similar decrease over time was also observed by other groups in initial earlier studies with $[^{18}F]$Fluoride performed in a limited number of patients.[23,24] Despite a general decrease in tissue metabolism over time as the bone heals, levels at these later time points are still higher compared to normal uninjured limb bones. This same observation was made previously by Piert *et al.*[24] in the study of bone metabolism after hip augmentation surgery with cancellous grafts. The authors suggested that the increased bone activity after a mean follow-up of 2 years was the result of continued mechanical stress on the normal surrounding bone. Unlike the ultimate decrease in the graft-to-normal metabolic rates observed during the normal healing, non-union in a study by Brenner *et al.*[33] resulted in the gradual constant increase of both K_i and K_{pat}. These changes were explained by increased blood flow, vascular permeability, and bone turnover in the presence of inflammation. In their short study, Blokhuis *et al.* demonstrated utility of PET imaging for the quantification of fractured bone healing. They showed a more rapid increase of a net ^{18}F uptake in the damaged area of normally healing tibia as compared to the fractured tibia resulting in a non-union.[26] The difference was especially marked at an early stage of healing. In summary, [^{18}F]Fluoride PET imaging has provided a sensitive approach to monitoring the early patterns of fractured bone healing.

2.4 *Advantages and limitations*

PET has several advantages that stand it apart from other molecular imaging modalities. Above all, its high sensitivity allows for detection of isotope quantities down to 10^{-10} moles. Additionally, absolute quantitative measurements enable precise calculations of radiolabeled substances distributed throughout the body. Furthermore, the tomographic character of PET facilitates image analysis and object localization in any body compartment at any depth. Combining PET with anatomical imaging modalities such as MRI and CT can further enhance this capability. However, as with any imaging modality with exceptional depth of penetration and a large field of view (>16 cm and 8 cm for clinical and preclinical scanners, respectively), PET provides a relatively low spatial resolution (starting with 4–5 mm and 1–2 mm for clinical and preclinical scanners, respectively), which makes PET best suited for tissue/organ imaging.

Although PET is based on administration of radioactive isotopes to the patient, highly sensitive and fast detectors can significantly minimize radioactive exposure of the patients and personnel by enabling the use of low doses of radiolabels. The cost of PET analysis and its dependency on the radioisotope supply have long been viewed as a shortcoming of the method limiting its affordability only to large medical centers with their own cyclotrons. Whereas the former issue still remains important, the latter has been alleviated with the widespread availability of commercial isotope suppliers. The non-invasive nature of PET and rapid analysis time (minutes to hours) offer substantial advantages in the clinic.

Background signal and non-specific uptake can significantly deteriorate image contrast and complicate PET image analysis and data interpretation. To address this challenge, post-acquisition correction methods (e.g. background subtraction) can be applied; nevertheless, more specific and efficient

PET probes are of high demand. Overall, despite limitations to be addressed by improved scanner design and radiochemistry, PET is a promising modality for quantitative imaging of wound metabolism, and can be particularly instrumental in the applications where microscopy-based methods are not feasible.

3. Hyperspectral Imaging

HSI technology has been utilized in the clinic to monitor and diagnose a variety of pathological conditions ranging from retinal disease to carcinomas,[76,77,78,79] and has proven itself to be a valuable tool in cutaneous wound healing research. Similar to PET, medical HSI research is based on the principle that changes in the local metabolism of the wound site precede changes that can be noticed anatomically or physiologically by a clinician.[80] Unlike PET, HSI has substantially lower operating costs and does not require the production of radiotracers from cyclotrons. As a method of optical spectroscopy, HSI generates contrast based on the absorbance of natural chromophores native to the skin, which can be used to monitor metabolic changes surrounding wounds.

Among the intrinsic chromophores found in the skin, oxyhemoglobin and deoxyhemoglobin are some of the most valuable to wound healing research as they can be used to measure levels of oxygen saturation, oxygenated hemoglobin content, deoxygenated hemoglobin content, and total hemoglobin. Oxy- and deoxyhemoglobin can be discriminated spectrally because deoxyhemoglobin has a single peak absorption of 554 nm while oxyhemoglobin has two local maxima at 542 nm and 578 nm.[81] This source of contrast has long been used in the clinic since the advent of pulse oximetry, which provides oxygen saturation readings at a single spatial point.[80] Given the spatial heterogeneity of most cutaneous wounds, imaging technologies such as HSI, capable of identifying local variability in the microcirculation and metabolism of the whole sample are critical. Measures of tissue oxygen obtained from oxygenated and deoxygenated hemoglobin are especially relevant in this wound healing environment where oxygen plays a critical role in immune defense, cell migration, and proliferation.[82,83]

3.1 *Hyperspectral imaging of chronic wounds*

Oxygen saturation is an important measure of the microvascular function and has been shown to be an useful factor in the assessment of diabetic foot ulcer sites.[84] Typically in the clinic, diabetic ulcer size is monitored over a

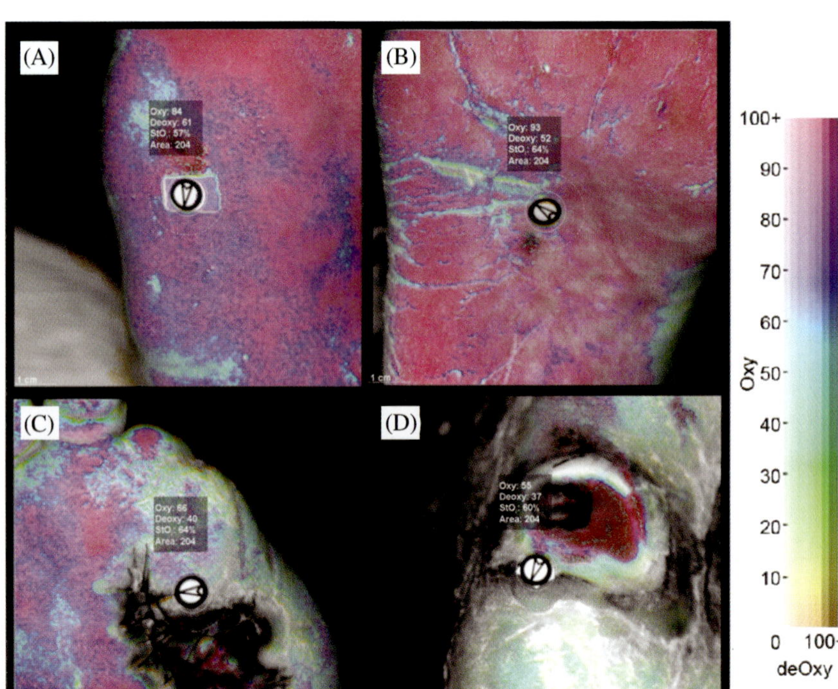

Figure 2. Hyperspectral images of plantar midfoot wounds. Healing wounds (A and B) demonstrate higher oxy- and deoxyhemoglobin concentrations than non-healing wounds (C and D). Images courtesy of Dr. Dimitrios Baltzis from the laboratory of Dr. Aristidis Veves.

4-week period to evaluate healing potential. This method of 4-week size reduction monitoring has a positive predictive value (PPV) of 58% and a negative predictive value (NPV) of 91% in the prediction of natural wound closure.[85,86] Using measurements of oxyhemoglobin and deoxyhemoglobin content, oxygen saturation can be mapped and healing indices can be calculated using linear discriminant analysis that separates healing and non-healing sites. Based on these indices of measured oxygenated and deoxygenated hemoglobin, HSI has a PPV ranging from 93% to 96% and a NPV of 86%[85,86] (Figure 2). These values correspond to a HSI index sensitivity that ranges between 86% and 93% and specificity between 86% and 88%.[85,86] However, patients with osteomyelitis or extensive callousing around the wound site proved difficult to image with HSI, and were accordingly removed from the sample pool in the previous work.[86] HSI has also demonstrated that patients suffering from diabetes have low skin oxygenation and impaired muscle metabolism in their feet relative to non-diabetics,[87] which may provide some prognostic value in evaluating diabetic wound healing. HSI can be performed in a few minutes over a wide range of wound locations and sizes in a single visit, and thus holds promise as a supplementary technique to the standard practice for evaluating diabetic foot ulcers and other chronic wounds.

3.2 *Hyperspectral imaging of cutaneous burns*

The diagnostic utility of HSI has also been investigated in the assessment of burn wounds. Accurate diagnosis of burn depth and severity is key to determining the appropriate course of action to promote healing. Burn wounds are commonly divided into superficial burns with minimal dermal damage and full thickness burns that will not heal without surgical intervention.[88] Early research with HSI used predictive models that sought to determine the potential for healing as a function of the thickness of the burn through the dermal layer, the volume fraction of blood in the tissue below the burn injury, and the oxygen saturation of perfusing blood in the wound site.[89] At three days post burn, HSI demonstrated a 91% PPV in the prediction of healing and 85% NPV, which was more accurate than predictions by three experienced burn surgeons with predictive values of 76% and 67%, respectively.[89] Since then, both point-based optical assessments[90] and NIR spectroscopy[91] have demonstrated that superficial and full thickness burns can be reliably discerned using measures of oxygen saturation and local hemodynamics at a sensitivity of 90% and specificity of 83%.[91] Regardless, there is a greater clinical need to differentiate superficial-partial thickness and deep partial-thickness burns.[88,90] These partial thickness burns are characterized by the extent of dermal damage, which determines the need for surgical intervention. However, these burns have yet to be accurately diagnosed through any spectroscopic approach.[88] Additionally, the eschar that forms from the necrotic epidermis and dermis may display stronger wavelength-dependent scattering than the uninjured dermis.[92] Removal or debridement of the eschar is suggested before optical spectroscopy of the wound to reduce these effects.[90]

3.3 *Other considerations for HSI of wounds*

Another prominent chromophore in the skin is melanin, which absorbs light within the 350–1200 nm range.[93] Given this broad range of absorption, melanin has the potential to interfere with spectral measurements of oxygenated hemoglobin and deoxygenated hemoglobin.[81] By modeling the skin as a semi-infinite medium, melanin can be accounted for using the modified Beer–Lambert Law in which the total absorption is the sum of the absorptions from the dominant chromophores in the skin — oxyhemoglobin, deoxyhemoglobin, and melanin — as well as the scattering by the tissue.[94] Alternatively, a two-layer optical skin model has also been developed that can simultaneously determine oxygen saturation, blood volume fraction, melanin concentration, and the tissue scattering coefficient.[95] In this model, an epidermal layer is modeled as a plane-parallel slab where absorption is dominated by melanin and a dermal layer is modeled as a semi-infinite medium in which absorption is a function of the blood volume fraction and oxy–deoxyhemoglobin concentrations. Second derivative spectroscopy has likewise been shown to demonstrate reliable tissue oxygenation readings independent of skin color.[96] Regardless of the method used, once the chromophores and the scattering properties of the tissue are taken into account, HSI systems can create wound site mappings of local oxygen saturation.[97]

3.4 Advantages and limitations

HSI offers many practical advantages in characterizing wound metabolism and identifying pathological states. As a non-invasive and non-contact method, it can be utilized in a variety of situations from the direct inspection of wounds to integration into other diagnostic methods such as wound site histopathology.[98] Because HSI relies on endogenous chromophores, it is limited to the finite number of chromophores naturally present *in vivo*. Additionally, HSI has a low penetration depth (approximately 3.57 mm at 850 nm and 0.48 mm at 550 nm[97]) compared to other clinical imaging modalities, such as PET, and also lacks any depth resolution. Despite these limitations, HSI has shown promising results in the early detection of wound chronicity and the evaluation of burns. This technique has the ability to provide relatively high lateral spatial resolution images of oxygenation, between 1 μm/pixel and 10 μm/pixel, over a large sample area within the span of a few minutes, which makes HSI particularly amenable to clinical settings in which quantitative metabolic information could be provided in real-time.[98,99] While most current HSI approaches collect information on oxy- and deoxyhemoglobin, measurements of water content and bilirubin, a product of bruising, also may provide a more detailed characterization of wound status in future work.[81] Additionally, collection of data in the mid-infrared field to monitor vibrational and rotational frequencies of biomolecules and the development of multi-modal imaging systems with HSI may hold increased diagnostic potential in future wound healing applications.[97]

4. Endogenous Fluorescence Measurements of Wound Metabolism

While HSI is capable of measuring changes in the oxygen content of wounds, fluorescence imaging has been used to assess cellular metabolism without the need for exogenous dyes or stains in a variety of tissues,[100,101] including cutaneous wound sites.[102–104] Early microfluorometry work[105] led to the discovery that nicotinamide adenine dinucleotide (NADH) and flavin adenine dinucleotide (FAD) naturally fluoresce under ultraviolet light.[105,106] NADH and FAD are electron carriers ubiquitously found throughout most metabolic pathways. As glucose metabolites are broken down during glycolysis and the citric acid cycle, energy is transferred to NADH and $FADH_2$ in the form of electrons. NADH and $FADH_2$ provide a key link in transferring these electrons to the electron transport chain, where that energy is ultimately harnessed to produce ATP. Upon donating their electrons, NADH and $FADH_2$ revert to their oxidized forms, NAD^+ and FAD. Because only the FAD and NADH forms are autofluorescent, ratiometric fluorescence measurements can provide a means of quantifying cell reduction-oxidation (i.e. redox) state.[100,107,108] NADH and FAD can be distinguished from each other because the emission of FAD is red-shifted relative to NADH, and FAD is preferentially excited with longer wavelengths (> 400 nm one-photon; > 800 nm two-photon).[109,110] An optical redox ratio of FAD/(NADH + FAD) autofluorescence has been shown to correlate with intracellular cofactor concentration ratios[108,111] in a variety of cell types and provides a relative measure of glucose catabolism relative to ATP production.

4.1 *Two-photon excited fluorescence microscopy*

While early detection methods of FAD and NADH relied on single-photon excitiaton,[105,107] two-photon excited fluorescence (TPEF) imaging has many advantages over standard fluorescence imaging techniques and is well-suited for characterizing 3D epithelial tissues.[112] TPEF relies on the simultaneous absorption of two photons (of half the energy and twice the wavelength) to bring molecules to an excited state. As a result, near infrared (rather than UV) light can be used to excite NADH and FAD, resulting in deeper imaging penetration depths. Since the simultaneous absorption of two photons by a molecule is limited, for all practical purposes, to the focal plane, TPEF provides an inherent depth sectioning ability, enabling more efficient light collection and resulting in reduced photodamage. The ability to section tissue with TPEF was demonstrated in guinnea pig cutaneous wound models using fluorescence of NADH, FAD, and collagen over a period of 28 days.[103] The endogenous fluorophores allowed for differentiation of wound site features such as collagen fibers, blood vessels, and inflammatory cells comparable to standard histological sections.[103] The metabolism during the initial phases of wound healing has also been monitored through an FAD/NADH redox ratio in rat models.[113] Mokrý *et al.* observed increases in the redox ratio during periods of heightened vascular flow such as in inflammatory vasodilation after 15–65 min post wounding, and when circulation increased due to angiogenesis after 3–4 days.[113]

Another benefit of TPEF is the ability to incorporate second harmonic generation imaging, which can be performed simultaneously to obtain information on collagen organization within the wound site.[114,115] Second harmonic generation imaging has been used in conjunction with NADH autofluorescence during studies of thermal wound healing in human skin-equivalent tissue models.[116] Using measures of metabolic activity and collagen deposition, dermal fibroblasts were observed to infiltrate the wound site and begin remodeling within 2 days post injury.[116] In similar studies of wound healing, second harmonic generation has shown that tissue models with keloid-derived fibroblasts lead to more rapid rates of collagen deposition than normal fibroblasts,[104] but direct measurements of cell metabolism through TPEF in these types of studies remain largely unexplored.

4.2 *Endogenous fluorescence as a measure of wound metabolism*

Despite the dynamic metabolic requirements of different cell types during processes such as phagocytosis, migration, proliferation, and ECM remodeling, only a few studies have specifically measured NADH and FAD autofluorescence changes during cutaneous wound healing. Unstained frozen wound sections from a mouse model of full-thickness skin wound healing exhibit distinct regional variability in optical redox ratio measurements (Figure 3). The proliferating keratinocytes at the wound edge demonstrate a significantly lower redox ratio than the underlying granulation tissue in the wound bed.[102] The biosynthetic demands of proliferation may result in a shift toward aerobic glycolysis rather than oxidative phosphorylation,[117] which causes an increase in NADH concentrations relative to FAD.

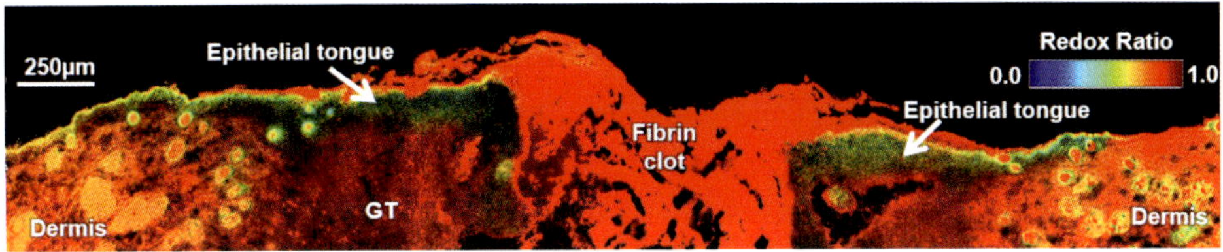

Figure 3. Optical redox ratio map of an excisional wound at day 10 demonstrates distinct differences in wound metabolism between the proliferating epithelial tongue and surrounding stromal tissue.

Interestingly, the keratinocyte redox ratio increases as the migrating tip of the epithelial tongue enters the wound bed, suggesting a sensitivity to the transition from keratinocyte hyperproliferation to migration. Furthermore, the epidermis of diabetic wounds at day 10 post wounding remained significantly lower than non-diabetic controls.[102] This is consistent with the continued hyperproliferation of keratinocytes and migration deficiencies in chronic wounds.[118] The elevated redox ratio within the wound bed (Figure 3) may be related to increased ATP production and consumption by migrating inflammatory cells and fibroblasts, as well as increases in specialized functions such as phagocytosis and the release of reactive oxygen species.[3]

In addition to optical redox ratio measurements, fluorescence lifetime imaging microscopy (FLIM) can be employed to distinguish free and bound NADH by measuring the time the molecule spends in the excited state.[119,120] Decreasing lifetimes have been observed for cells in conditions of hypoxia, which may be helpful in characterizing the wound microenvironment.[101,121] FLIM has already been used in the creation of multimodal systems that characterize fluorescence of engineered and native skin tissue utilizing NADH and FAD[122,123] with future directions looking into the classification of wound repair and regeneration.[122]

4.3 *Advantages and limitations*

Compared to other non-invasive metabolic imaging modalities (i.e. PET and HSI), TPEF imaging uniquely offers microscale resolution in 3D, which enables the characterization of metabolic heterogeneity through different skin layers and wound regions. Initial studies of cell redox state through NADH and FAD autofluorescence at the wound edge indicate distinct metabolic profiles between stromal and epithelial cell types (Figure 3). The high lateral and axial resolution of TPEF comes at the cost of limited imaging penetration depths (maximum of 0.5–1 mm), which may be insufficient for the complete characterization of deep wounds. Furthermore, the smaller field of view of most objective lenses used in TPEF imaging makes imaging of the entire wounds somewhat time-consuming. Compared to HSI, the initial cost of a TPEF system (which typically requires a tunable Ti: Sapphire laser) is significantly greater. Nonetheless, its unique insight into microscale wound metabolism and non-invasive nature, make TPEF a promising tool for wound healing applications.

In addition to NADH and FAD autofluorescence imaging, other endogenous fluorophores found in bacteria may offer future diagnostic opportunities, and handheld single-photon fluorescent systems have already seen trial use in the clinical monitoring of wound site bioburden.[124] The continued development of handheld fluorescence imaging systems for the clinic[125] and the refinement of quantitative optical biomarkers for delayed wound healing using endogenous NADH and FAD autofluorescence will provide exciting new approaches to guide wound care product development and assist in the clinical management of ulcers.

5. Future Directions and Considerations for Imaging Wound Metabolism

PET, HSI, and TPEF imaging collectively offer a wealth of quantitative metabolic information for characterizing different phases of the wound healing process. Given the lack of non-invasive quantitative biomarkers available to the wound healing research community beyond wound closure, further development and validation of these imaging methods is certainly warranted. PET offers excellent sensitivity and specificity to certain aspects of metabolism through the use of radiotracers. However, the production of these short-lived radionuclides through a cyclotron limits the use of this imaging modality in some rural areas and also contributes to its high operating costs. Unlike PET, HIS and TPEF rely on intrinsic sources of contrast and consequently have relatively low operating costs. However, additional work is needed to understand and remove the potential effects of other naturally occurring chromophores such as melanin, keratin, and lipofuscin during optical measurements of wound metabolism prior to their widespread use in the clinic.

Clinical adoption of these imaging modalities for wound healing monitoring may likely be dependent on the application. PET has clearly demonstrated its value in monitoring different aspects of bone healing through [^{18}F] Fluoride and [^{18}F]FDG imaging. In contrast, clinical applications of HSI and

TPEF imaging in the healing process will likely be limited to epithelial wounds and applications, which only require light penetration of < 1 mm. While lacking the depth of penetration of clinical imaging modalities, TPEF methods provide superior, subcellular image resolution compared to PET, which is critical in discriminating the activity of different cell types and heterogeneity throughout the wound microenvironment. Multi-modal imaging, combining structural and morphological information with the metabolic imaging techniques described in this chapter, may help to minimize the limitations of specific individual techniques and provide a more complete understanding of the wound healing process.

References

1. Diegelmann, R. F. and Evans, M. C. Wound healing: An overview of acute, fibrotic and delayed healing. *Front Biosci* **9**, 283–289 (2004).
2. Stossel, T. P. and Hartwig, J. H. Interactions of actin, myosin, and a new actin-binding protein of rabbit pulmonary macrophages. II. Role in cytoplasmic movement and phagocytosis. *J Cell Biol* **68**, 602–619 (1976).
3. Kominsky, D. J., Campbell, E. L. and Colgan, S. P. Metabolic shifts in immunity and inflammation. *J Immunol* **184**, 4062–4068 (2010).
4. Oren, R., Farnham, A. E. Saito, K. Milofsky, E. and Karnovsky, M. L. Metabolic patterns in three types of phagocytizing cells. *J Cell Biol* **17**, 487–501 (1963).
5. Tandara, A. A. and Mustoe, T. A. Oxygen in wound healing — more than a nutrient. *World J Surg* **28**, 294–300 (2004).
6. de Oliveira-Junior, E. B. Bustamante, J. Newburger, P. E. and Condino-Neto, A. The human NADPH oxidase: Primary and secondary defects impairing the respiratory burst function and the microbicidal ability of phagocytes. *Scand J Immunol* **73**, 420–427 (2011).
7. Broughton, G. Janis, J. E. and Attinger, C. E. The basic science of wound healing. *Plast Reconstr Surg* **117**, 12S–34S (2006).
8. Singer, A. and Clark, R. Cutaneous wound healing. *N Engl J Med* **341**, 738–746 (1999).
9. Siddiqui, A. *et al.* Differential effects of oxygen on human dermal fibroblasts: Acute versus chronic hypoxia. *Wound Repair Regen* **4**, 211–218 (1996).
10. Enoch, S. and Leaper, D. J. Basic science of wound healing. *Surgery* **23**, 37–42 (2006).
11. Brem, H. Balledux, J. Bloom, T. Kerstein, M. D. and Hollier, L. Healing of diabetic foot ulcers and pressure ulcers with human skin equivalent: A new paradigm in wound healing. *Arch Surg* **135**, 627–634 (2000).

12. Falanga, V. Wound healing and its impairment in the diabetic foot. *Lancet* **366**, 1736–1743 (2005).

13. Bergan, J. J. *et al.* Chronic venous disease. *N Engl J Med* **355**, 488–498 (2006).

14. Takase, S. Schmid-Schönbein, G. and Bergan, J. J. Leukocyte activation in patients with venous insufficiency. *J Vasc Surg* **30**, 148–156 (1999).

15. Vincent, A. M. Russell, J.W. Low, P. and Feldman, E. L. Oxidative stress in the pathogenesis of diabetic neuropathy. *Endocr Rev* **25**, 612–628 (2004).

16. Guo, S. and Dipietro, L. A. Factors affecting wound healing. *J Dent Res* **89**, 219–229 (2010).

17. Woo, K. Ayello, E. A. and Sibbald, R. G. The edge effect: Current therapeutic options to advance the wound edge. *Adv Skin Wound Care* **20**, 99–117, 118–119 (2007).

18. Diegelmann, R. F. Cohen, I. K. and McCoy, B. J. Growth kinetics and collagen synthesis of normal skin, normal scar and keloid fibroblasts *in vitro*. *J Cell Physiol* **98**, 341–346 (1979).

19. Ehrlich, H. P. *et al.* Morphological and immunochemical differences between keloid and hypertrophic scar. *Am J Pathol* **145**, 105–113 (1994).

20. Phelps, M. E. Hoffman, E. J., Mullani, N. A. and Ter-Pogossian, M. M. Application of annihilation coincidence detection to transaxial reconstruction tomography. *J Nucl Med* **16**, 210–224 (1975).

21. Phelps, M. E. PET: The merging of biology and imaging into molecular imaging. *J Nucl Med* **41**, 661–681 (2000).

22. Dyke, J. P. and Aaron, R. K. Noninvasive methods of measuring bone blood perfusion. *Ann N Y Acad Sci* **1192**, 95–102 (2010).

23. Berding, G. *et al.* Evaluation of the incorporation of bone grafts used in maxillofacial surgery with [18F]fluoride ion and dynamic positron emission tomography. *Eur J Nucl Med* **22**, 1133–1140 (1995).

24. Piert, M. *et al.* Allogenic bone graft viability after hip revision arthroplasty assessed by dynamic [18F]fluoride ion positron emission tomography. *Eur J Nucl Med* **26**, 615–624 (1999).

25. Schliephake, H. Berding, G. Knapp, W. H. and Sewilam, S. Monitoring of graft perfusion and osteoblast activity in revascularised fibula segments using [18F]-positron emission tomography. *Int J Oral Maxillofac Surg* **28**, 349–355 (1999).

26. Blokhuis, T. J. *et al.* Quantitative assessment of fracture healing using positron emission tomography. *Eur J Nucl Med Mol Imaging* **30**, 329–330 (2003).

27. Kelly, P. J. Pathways of transport in bone. In: Shepherd, J. T. Abboud, F. M. and Geiger, S. R. (eds.), *Handbook of Physiology*, Vol. 3. American Physiological Society, pp. 371–396 (1983).

28. Wang, Y. X. *et al.* Reduced bone perfusion in proximal femur of subjects with decreased bone mineral density preferentially affects the femoral neck. *Bone* **45**, 711–715 (2009).

29. Griffith, J. F. *et al.* Compromised bone marrow perfusion in osteoporosis. *J Bone Miner Res* **23**, 1068–1075 (2008).

30. Hawkins, R. A. *et al.* Evaluation of the skeletal kinetics of fluorine-18-fluoride ion with PET. *J Nucl Med* **33**, 633–642 (1992).

31. Piert, M. *et al.* Blood flow measurements with [(15)O]H2O and [18F]fluoride ion PET in porcine vertebrae. *J Bone Miner Res* **13**, 1328–1336 (1998).

32. Piert, M. *et al.* Assessment of porcine bone metabolism by dynamic. *J Nucl Med* **42**, 1091–1100 (2001).

33. Brenner, W. Vernon, C. Conrad, E. U. and Eary, J. F. Assessment of the metabolic activity of bone grafts with (18)F-fluoride PET. *Eur J Nucl Med Mol Imaging* **31**, 1291–1298 (2004).

34. Schiepers, C. Broos, P. Miserez, M. Bormans, G. and De Roo, M. Measurement of skeletal flow with positron emission tomography and 18F-fluoride in femoral head osteonecrosis. *Arch Orthop Trauma Surg* **118**, 131–135 (1998).

35. Niu, G. and Chen, X. PET Imaging of Angiogenesis. *PET Clin* **4**, 17–38 (2009).

36. Laking, G. R. and Price, P. M. Positron emission tomographic imaging of angiogenesis and vascular function. *Br J Radiol* **76**(1), S50–S59 (2003).

37. Ellingson, B. M., Salamon, N. and Holly, L. T. Imaging techniques in spinal cord injury. *World Neurosurg* **82**, 1351–1358 (2014).

38. Cai, W. *et al.* PET of vascular endothelial growth factor receptor expression. *J Nucl Med* **47**, 2048–2056 (2006).

39. Rodriguez-Porcel, M. *et al.* Imaging of VEGF receptor in a rat myocardial infarction model using PET. *J Nucl Med* **49**, 667–673 (2008).

40. Cai, W. *et al.* Positron emission tomography imaging of poststroke angiogenesis. *Stroke* **40**, 270–277 (2009).

41. Willmann, J. K. *et al.* Monitoring of the biological response to murine hindlimb ischemia with 64Cu-labeled vascular endothelial growth factor-121 positron emission tomography. *Circulation* **117**, 915–922 (2008).

42. Collingridge, D. R. *et al.* The development of [(124)I]iodinated-VG76e: A novel tracer for imaging vascular endothelial growth factor *in vivo* using positron emission tomography. *Cancer Res* **62**, 5912–5919 (2002).

43. Jayson, G. C. *et al.* Molecular imaging and biological evaluation of HuMV833 anti-VEGF antibody: Implications for trial design of antiangiogenic antibodies. *J Natl Cancer Inst* **94**, 1484–1493 (2002).

44. Nagengast, W. B. *et al. In vivo* VEGF imaging with radiolabeled bevacizumab in a human ovarian tumor xenograft. *J Nucl Med* **48**, 1313–1319 (2007).

45. Cai, H. and Conti, P. S. RGD-based PET tracers for imaging receptor integrin alphav beta3 expression. *J Labelled Comp Radiopharm* **56**, 264–279 (2013).

46. Tateishi, U. Oka, T. and Inoue, T. Radiolabeled RGD peptides as integrin alpha(v)beta3-targeted PET tracers. *Curr Med Chem* **19**, 3301–3309 (2012).

47. Almutairi, A. *et al.* Biodegradable dendritic positron-emitting nanoprobes for the noninvasive imaging of angiogenesis. *Proc Natl Acad Sci USA* **106**, 685–690 (2009).

48. Krohn, K. A. Link, J. M. and Mason, R. P. Molecular imaging of hypoxia. *J Nucl Med* **49**(2), 129S–148S (2008).

49. Chapman, J. D. Hypoxic sensitizers — implications for radiation therapy. *N Engl J Med* **301**, 1429–1432 (1979).

50. Lee, S. T. and Scott, A. M. Hypoxia positron emission tomography imaging with 18f-fluoromisonidazole. *Semin Nucl Med* **37**, 451–461 (2007).

51. Basu, S. *et al.* Positron emission tomography as a diagnostic tool in infection: Present role and future possibilities. *Semin Nucl Med* **39**, 36–51 (2009).

52. Calder, P. C. Dimitriadis, G. and Newsholme, P. Glucose metabolism in lymphoid and inflammatory cells and tissues. *Curr Opin Clin Nutr Metab Care* **10**, 531–540 (2007).

53. Fu, Y. Maianu, L. Melbert, B. R. and Garvey, W. T. Facilitative glucose transporter gene expression in human lymphocytes, monocytes, and macrophages: A role for GLUT isoforms 1, 3, and 5 in the immune response and foam cell formation. *Blood Cells Mol Dis* **32**, 182–190 (2004).

54. Phelps, M. E. *et al.* Tomographic measurement of local cerebral glucose metabolic rate in humans with (F-18)2-fluoro-2-deoxy-D-glucose: Validation of method. *Ann Neurol* **6**, 371–388 (1979).

55. Fischman, A. J. and Alpert, N. M. FDG-PET in oncology: There's more to it than looking at pictures. *J Nucl Med* **34**, 6–11 (1993).

56. Odekerken, J. C. Brans, B. T. Welting, T. J. and Walenkamp, G. H. (18) F-FDG microPET imaging differentiates between septic and aseptic wound healing after orthopedic implant placement: A longitudinal study of an implant osteomyelitis in the rabbit tibia. *Acta Orthop* **85**, 305–313 (2014).

57. Chacko, T. K. Zhuang, H. Stevenson, K. Moussavian, B. and Alavi, A. The importance of the location of fluorodeoxyglucose uptake in periprosthetic infection in painful hip prostheses. *Nucl Med Commun* **23**, 851–855 (2002).

58. Reinartz, P. *et al.* Radionuclide imaging of the painful hip arthroplasty: Positron-emission tomography versus triple-phase bone scanning. *J Bone Joint Surg Br* **87**, 465–470 (2005).

59. Zhuang, H. *et al.* The promising role of 18F-FDG PET in detecting infected lower limb prosthesis implants. *J Nucl Med* **42**, 44–48 (2001).

60. Pill, S. G. *et al.* Comparison of fluorodeoxyglucose positron emission tomography and (111)indium-white blood cell imaging in the diagnosis of periprosthetic infection of the hip. *J Arthroplasty* **21**, 91–97 (2006).

61. Yamada, S. Kubota, K. Kubota, R. Ido, T. and Tamahashi, N. High accumulation of fluorine-18-fluorodeoxyglucose in turpentine-induced inflammatory tissue. *J Nucl Med* **36**, 1301–1306 (1995).

62. Ozawa, T. *et al.* Accumulation of glucose in keloids with FDG-PET. *Ann Nucl Med* **20**, 41–44 (2006).

63. Zhuang, H. M. *et al.* Do high glucose levels have differential effect on FDG uptake in inflammatory and malignant disorders? *Nucl Med Commun* **22**, 1123–1128 (2001).

64. Maitra, S. R. Wojnar, M. M. and Lang, C. H. Alterations in tissue glucose uptake during the hyperglycemic and hypoglycemic phases of sepsis. *Shock* **13**, 379–385 (2000).

65. Zhuang, H. and Alavi, A. 18-fluorodeoxyglucose positron emission tomographic imaging in the detection and monitoring of infection and inflammation. *Semin Nucl Med* **32**, 47–59 (2002).

66. Blumer, S. L. *et al.* Cutaneous and subcutaneous imaging on FDG-PET: Benign and malignant findings. *Clin Nucl Med* 34, 675–683 (2009).

67. Carter, E. A. *et al.* Metabolic alterations in muscle of thermally injured rabbits, measured by positron emission tomography. *Life Sci* **61**, 39–44 (1997).

68. Pawlik, T. M. *et al.* Acne vulgaris: False-positive finding on integrated 18F-FDG PET/CT in a patient with melanoma. *AJR Am J Roentgenol* **187**, W117–W119 (2006).

69. El-Haddad, G. Chamroonrat, W. Houseni, M. Alavi, A. and Zhuang, H. Beware of mosquitoes: The first instance of a mosquito bite detected by fluorodeoxyglucose positron emission tomography. *Pediatr Dermatol* **24**, 344–345 (2007).

70. Yen, R. F. Shun, C. T. Pan, M. H. Tsai, Y. C. and Wu, Y. W. FDG PET manifestation of lobular panniculitis. *Clin Nucl Med* **29**, 442–443 (2004).

71. Kerrou, K. *et al.* [18F]-FDG uptake in soft tissue dermatome prior to herpes zoster eruption: An unusual pitfall. *Ann Nucl Med* **15**, 455–458 (2001).

72. Golberg, A. *et al.* Skin rejuvenation with non-invasive pulsed electric fields. *Sci Rep* **5**, 10187 (2015).

73. Tehrani, O. S. and Shields, A. F. PET imaging of proliferation with pyrimidines. *J Nucl Med* **54**, 903–912 (2013).

74. Patlak, C. S. Blasberg, R. G. and Fenstermacher, J. D. Graphical evaluation of blood-to-brain transfer constants from multiple-time uptake data. *J Cereb Blood Flow Metab* **3**, 1–7 (1983).

75. Patlak, C. S. and Blasberg, R. G. Graphical evaluation of blood-to-brain transfer constants from multiple-time uptake data. Generalizations. *J Cereb Blood Flow Metab* **5**, 584–590 (1985).

76. Mordant, D. *et al.* Spectral imaging of the retina. *Eye* **25**, 309–320 (2011).

77. Ramella-Roman, J. *et al.* Measurement of oxygen saturation in the retina with a spectroscopic sensitive multi aperture camera. *Opt Express* **16**, 6170–6182 (2008).

78. Akbari, H. Uto, K. Kosugi, Y. Kojima, K. and Tanaka, N. Cancer detection using infrared hyperspectral imaging. *Cancer Sci* **102**, 852–857 (2011).

79. Ferris, D. *et al.* Multimodal hyperspectral imaging for the noninvasive diagnosis of cervical neoplasia. *J Low Genit Tract Dis* **5**, 65–72 (2001).

80. Kellicut, D. *et al.* Emerging technology: Hyperspectral imaging. *Perspect Vasc Surg Endovasc Ther* **16**, 53–57 (2004).

81. Yudovsky, D. Nouvong, A. and Pilon, L. Hyperspectral imaging in diabetic foot wound care. *J Diabetes Sci Technol* **4**, 1099–1113 (2010).
82. Rodriguez, P. Felix, F. Woodley, D. and Shim, E. The role of oxygen in wound healing: A review of the literature. *Dermatologic Surg* **34**, 1159–1169 (2008).
83. Schreml, S. *et al.* Oxygen in acute and chronic wound healing. *British J Dermatol* **163**, 257–268 (2010).
84. Zimny, S. Dessel, F. Ehren, M. Pfohl, M. and Schatz, H. Early detection of microcirculatory impairment in diabetic patients with foot at risk. *Diabetes Care* **24**, 1810–1814 (2001).
85. Khaodhiar, L. *et al.* The use of medical hyperspectral technology to evaluate microcirculatory chaves in diabetic foot ulcers and to predict clinical outcomes. *Diabetes Care* **30**, 903–910 (2007).
86. Nouvong, A. *et al.* Evaluation of diabetic foot ulcer healing with hyperspectral imaging of oxyhemoglobin and deoxyhemoglobin. *Diabetes Care* **32**, 2056–2061 (2009).
87. Greenman, R. *et al.* Early changes in the skin microcirculation and muscle metabolism of the diabetic foot. *Lancet* **366**, 1711–1717 (2005).
88. Kaiser, M. Yafi, A. Cinat, M. Choi, B. and Durkin, A. Noninvasive assessment of burn wound severity using optical technology: A review of current and future modalities. *Burns* **37**, 377–386 (2011).
89. Afromowitz, M. Callis, J. Heimbach, D. Desoto, L. and Norton, M. Multispectral imaging of burn wounds — a new clinical instrument for evaluating burn depth. *IEEE Trans Biomed Eng* **35**, 842–850 (1988).
90. Cross, K. *et al.* Clinical utilization of near-infrared spectroscopy devices for burn depth assessment. *Wound Repair Regen* **15**, 332–340 (2007).
91. Sowa, M. G. *et al.* Classification of burn injuries using near-infrared spectroscopy. *J Biomed Opt* **11**, 054002 (2006).
92. Afromowitz, M. Vanliew, G. and Heimbach, D. Clinical-evaluation of burn injuries using an optical reflectance technique. *IEEE Trans Biomed Eng* **34**, 114–127 (1987).
93. Anderson, R. and Parrish, J. The optics of human-skin. *J Invest Dermatol* **77**, 13–19 (1981).
94. Sassaroli, A. and Fantini, S. Comment on the modified Beer-Lambert law for scattering media. *Phys Med Biol* **49**, N255–N257 (2004).
95. Yudovsky, D. and Pilon, L. Rapid and accurate estimation of blood saturation, melanin content, and epidermis thickness from spectral diffuse reflectance. *Appl Opt* **49**, 1707–1719 (2010).
96. Xu, R. X. *et al.* Dual-mode imaging of cutaneous tissue oxygenation and vascular function. *J Vis Exp* (2010).
97. Lu, G. and Fei, B. Medical hyperspectral imaging: A review. *J Biomed Opt* **19** (2014).
98. Shah, S. *et al.* Cutaneous wound analysis using hyperspectral imaging. *Biotechniques* **34**, 410–+ (2003).

99. Kong, S. Martin, M. and Vo-Dinh, T. Hyperspectral fluorescence imaging for mouse skin tumor detection. *ETRI J* **28**, 770–776 (2006).

100. Georgakoudi, I. and Quinn, K. P. Optical imaging using endogenous contrast to assess metabolic state. *Annu Rev Biomed Eng* **14**, 351–367 (2012).

101. Mayevsky, A. and Rogatsky, G. G. Mitochondrial function *in vivo* evaluated by NADH fluorescence: From animal models to human studies. *Am J Physiol Cell Physiol* **292**, C615–C640 (2007).

102. Quinn, K. *et al.* Diabetic wounds exhibit distinct microstructural and metabolic heterogeneity through label-free multiphoton microscopy. In: *Department of Biomedical Engineering*, T. U., (ed.), (2015).

103. Navarro, F. A. *et al.* Two-photon confocal microscopy: A nondestructive method for studying wound healing. *Plast Reconstr Surg* **114**, 121–128 (2004).

104. Torkian, B. A. *et al.* Modeling aberrant wound healing using tissue-engineered skin constructs and multiphoton microscopy. *Arch Facial Plast Surg* **6**, 180–187 (2004).

105. Chance, B. and Thorell, B. Localization and kinetics of reduced pyridine nucleotide in living cells by microfluorometry. *J Biol Chem* **234**, 3044–3050 (1959).

106. Scholz, R. Thurman, R. G. Williamson, J. R. Chance, B. and Bücher, T. Flavin and pyridine nucleotide oxidation-reduction changes in perfused rat liver. I. Anoxia and subcellular localization of fluorescent flavoproteins. *J Biol Chem* **244**, 2317–2324 (1969).

107. Chance, B. Schoener, B. Oshino, R. Itshak, F. and Nakase, Y. Oxidation–reduction ratio studies of mitochondria in freeze–trapped samples. NADH and flavoprotein fluorescence signals. *J Biol Chem* **254**, 4764–4771 (1979).

108. Quinn, K. P. *et al.* Quantitative metabolic imaging using endogenous fluorescence to detect stem cell differentiation. *Sci Rep* **3**, 3432 (2013).

109. Huang, S. Heikal, A. A. and Webb, W. W. Two-photon fluorescence spectroscopy and microscopy of NAD(P)H and flavoprotein. *Biophys J* **82**, 2811–2825 (2002).

110. Xu, C. Zipfel, W. Shear, J. B., Williams, R. M. and Webb, W. W. Multiphoton fluorescence excitation: New spectral windows for biological nonlinear microscopy. *Proc Natl Acad Sci USA* **93**, 10763–10768 (1996).

111. Varone, A. *et al.* Endogenous two-photon fluorescence imaging elucidates metabolic changes related to enhanced glycolysis and glutamine consumption in precancerous epithelial tissues. *Cancer Res* **74**, 3067–3075 (2014).

112. Denk, W. Strickler, J. H. and Webb, W. W. Two-photon laser scanning fluorescence microscopy. *Science* **248**, 73–76 (1990).

113. Mokrý, M. *et al. In vivo* monitoring the changes of interstitial pH and FAD/NADH ratio by fluorescence spectroscopy in healing skin wounds. *Photochem Photobiol* **82**, 793–797 (2006).

114. Campagnola, P. J. and Loew, L. M. Second-harmonic imaging microscopy for visualizing biomolecular arrays in cells, tissues and organisms. *Nat Biotechnol* **21**, 1356–1360 (2003).
115. Williams, R. M. Zipfel, W. R. and Webb, W. W. Multiphoton microscopy in biological research. *Curr Opin Chem Biol* **5**, 603–608 (2001).
116. Yeh, A. T. *et al.* Imaging wound healing using optical coherence tomography and multiphoton microscopy in an *in vitro* skin-equivalent tissue model. *J Biomed Opt* **9**, 248–253 (2004).
117. Vander Heiden, M. G. Cantley, L. C. and Thompson, C. B. Understanding the Warburg effect: The metabolic requirements of cell proliferation. *Science* **324**, 1029–1033 (2009).
118. Brem, H. *et al.* Molecular markers in patients with chronic wounds to guide surgical debridement. *Mol Med* **13**, 30–39 (2007).
119. Ghukasyan, V. and Kao, F. Monitoring cellular metabolism with fluorescence lifetime of reduced nicotinamide adenine dinucleotide. *Journal of Physical Chemistry C* **113**, 11532–11540 (2009).
120. Lakowicz, J. R. Szmacinski, H. Nowaczyk, K. and Johnson, M. L. Fluorescence lifetime imaging of free and protein-bound NADH. *Proc Natl Acad Sci USA* **89**, 1271–1275 (1992).
121. Bird, D. K. *et al.* Metabolic mapping of MCF10A human breast cells via multiphoton fluorescence lifetime imaging of the coenzyme NADH. *Cancer Res* **65**, 8766–8773 (2005).
122. Zhao, Y. *et al.* Integrated multimodal optical microscopy for structural and functional imaging of engineered and natural skin. *J Biophotonics* **5**, 437–448 (2012).
123. Park, J. *et al.* A dual-modality optical coherence tomography and fluorescence lifetime imaging microscopy system for simultaneous morphological and biochemical tissue characterization. *Biomed Opt Express* **1**, 186–200 (2010).
124. DaCosta, R. S. *et al.* Point-of-care autofluorescence imaging for real-time sampling and treatment guidance of bioburden in chronic wounds: First-in-human results. *PLoS One* **10**, e0116623 (2015).
125. Konig, K. and Riemann, I. High-resolution multiphoton tomography of human skin with subcellular spatial resolution and picosecond time resolution. *J Biomed Opt* **8**, 432–439 (2003).

9. Functional Skin Substitutes — The Intersection of Tissue Engineering and Biomaterials

Kevin Dooley*, Julie Devalliere* and Basak Uygun

Center for Engineering in Medicine, Massachusetts General Hospital, Harvard Medical School, and Shriners Hospitals for Children 51 Blossom St. Boston, Massachusetts, USA

Abstract

In recent years, a number of skin substitutes have been developed for the treatment of acute and chronic cutaneous wounds in order to provide sterile wound coverage and stimulate tissue repair. Advancements in tissue engineering and biomaterials have led to an array of products designed to target several facets of the regeneration process. Some of these products have been rigorously evaluated in clinical settings, producing variable results. Cutting-edge fabrication techniques, stem cell technology, growth factor therapies, and genetic engineering are poised to overcome limitations associated with current products on the market. Due to the enormous financial burden cutaneous wound management places on healthcare systems worldwide, there is a significant demand for treatment strategies that are cost effective, easy to use, and result in accelerated and cosmetically pleasing wound closure.

1. Introduction

The skin is a complex, multifunctional organ primarily responsible for creating a physical barrier between internal organs and the environment in order to protect the body from trauma, radiation, pathogens, and other potentially

---

*Both authors contributed equally to this work.

harmful agents. It is also intimately involved with thermoregulation, fluid balance, vitamin D production, and sensation.[1,2] Any disruption to the structural integrity of the skin requires an immediate and efficient response to repair damaged tissue and regenerate the barrier. The wound healing response is a complex and dynamic process that relies on a coordinated effort from an array of cell types along with protein and chemical mediators to restore skin function. Several pathologies, medications, and other systemic variables can lead to an impaired response resulting in a non-healing wound characterized by a pathological state of inflammation. Non-healing surgical wounds, venous ulcers, diabetic ulcers, and pressure sores account for the vast majority of such wounds.[3,4] Chronic wounds represent an enormous burden on healthcare systems worldwide with a recent study conservatively estimating a \$50 billion annual price tag in the US alone.[3] Traditionally, treatment options for such wounds focus mainly on sterilization, fluid retention, nutrition, and other externally controllable factors. The ineffectiveness of these traditional therapeutic approaches have pushed researchers and clinicians to investigate alternative, targeted treatment strategies. Recently, a number of bioengineered skin substitutes have been developed to serve as provisional scaffolds and barriers to stimulate physiologic wound healing. Scaffolds constructed from natural polymers, synthetic materials, and decellularized tissues have been further engineered to incorporate growth factors, stem cells, and other biologically active moieties to address several facets of an impaired wound healing cascade. The focus of this chapter serves to provide an overview of currently available technologies and an assessment of their efficacy in clinical trials. Significant hurdles associated with the implementation of these skin substitutes as well as strategies to address these challenges will be discussed.

2. Skin Architecture and Wound Healing Basics

The skin is structurally organized into two mutually dependent layers, the epidermis and the dermis, which sit on top of a layer of subcutaneous fatty tissue. The epidermis is a stratified squamous epithelium composed primarily of keratinocytes (~95%) with Langerhans cells, melanocytes, and Merkel cells making up the rest (~5%). This top layer is on average 100 μm thick, but can vary considerably from 50 μm on the eyelids to 1 mm on the palms and soles.[5,6] Sweat glands, sebaceous glands, and hair follicles, though housed primarily in the dermis and subcutaneous layers, are considered epidermal appendages as they originate as downgrowths in the epidermis during fetal development.[7,8] A thin basement membrane, composed of specialized extracellular matrix (ECM) separates the epidermal and dermal compartments and provides mechanical support for the adhesion of the two layers. Additionally, it plays a key role in trafficking cells, metabolites, growth factors, and other biologically active domains between the two layers, especially during cutaneous wound healing.[9,10]

The dermis is primarily responsible for conferring support, elasticity, and tensile strength to the skin and is largely composed of fibrous and amorphous ECM with collagen being the most abundant protein. It is organized into two sections: the superficial papillary layer and the deep reticular layer with the distinction mainly based on cell and vasculature density as well as connective tissue organization. The papillary dermis forms conical projections which protrude upward and interdigitate with the rete pegs in the epidermis. This significantly increases the adhesive surface area between the two layers, providing additional support. The papillary layer is composed of fibroblasts, mast cells, dermal dendrocytes, capillaries, and connective tissues made up of small-diameter collagen fibrils. In contrast, the reticular dermis is characterized by large-diameter collagen fibers spatially organized into bundles which lie parallel to the skin surface. This thick, elastic network of collagen is responsible for the robust mechanical properties of the skin. Populations of fibroblasts and other resident dermal cell types are less dense in the reticular dermis compared to the papillary layer. The transition from the dermal layer to the subcutaneous layer is characterized by an abrupt shift in largely fibrous connective tissue to fatty adipose tissue.[5,11]

In order to more effectively tailor treatment protocols to address a disruption in the continuity of the skin, the underlying molecular mechanisms of the wound healing cascade must be understood. Classically, wound healing is divided into four overlapping, interdependent phases: hemostasis, inflammation, proliferation and remodeling, with each phase defined by the infiltrating cell types, growth factors, and chemical gradients present in the wound site. Immediately following a disruption to the dermal layer and extravasation of blood into the wound site, several physiological processes are set into motion to initiate wound healing. Vascular smooth muscle contraction and subsequent vasoconstriction occurs within the endothelium to curb blood loss.[12–14] Circulating platelets bind to exposed collagen fibrils in the ECM and degranulate, promoting further platelet aggregation and eventually creating a plug. An elegant coagulation cascade is triggered by the interaction of circulating activated factor VII with subendothelial tissue factor leading to a fibrin-rich clot and hemostasis.[15,16] The hemostatic clot serves as a provisional scaffold for cellular migration into the wound bed.[17]

The inflammatory phase is observed within hours post injury and is principally characterized by vasodilation, increased capillary permeability, and infiltration of leukocytes into the damaged tissue.[12,18] Neutrophils are the first pro-inflammatory cells to arrive at the wound and are primarily responsible for eliminating pathogens and breaking down damaged matrix proteins. Macrophages, derived from extravasated monocytes, arrive next to assist in sterilizing and debriding the wound bed and removing expended neutrophils. They are also responsible for releasing an array of growth factors and cytokines including, epidermal growth factor (EGF), transforming growth factor α and β (TGF-α, β), fibroblast growth factors (FGF), and vascular endothelial growth factor (VEGF), which facilitate the transition to the proliferative phase.

Once the wound is cleared of cellular debris and fragmented matrix proteins, the healing process enters the proliferation stage where ECM is deposited, new dermal tissue is vascularized, and the epithelium is reconstructed. Fibroblasts, both resident and recruited, begin synthesizing a provisional matrix composed largely of Type III collagen, glycosaminoglycans, and fibronectin in order to reestablish tissue continuity and provide structural support for further cellular influx. The microvasculature in the wound must be reconstructed to supply the developing tissue with oxygen and other blood constituents. VEGF, FGF-2, and TNF-α are responsible for tightly regulating endothelial cell migration, proliferation and tubule formation. This newly formed, vascularized connective tissue at the base of the wound bed composed of fibroblasts, macrophages and other matrix proteins is called granulation tissue. At the surface of the wound, the re-epithelization process begins within 24 h post injury to prevent fluid loss and reestablish a protective barrier. Proliferative epithelial progenitors located in the bulge region of hair follicles are released and proliferate from the edges of the wound inward, providing coverage. These cells continue to proliferate and differentiate into keratinocytes to form the stratified epidermal layers.[12]

The remodeling phase primarily serves to break down, reorganize, and strengthen the newly formed network of cells, matrix proteins, and blood vessels. The haphazardly constructed collagen framework produced in the proliferative phase is degraded by matrix metalloproteinases (MMPs) secreted by epithelial cells and fibroblasts. Highly cross-linked and spatially well-organized layers of collagen are deposited with the appropriate proportion of Type III collagen (10–20%). As this process of collagen maturation continues, the tensile strength of the newly formed skin increases from about 30% at 3 weeks to 80% at 3 months post injury, as compared to intact skin.[12] The majority of macrophages, fibroblasts and unnecessary vasculature are cleared from the injury site by apoptosis. Over time, cellular and metabolic functions in the wound site return to normal with the mature wound being largely acellular and avascular.[19,20]

3. Impaired Wound Healing and Treatment Options

For proper cutaneous healing, the dynamic process outlined above must proceed for the appropriate amount of time and intensity at each stage. There are several factors that can interfere with the initiation, regulation, and/or termination of each stage leading to an impaired wound healing response. These factors can be generally categorized as local, pertaining to the wound site itself, or systemic, pertaining to overall health of the affected individual.[13] First off, the severity and location of the primary injury can have a significant impact on healing. A recent study examining diabetic foot ulcer treatment demonstrated a substantial difference between healing rates of ulcers located on the heel, midfoot, or toe of the patient.[21] In addition, improper oxygenation, insufficient vascular supply, and bacterial infection can locally contribute to delayed or improper wound healing.[22] Systemic variables such as age, gender, stress, obesity, medication, and disease profile can all adversely affect the healing response as well.

3.1 *Acute versus chronic wounds*

Human wounds can be classified as acute or chronic based on the nature of the repair process. Acute wounds are mostly healable within 8–12 weeks and are caused by mechanical injuries such as abrasions, avulsions, incisions or lacerations.[23] Burns and chemical injuries represent another category of acute wounds caused by exposure to extreme heat, irradiation, electrical shock, or corrosive chemicals. Wounds can be further categorized by the number of skin layers and area affected as illustrated by the classification of burns: first degree burns are superficial and limited to the epidermis, second degree burns affect a superficial portion of the dermis in addition to the epidermis and are referred to as partial-thickness wounds, whereas third and fourth degree burns are considered full-thickness wounds and affect all epidermal and dermal layers, extending down to the subcutaneous tissue, muscle, or bone.[24] While the ability of human skin to repair itself after a minor injury is a remarkable process, any loss of full-thickness skin more than 4 cm in diameter will require skin grafting as these wounds cannot re-epithelialize. Most cases result in extensive scarring, mobility limitations and severe cosmetic deformities.[25]

Unlike acute wounds, the healing process for chronic wounds usually exceeds 12 weeks, requiring frequent hospital visits in addition to high recurrence rates. While acute wounds proceed through the normal stages of healing, chronic wounds generally do not follow an orderly process of regeneration and repair.[26] Physiologically, these wounds are in a perpetual state of inflammation caused by the over-infiltration of neutrophils in the wound site.[27] They release an excessive amount of MMPs which break down collagen and other matrix proteins faster than fibroblasts can deposit them. This, in turn, leads to an uncontrolled pro-inflammatory feedback loop and inhibits progression to the proliferation phase.[28]

The majority of impaired wound healing cascades manifest in the form of chronic ulcers. These wounds can generally be assigned to one of three clinical categories: leg ulcers, diabetic foot ulcers or pressure ulcers also known as bed sores.[29] Leg ulcers can have a variety of etiologies. Venous ulcers are the most common, often resulting from dysfunction of valves in veins of the lower leg that normally prevent the backflow of venous blood. Arterial insufficiency and diabetes can lead to arterial blockage causing tissue ischemia that contributes to the development of leg ulcers or necrosis.[30] Patients with diabetes are prone to leg ulcers due to their disease features that include neuropathy, poor circulation, and reduced response to infection. Pressure ulcers, characterized by tissue ischemia and necrosis,[31] are common among patients in long-term care settings, but patients hospitalized for short-term care or in home settings with reduced mobility are also at risk.[32] Other manifestations of abnormal wound healing responses include keloid and hypertrophic scar formation, resulting from pathological collagen deposition in the dermal layer.[33–35]

3.2. *Traditional wound management*

In the US alone, approximately 6.5 million people suffer from chronic wounds, most of which are attributed to some form of non-healing ulcer or surgical wound.[36] The costs associated with the treatment of chronic wounds has been conservatively estimated at $50 billion dollars annually in a 2012 study.[3] These figures are only predicted to rise as the population begins to age and causative pathologies like diabetes and obesity become more prevalent worldwide. There are currently several treatment options available to clinicians which primarily use some sort of occlusive dressing in order to retain a moist, sterile wound environment. Several studies have confirmed the importance of pressure relief for the healing of diabetic ulcers; compression therapy is still considered the standard treatment for venous ulcers as well.[37] Depending on the state of the injury a variety of barrier products, gauzes, films, and hydrogels have been developed to help stem pathologic inflammation and stimulate physiologic healing. However, there is no consensus among physicians as to which products are superior in facilitating wound healing, while remaining cost effective as there is a substantial lack of quantitative clinical data.[38] Wound debridement has also proven to be beneficial in chronic wound management. Necrotic tissue, bacterial films, and eschar can be removed surgically or enzymatically in an attempt to push the wound to an acute state.[39] Other variables such as nutrition,[40] alcohol consumption,[22] smoking habits,[41] and prescribed medications[42] can be managed appropriately to create a more hospitable healing environment.

Outside of managing externally controllable factors, a handful of other approaches are available for treating chronic wounds. It has been known for decades that topically applied growth factors can accelerate the wound healing process. However, due to the highly proteolytic nature of the wound microenvironment these growth factors are swiftly degraded. Repeated application is required to achieve the desired therapeutic effect thereby inflating treatment costs. Currently, only one such treatment consisting of human platelet-derived growth factor (PDGF) is approved by the Food and Drug Administration (FDA) for neuropathic ulcers. Reperfusion via infrapopliteal endovascular revascularization, a minimally invasive procedure, has shown promising results for the treatment of diabetic foot ulcers in clinical studies.[43,44] Non-invasive low-level laser therapy (LLLT), or phototherapy, uses light emitting diodes to stimulate cellular processes involved in wound healing. Recent findings have shown LLLT to enhance migration, proliferation, and re-epithelialization in animal models.[45–47]

For trauma extending deep into the dermis surgical skin grafting is often required, particularly in severe burn injury. Biological grafts can be categorized based on tissue thickness and donor origin. Split thickness skin grafts (STSG) are considered the standard of care for defect coverage in such injuries and are composed of an intact epidermal layer and a fractional slice of the dermal layer. Thin tissue slices are typically harvested from the arm, thigh, back, or abdomen leaving behind skin appendages (hair follicles, sweat glands, sebaceous glands) and the reticular dermis, which typically allow the donor site to re-epithelialize and remain functional. STSGs are commonly used to cover large skin defects and can be meshed to increase their surface coverage, but often times result in less than satisfactory scarring, especially in children.[48,49] Alternatively, full thickness skin grafts (FTSG) composed of the entire epidermal and dermal layers tend to produce more aesthetically acceptable results and are commonly used to address injuries to the face and neck. However, only small grafts are practical given that the constituents required for healing and re-epithelization are removed from the donor site. Furthermore, FTSGs require suitable vascularization in the surrounding tissue in order to be successfully integrated.[1,50]

The origin of the donor tissue plays a key role in the successful outcome of a skin graft procedure. Autologous grafts, or autografts, refer to the transfer of tissue from one site to another on the same individual. While autografts do not elicit an immune response, scar tissue is created at both the donor and transplant site. Allografts, or homografts, refer to the transfer of tissue from one individual to another of the same species. These are typically used in cases of severe burn injury where autograft donor tissue is not available. The transplanted tissue is most commonly isolated from a cadaveric source. Xenografts describe the transfer of tissue isolated from a different species, typically porcine. Both allografts and xenografts will eventually undergo immunogenic rejection, anywhere from the third to seventh day post transplantation, whereas autografts have the potential to become fully integrated.[13,50] Other ethical and safety issues, particularly disease transmission, are associated with allografts and xenografts as well.[2,51]

The sheer complexity of the wound healing process coupled with the enormous burden acute and chronic wound management places on healthcare systems has led to the development of an extraordinary amount of treatment options and protocols. However, most of these passive treatments lead to debilitating scar formation and require lengthy hospital stays. Furthermore, skin grafting procedures must be performed under anesthesia by extensively trained surgeons and are highly dependent on the availability of donor tissue, which may not be an option for the 11 million people suffering from severe burn injury each year.[52] In addition to wound contracture and pigmentation loss, insufficient coverage by STSGs can lead to bacterial sepsis, which is the leading cause of mortality in burn patients following initial stabilization.[53] These shortcomings have pushed scientists and clinicians to explore alternative approaches. Tissue engineering principles and advanced biomaterials have opened the door to a class of bioengineered skin substitutes for the treatment of acute and chronic wounds.

4. Engineered Skin Substitutes

The ultimate goal for tissue engineered skin substitutes is to rapidly culture autologous full-thickness skin grafts *in vitro* containing all the skin appendages for non-immunogenic, scarless incorporation into a wound site. While this lofty goal is still far off, several strides have been made in the tissue engineering and biomaterials fields which have translated to a myriad of technologies for functional skin replacement. These products aim to address the problems outlined in the previous section by providing immediate defect coverage, moisture retention, and protection from invading microorganisms in hopes of stimulating physiologic healing. Despite these recent advances, there appears to be no consensus as to which products are superior for treating wounds extending deep into the dermis. With such a diverse catalogue of skin substitutes available, it is important to rigorously evaluate manufacturing procedures, storage conditions, tissue origin/preparation, and biological activity as these properties can all affect the wound healing response. For the sake of this chapter, we have generally categorized these bioengineered skin equivalents based on whether or not living cells are incorporated, with further classification into the skin layers targeted for regeneration.

4.1. *Acellular scaffolds and coverage technologies*

There are several forms of acellular scaffolds available to clinicians for use in treatment of burn injury and chronic ulcers, ranging from artificial or natural polymers to decellularized matrices derived from human tissue. These scaffolds are typically cheaper to manufacture and store when compared to cell-based therapies, and can be used "off the shelf". Additionally, depending on their mechanism of action, some products may circumvent time consuming and costly clinical trials by moving through the FDA's premarket approval process, where they are considered medical devices. Even though cells are not directly incorporated, these scaffolds can (1) use their physical characteristics, such as stiffness, pore size, and hydrophobicity, and/or (2) be supplemented with biologically active moieties, in order to influence proximal cellular behavior and the wound healing cascade.[54]

Acellular polymeric wound dressings are routinely used in the clinical setting to achieve rapid epidermal coverage of skin defects, especially when donor tissue is not available for STSGs. These treatments aim to achieve a moist, sterile wound environment, while simultaneously allowing for sufficient oxygen diffusion across the barrier. A staggering array of such products are being currently developed in the laboratory in addition to those already approved by the FDA for clinical use. These products can generally be broken down by their composition: naturally occurring versus. non-biological. A range of naturally occurring polysaccharides including cellulose, dextran, chitin/chitosan, and alginate have all been used as starting materials to create epidermal wound dressings. Proteins and peptides including collagen, keratin, silk fibroin, and fibrin have been explored as potential building blocks as well. Non–biological materials such as silicone, polyurethane, and poly-lactic acid have been manufactured into membranes, foams, and nanofibers to provide epidermal coverage. Several in-depth reviews have been published on the acellular materials used for wound dressings.[55–62] Table 1 provides a

Table 1. Acellular scaffolds and wound dressings.

Type	Material	Product and Notes	Reference
Epidermal			
Polysaccharide	Alginate	Alginate/PVP hydrogels loaded with nanosilver demonstrate enhanced antimicrobial activity, fluid absorption capacity, and moisture permeability necessary for wound dressings.	173
		Alginate/chitin biocomposite fibers accelerate wound closure with single application in rats.	174
	Cellulose	Bacterial cellulose–chitosan membranes re-epithelialize skin defects faster than Tegaderm® in rat model.	175
		Crystalline cellulose dressing Veloderm® proves to be a safe and effective dressing in management of STSG donor site injury in burn and plastic surgery patients.	176
		Pure cellulose dressing Cuticell® Epigraft reduces pain and enhances re-epithelialization in donor areas for STSGs in clinical study.	177
	Chitin/Chitosan	Chitin/nanosilver composite hydrogel shows enhanced antimicrobial activity.	178
		Nonwoven chitin fabrics enhance wound healing in rabbit model.	179
		Electrospun chitosan-based fibrous mats containing chitin nanocrystals create flexible, strong materials for wound dressing application.	180
		Chitosan, PVP, TiO_2 blended materials demonstrate antimicrobial activity, biocompatibility and accelerate excisional wound healing in rats.	181
Protein	Collagen	Antibiotic loaded collagen matrices exhibit favorable biocompatibility and mechanical properties. Further *In vivo* studies are required in an impaired wound healing model.	182

			90% collagen 10% alginate mix (FIBRACOL™ PLUS) provides structural support and enhances re-epithelialization at donor sites for STSGs.	183
			Biobrane® is composed of a nylon mesh coated with porcine collagen. A semi-permeable silicone layer is applied on top of the mesh and is used for superficial or partial thickness burns.	111
		Silk	Silk sericin-releasing wound dressing is less adhesive than a clinical standard (Bactigras®) and significantly reduces pain at STSG donor sites	184
		Fibrin	Fibrin bio-matrix conjugated with PDGF was used to secure autologous STSGs in a third degree porcine burn model and showed to accelerate and improve wound healing	185
	Non-biological	Silicone	Microporous silicone bilayer membranes prevent bacterial invasion, promote cell adhesion and proliferation, and enhance re-epithelialization in full-thickness mouse models.	186
		Polyurethane	Polyurethane foam infused with silver-hydrozyapatite promote wound contraction, re-epithelialization, and collagen deposition in an infected wound rat model.	187
		Polylactic acid	Polylactic acid nanofibers loaded with curcumin enhance the rate of wound closure in mouse.	188
Dermal				
	Polysaccharide	Carboxy-methylcellulose	PDGF infused in sodium carboxymethylcellulose gel (Regranex®) boosts granulation tissue formation and reduces wound closure time of chronic diabetic neuropathic ulcers.	189

(Continued)

Table 1. (*Continued*)

Type	Material	Product and Notes	Reference
Protein	Collagen	Collagen/hyaluronic acid dressing infused with EDF and vitamin C derivative promotes granulation tissue formation and angiogenesis in a diabetic mouse model.	190
	Fibrin	Fibrin gels laden with keratinocyte growth factor (KGF) conjugated elastin, like polypeptide (ELP) nanoparticles enhance re-epithelialization and granulation tissue formation in a diabetic mouse model.	191
		Fibrin/poly(ether)urethane–polydimethylsiloxane scaffolds mixed with platelet lysate promote granulation tissue formation, collagen deposition, and re-epithelialization in diabetic mice.	192
	Silk	Silk fibroin scaffold loaded with antibiotic gelatin microspheres serve as a dermal regeneration template for full-thickness burns. Persistent release of antibiotic staves off infection from several pathogens.	193
Decellularized tissue	Allogeneic skin	AlloDerm® Regenerative Tissue Matrix provides provisional matrix for cellular infiltration and granulation tissue formation.	194
		GraftJacket® Regenerative Tissue Matrix is specifically designed for the treatment of chronic wounds.	145
	Porcine small intestine submucosa	OASIS® Wound Matrix outperforms moist wound dressings for the treatment of venous ulcers.	195
	Porcine urinary bladder	Matristem® ECM material has shown promising clinical results for the treatment of chronic wounds in previously irradiated tissue.	196
	Fetal bovine dermis	PriMatrix® Dermal Repair Scaffold has been successfully used as a treatment regimen for diabetic foot ulcers.	197

Dermo-epidermal

Synthetic/ Protein	Silicone/Collagen	INTEGRA® Dermal Regeneration Template is composed of an epidermal substitute layer made of silicone and a dermal layer composed of a porous matrix of cross-linked bovine tendon collagen and glycosaminoglycan. It is one of the most frequently used templates for skin regeneration.	96
Protein/ Protein	Silk/Gelatin	Wax-coated silk fibroin woven fabric is used as a non-adhesive top layer to retain moisture. A bioactive bottom layer composed of silk fibroin/gelatin/sericin is used to support cell adhesion and collagen deposition. The bi-layered construct promotes healing in full-thickness wounds in rat when compared to Tegaderm alone.	198
	Gelatin/ECM	Gelatin hydrogel top layer is infused with EGF conjugated microspheres. Tissue engineered ECM serves as a provisional dermal scaffold. The bi-layered material induces cellular migration and proliferation as well as re-epithelialization in rat model.	199

summary of some of the most recently developed and clinically successful technologies for acellular wound coverage and regeneration.

For non-superficial wounds that extend deep into the dermal layer, a more robust scaffold is required to help promote cellular influx and dermal regeneration to prepare the wound for STSGs or other coverage mechanisms. Again, polysaccharide, and protein, based templates have been explored as potential scaffolds. However, the most widely implemented form of dermal regeneration templates is derived from decellularized tissues. Allogeneic (cadaver) and xenogeneic (porcine, bovine) tissues can be harvested and chemically and/or enzymatically treated to remove all cellular components of the graft. Connective tissues, vascular templates, and other ECM components are left behind which help stimulate cellular migration and proliferation in a properly debrided wound bed. There are significant risks of graft rejection or immunogenic response associated with these technologies, as is the case for any non-autogenic substitute. More complex acellular skin substitutes are available that target both epidermal and dermal regeneration simultaneously. Typically, these products are composed of a thin, semi-permeable top layer coupled with an ECM-like dermal analog to act as a provisional scaffold. One of the most clinically successful bi-layered acellular products is INTEGRA® Dermal Regeneration Template, and will be discussed in detail in Section 5.

4.2 Cellularized scaffolds and coverage technologies

Compared to the vast number of acellular products either approved by the FDA or currently in development in research laboratories, there are far fewer technologies available which directly incorporate living cells. Often times, living cells are seeded into scaffolds similar to those described in the acellular section in hopes of amplifying the clinical efficacy. However, there is an inherent tradeoff associated with the cellularized scaffold approach; the benefits, which include *in situ* production of growth factors and cytokines for the subsequent recruitment of regenerative host cells directly to the wound bed, must be weighed against the inflated cell culture costs, time consuming production requirements, truncated shelf-lives (on the order of hours to days), and complicated handling and application protocols.[63] Technologies that incorporate living cells are also subject to much more stringent regulatory standards and approval procedures.[54]

Most commercial cellular epidermal substitutes are composed of autologous keratinocytes obtained via skin biopsy and expanded *ex vivo*. Cultured epidermal autografts, under the tradename Epicel®, can be expanded

10,000-fold from a 3 cm^2 tissue sample over the course of 4 weeks. This technology is FDA approved for the treatment of severe burn injury covering at least 30% of the total body surface area.[13,63] The outer root sheath of hair follicles has been explored as a potential source of keratinocytes for *ex vivo* expansion. Cells obtained from this source have superior proliferative capacity as they are thought to be the immediate progeny of pluripotent stem cells. Tissue engineered autologous epidermal sheets derived from these cells (EpiDex®) have been clinically evaluated for the treatment of recalcitrant venous leg ulcers.[64–66] Non-cultured autologous and allogeneic keratinocyte suspensions are also commercially available for use as epidermal substitutes.[67–69] More recently, genetically modified immortalized allogeneic keratinocytes reprogrammed to overexpress EGF have been shown to improve epidermal morphogenesis and accelerate wound healing in a rat burn model.[70] While most similar studies are still in the experimental phase, the advent of genome editing and induced pluripotent stem cell (ips) technology has the potential to revolutionize bioengineered skin substitutes.

Cellularized scaffolds which specifically target the dermal layer are typically composed of a collagen or similar biodegradable framework infused with allogeneic fibroblasts. These cells deposit matrix proteins, growth factors, cytokines, and other molecules necessary to stimulate cellular influx and dermal regeneration. Dermagen®, Dermagraft®, ICX-SKN, and Transcyte® (no longer commercially available) all operate using this same basic framework, with slight differences in processing techniques and cell sources. These technologies are most commonly used for temporary coverage of full and partial thickness burns as well as treatment alternatives for diabetic foot ulcers.[63,71–75] *In situ* self-assembling scaffolds offer an interesting alternative to the products described above which necessitate time consuming fabrication and cryopreservation requirements. Recently, a partially cross-linked hydrogel–collagen biohybrid platform was used to deliver genetically modified fibroblasts which overexpress indoleamine 2,3-dioxygenase to full-thickness wounds in a rabbit fibrotic ear model. A significant reduction in scar elevation as well as improved overall healing quality was observed when compared to control groups.[76] Other potential therapeutic strategies which deliver cells via *in situ* forming gels have been reported.[77,78]

The most complex cellularized skin equivalents are designed to resemble native skin architecture. This is typically accomplished by culturing a layer of autologous or allogeneic keratinocytes on top of a fibroblast infused dermal analog composed of collagen or fibrin, thereby simultaneously targeting dermal and epidermal regeneration upon transplantation. Products using this approach include Stratagraft®, OrCel® (no longer commercially available),

PermaDerm™, and Apligraf®, the last of which has been extensively characterized in the clinic.[79–86] Laser-assisted bioprinting and 3D freeform fabrication have recently been explored as alternative production techniques for bi-layered skin substitutes and have shown promising results in animal models.[87–89] These methods offer precise control over the size, shape, and cellular composition of the 3D tissue substitutes. In addition to addressing dermal and epidermal tissue repair, research groups have begun to focus on reconstituting cutaneous appendages during the regeneration process. Using an engineered bi-layered dermo-epidermal substitute populated with murine dermal papilla cells and human keratinocytes, chimeric hair follicles have been successfully generated in an athymic mouse model.[90]

Despite the multitude of products approved by the FDA for clinical use and the emerging technologies currently in development in the laboratory, several challenges remain to be addressed. Most of the dermal and dermo-epidermal products discussed above suffer from insufficient vascularization, cosmetically unacceptable scar formation and pigmentation, and expensive production requirements. Advances in genetic engineering, programmable biomaterials, and cell printing are poised to overcome some of these limitations. In the meantime, it is important to fully understand the products being currently used in the clinic for chronic wound and burn injury. Sections will focus on the clinical successes and shortcomings of these technologies.

5. Clinical Trials

Surgical interventions, burns, and chronic ulcers constitute the main clinical conditions in which wound healing plays a critical role. More than 50 million inpatient surgical procedures were performed in the US in 2014 (Center for Disease Control and Prevention) with more than one-third resulting in hypertrophic scarring or keloid formation requiring wound management. Burn injuries remain the most common cause of significant skin loss, which accounts for approximately 500,000 hospital emergency visits per year in the US (American Burn Association, 2011). In addition to acute wounds, approximately 6.5 million people in the US suffer from ulcers resistant to conventional healing which places an enormous burden on the healthcare system, and is only predicted to get worsen in the coming years.[3,22,36] Engineered skin substitutes are designed to restore skin barrier function by increasing or restoring the wound healing abilities. The treatment of skin lesions is a critical issue in healthcare and several parameters affecting the healing process such as the wound type, the wound depth, and the level of the exudate have to be considered.[22] An ideal wound therapy must induce rapid healing, with minimum pain, discomfort and scarring to the patient. In this section, the most relevant and recent clinical trials using biological skin substitutes will be reviewed and their advantages or drawbacks will be discussed.

5.1 *Acellular scaffolds*

5.1.1 *Synthetic scaffolds*

The most basic of the skin substitutes are synthetic, acellular materials designed to act primarily as barriers to fluid loss and microbial contamination. In this section, we will illustrate this engineering approach by two examples of products that have been widely used and documented for coverage of excised burn wounds: INTEGRA® Dermal Regeneration Template (INTEGRA LifeSciences Corporation, Plainsboro, NJ) and Biobrane® (Smith & Nephew global product, Andover, MA).

INTEGRA® is the most studied of the biosynthetic templates and consists of two layers: a dermal substitute made of porous bovine collagen and chondroitin-6-sulfate glycosaminoglycan coupled with an epidermal substitute made of a semipermeable silicone polymer. The dermal layer provides a matrix for the migration of native wound bed cells, mostly fibroblasts. Sequential remodeling of the collagen by infiltrating cells leads to the degradation of the bovine collagen and replacement by newly synthesized collagen. The silicone layer provides shear strength and flexibility and creates a less permeable layer to control fluid loss from the wound site. This layer is removed upon vascularization of the dermis and a STSG is placed on top to provide definitive soft-tissue. This technology was originally developed by Drs. Ioannis Yannas and John Burke, of the Massachusetts Institute of Technology and Shriners Burns Institute, respectively.[91–95]

INTEGRA® was approved by the FDA in 1996 to treat patients with severe burns or patients undergoing reconstructive surgery for contractures. Its use was further expanded for other indications such as chronic, surgical, and traumatic wounds in 2008. To date, the largest study addressing the safety and effectiveness of INTEGRA® for the treatment of deep partial-thickness thermal injuries is a 2003 prospective non-randomized clinical trial involving 216 burn injury patients who were treated at 13 burn care facilities in the US.[96] The primary objective was to determine the incidence of invasive infection associated with INTEGRA®-treated wounds as part of a post-approval study requested by the FDA. In a pivotal clinical trial conducted 20 years earlier, INTEGRA® was shown to be efficacious in treating

136 burn wounds on 106 patients but a higher infection rate was observed with INTEGRA®-grafted sites compared to autograft controls.[97] To re-evaluate this potential infection issue, 222 patients with life-threatening full-thickness and deep partial thickness thermal injuries were enrolled in a nonblinded observational study. INTEGRA® was applied to fresh, clean, surgically excised burn wounds and within 2 to 3 weeks, a thin epidermal autograft was placed on top of the regenerated dermal layer. The incidence of infection at INTEGRA®-treated sites was low compared to the pivotal clinical with only 3.1% of invasive infection and 13.2% of superficial infection. According to the authors, these good results were due in part to the implementation of standardized methods for wound site management with INTEGRA®. They also demonstrated a high take rate of the scaffold (76.2%) and the subsequent skin graft (87.7%). This post-approval study was the first to clearly establish that INTEGRA® is a safe and effective treatment modality in the hands of properly trained clinicians under conditions of routine clinical use.

However, the number of randomized, controlled studies analyzing the benefit of using INTEGRA® on burn wounds is very limited. In the earlier study conducted by Heimbach *et al.*, INTEGRA® was compared to autograft, allograft, xenograft, and synthetic dressing. They found that INTEGRA® was comparable in appearance and function, and formed less hypertrophic scarring. Patients preferred INTEGRA® as it caused less itching than control materials.[97] In a randomized pediatric clinical trial comparing INTEGRA® to allogeneic cadaver graft treatment of full-thickness burns, Branski *et al.* revealed improved scarring in terms of height, thickness, vascularity, and pigmentation of the skin (12 months and 18–24 months) in the INTEGRA® group.[98] Nevertheless, Lagus *et al.* did not find INTEGRA® to have better scar quality than STSGs or cellulose sponge dressings in their randomized controlled trial.[99] Additionally, only a modest inflammatory response was observed during the entire tissue repair process with the three treatments used in this study. Thus, the evidence of INTEGRA® superiority to STSG is mainly based on very heterogenic studies using subjective scar assessments (patient satisfaction/perception) and clinical experiences. An attempt at an objective scar assessment has been made in a few studies by using the Vancouver Scar Scale in conjunction with the range of motion assessment. In most of these studies, INTEGRA® was used for reconstructive surgery and has been found to improve both function and cosmetic outcome.[100–103] Skin regenerated using INTEGRA® is reported to be more pliable,[100,104] more elastic[105] and less contractile than STSG, allowing for a better range of motion.[100]

Despite the limited data analyzing its healing effect in a randomized setting, INTEGRA® is a valuable tool for the treatment of challenging wounds. INTEGRA® appears to be suitable not only in acute phase burn surgery, but also in the coverage of chronic ulcers,[106] of wounds complicated by exposed tendon and bone or joint tissues[107,108] as well as scar reconstructions or tumor extirpation defects.[109,110] However, the high cost and the need for two surgical procedures, which increases hospitalization and immobilization times before the final autograft, illustrate some of the disadvantages of INTEGRA®.

The second most widely used engineered wound dressing is Biobrane®. It is a synthetic dressing composed of an ultrathin, semipermeable silicone film with a nylon fabric partially imbedded into the film. The nylon mesh is made of a complex 3D structure of trifilament thread to which porcine dermal collagen has been chemically bound, providing an adherent surface for wound coverage. Similar to INTEGRA®, the silicone surface controls water vapor loss from the wound, however Biobrane® is intended as a temporary wound dressing and has to be removed either when the wound is healed or when autograft is available. Although Biobrane® has received FDA clearance in 1979, its design limits its use to superficial or partial thickness burn wounds. To date, the results from clinical trials are mixed. Pediatric studies have reported efficacy of Biobrane® for providing immediate wound coverage of superficial or partial thickness burn wounds.[111,112] Biobrane® was associated with improved healing and decreased hospitalization times in pediatric patients with second degree burns.[113] While multiple studies pointed out that Biobrane® might be associated with increased rates of infection in deeper partial thickness burns,[111,114] a recent retrospective case series showed that Biobrane® can serve as a reliable tool in the treatment of partial burns by maintaining a healthy wound bed with a low infection rate of 3.6%.[115] Data from comparative analysis of Biobrane® to other burn dressing options are also unclear. Biobrane® has been found to promote faster healing than silver sulfadiazine topical cream,[116,117] β-glucan collagen matrix[118] and Suprathel® dressing (BioMed Sciences, Allentown, PA).[119] Furthermore, Biobrane® has been shown to be superior to cadaveric allograft as a temporary skin substitute, both in terms of procedure time and associated cost.[120] However, separate studies showed no benefit of Biobrane® when compared to the cheaper Dressilk® (PREVOR, Valmondois, France)[121] or DuoDERM (ConvaTec, Bridgewater, NJ)[122] in partial thickness burn treatment. Giving these heterogeneous data, there is limited clinical evidence to opt for Biobrane® at this point and further studies are needed to validate its use.

This class of acellular synthetic scaffolds provides immediate coverage and a matrix for resident cell and vascular ingrowth. Depending on their design and composition, they act as temporary wound covers or permanent skin replacements. While they have demonstrated positive results for coverage of burn wounds in clinical settings, their efficacy is variable and future controlled studies are required. The comparison of these products in terms of clinical efficacy, convenience of use and cost is needed to develop consensual tissue-engineered materials utilization. Efforts to objectively compare skin substitutes and assess healing and scars are made with the development of new animal models in which skin substitutes can be tested under standardized conditions,[123] and the use of non-invasive physical techniques (ultrasonography, elastometry, oxygen-sensing electrode) to measure elasticity, extensibility and flexibility of the skin[124] or tissue oxygenation.[125]

5.1.2 *Decellularized scaffolds*

AlloDerm® Regenerative Tissue Matrix (LifeCell Corp., Bridgewater, NJ) is similar to INTEGRA® in that it is intended to provide immediate coverage of the wound and offers a matrix for resident cell infiltration and granulation tissue formation, but it is not a synthetic material. AlloDerm® is composed of human allograft skin that has been screened for absence of transmissible pathogens and then processed to remove all epidermal and dermal cells. AlloDerm® is grafted like a dermal autograft and covered with STSG. GraftJacket® Regenerative Tissue Matrix (Wright Medical Technology, Memphis, TN) is also an acellular matrix derived from human dermis with different proprietary processes to remove the cells and to deactivate or destroy pathogens. They are both regulated by the FDA as human tissue for transplantation. Due to this classification, they have not required extensive clinical trials for regulatory approval.

The first clinical studies reporting the use of AlloDerm® relate to the treatment of full-thickness burns.[126,127] In a multicenter clinical trial, Wainwright *et al.* have shown that thin split-thickness autografts combined with AlloDerm® performed similarly to thicker split-thickness autografts. Histology of the dermal matrix showed fibroblast infiltration, neovascularization, and neoepithelialization without evidence of rejection[127] In fact, as opposed to INTEGRA® that induces a foreign body reaction and giant cell infiltration since it is made of chemically cross-linked materials, AlloDerm® does not induce a nonspecific inflammatory response as all allogeneic cells have been cleared off the matrix.[128] This result was further confirmed by a pilot trial conducted by Sheridan *et al.* in 1998.[126] Despite these encouraging data in burn wound management, the use of AlloDerm® rapidly shifted to soft tissue injuries such as large traumatic open abdomen[129] or hernia surgery,[130] periodontal treatments including gingival recessions[131–133] and recently breast reconstruction.[134–136] Since reported in 2005,[137] the use of AlloDerm® in tissue expander breast reconstruction has rapidly become an accepted alternative to the total submuscular technique. However, recent retrospective studies have demonstrated significant limitations to the use of dermal substitutes in breast reconstruction, with a higher risk of seroma,[138–140] infection[138,141] and tissue expander/implant loss.[142,143] Prospective, randomized trials, such as the ongoing BREASTrial,[144] are needed to identify the risk of complications and compare outcomes after using AlloDerm® or other dermal substitutes.

GraftJacket® was commercialized more recently and is specifically designed for chronic wounds. In a prospective, randomized, multicenter study, GraftJacket® has been shown to double the probability of healing diabetic lower extremity wounds compared to standard of care therapies.[145] Other retrospective studies confirm GraftJacket® efficiency with a high healing rate of complex lower extremity wounds.[146,147]

A multitude of non-human biological scaffolds have emerged for treatment of chronic wounds and are currently cleared for clinical use. OASIS® Wound Matrix (Smith & Nephew global product) is derived from the submucosal layers of porcine jejunum and was shown to be more efficient than compression or wet-to-dry dressings for healing chronic lower extremity wounds.[148] Other such matrices include PriMatrix® (TEI Biosciences, Waltham, MA), Matristem® (ACell, Columbia, MD), and MatriDerm® (Medskin, Billerbeck, Germany) which are derived from fetal bovine dermis, porcine urinary bladder, and bovine collagen fibrils, respectively. Although all of these scaffolds have demonstrated benefits for patients, future randomized trials are needed to compare one product to another in a controlled clinical setting. A clear assessment of all dermal substitutes based on efficacy and cost will clarify the decision-making process for treating clinicians.

5.2 *Cellular scaffolds*

This class of skin substitutes is generally composed of composite dressings seeded with allogeneic or autogenic skin cells. The addition of cells attempts to provide an organized cell mass for immediate coverage, while providing new collagen, growth factors and cytokines to the wound bed, helping to reconstitute epidermal and/or dermal layers. Many cellular scaffolds are still under investigation and are not discussed in this section due to the lack of clinical data.

5.2.1 *Allogeneic cells*

TransCyte[®] (Advanced BioHealing, La Jolla, USA) is comprised of a scaffold similar to Biobrane[®] which is seeded with allogeneic fibroblasts cultured from newborn human foreskin. The fibroblasts secrete ECM components and growth factors over a period of 17 days. Before grafting, the cells in TransCyte[®] are destroyed to reduce the risk of immune response by using a freezing process intended to preserve the tissue matrix and growth factors. This processing has shown clinical benefits for local wound bed preparation since partial-thickness burn wounds re-epithelialize faster when compared to the Biobrane[®] scaffold alone.[116] Additionally, a multicenter randomized clinical study showed that TransCyte[®] is equivalent or superior to frozen human cadaver allograft for the temporary closure of excised burn wounds by improving STSG take.[149]

Dermagraft[®] (Organogenesis Inc., La Jolla, USA) is constructed from human neonatal foreskin-derived fibroblasts that are seeded onto a bioabsorbable polyglactin mesh that degrades by hydrolysis in 20–30 days. Unlike TransCyte[®], the fibroblasts from Dermagraft[®] are cryopreserved to maintain cell viability at the time of implantation. This material is licensed and mainly used for chronic diabetic foot ulcers[72,150,151] and venous ulcers[152] with superior effect than standard wound therapies (wet-to-dry dressings, compression therapy). Dermagraft[®] can also be used for burn treatment when combined with skin grafts with no evidence of rejection of the cultured allogeneic fibroblasts.[153]

An additional layer of complexity is found in Apligraf[®] (Organogenesis Inc.), which contains both viable allogeneic fibroblasts and keratinocytes. Neonatal fibroblasts are grown in a bovine Type I collagen matrix and combined with neonatal keratinocytes forming a confluent superficial layer, recapitulating the epidermis. To date, Apligraf[®] is one of the most advanced and sophisticated skin substitutes but also the most expensive FDA-approved product for tissue repair with a shelf-life of only 10 days. Apligraf[®] is licensed for the treatment of venous leg and diabetic foot ulcers, with three large, multicenter randomized studies supporting its clinical efficacy.[86,154,155] In these studies, application of Apligraf[®] as a bioactive wound dressing resulted in a higher healing rate of diabetic foot ulcers and venous ulcers when compared with currently available treatment (wet-to-dry dressing or compression therapy) and is not associated with any significant side effects. Furthermore, a retrospective cohort study found Apligraf[®] to be superior to topical recombinant PDGF treatments (Regranex[®] and Procuren[®]) for accelerating chronic wound healing.[156] A clinical trial also mentioned the use of Apligraf[®] for burn wounds treatment.[85] In this study, Waymack *et al.*

found Apligraft® to be a suitable and clinically effective treatment for burn wounds when applied over meshed autografts. In fact, addition of Apligraft® over meshed autograft showed better cosmetic and functional outcomes compared to meshed autografts alone or combined with an overlying allograft. Based on these strong clinical data, Apligraft® appears to be a powerful tool for wound healing management. It was shown to deliver ECM components to the wound, as well as interferons α and β, PDGF, interleukins 1, 6, and 8, thereby stimulating the wound healing process.[157] However, this product has no engraftment features and is used as a temporary biological dressing rather than a skin substitute. The survival of the Apligraft® allogeneic cells does not exceed 6 weeks *in vivo* limiting its action and its ability to deliver a definitive wound closure in full-thickness injuries.[158]

Other cellular scaffolds such as StrataGraft® (Stratatech Corp., Madison, WI), containing human dermal fibroblast and keratinocyte cell lines, or OrCel® (Ortec International Inc., New York, NY), made of fibroblasts and keratinocytes from the same neonatal foreskin, are other allogeneic products that have demonstrated significant results in a clinical setting. However, comparative analyses are still missing and future studies are needed to establish clear wound care guidelines based on efficacy of the product, wound type, and cost.

5.2.2 *Autologous cells*

The major drawback of the allogeneic cellular biomaterials is the risk of disease transfer and immunogenicity induced by fibroblasts and keratinocytes. To address this issue, strategies using autologous cells that are not rejected by the host and can potentially engraft permanently, were developed. The possibility to expand keratinocytes *ex vivo* has led to the emergence of several epidermal substitutes such as Epicel® (Vericel, Cambridge, MA) or Bioseed-S® (BioTissue TechnologiesGmbH, Feiburg, Germany) which are stratified cultured epithelial autografts or ReCell® (Avita medical, Perth, Australia), a device that allows the application of autologous subconfluent keratinocytes to the wound bed via an aerosol cell suspension. Despite a theoretical attractive technology, the value of cultured keratinocytes remains controversial. These products suffer from major disadvantages of long culture time, difficulties of handling and application, poor take rate, few long-term results, as well as high cost.[159]

6. Significant Limitations of Current Skin Substitutes

Despite some clinical success, none of the existing engineered skin substitutes provide an outcome consistently comparable to a full-thickness autograft.

As discussed in this section, there are several limitations of the current engineered skin substitutes and further improvement is necessary to safely replace the functional and anatomical properties of native skin.

6.1. *Safety and biocompatibility*

One of the major requirements for skin substitutes is that they must be safe for the patient, meaning they must not be toxic, immunogenic, or cause excessive inflammation. Materials for skin substitutes must induce the appropriate response within the host organism and influence the local biological environment. Any cultured cell material carries the risk of transmitting viral or bacterial infection. The use of autologous or allogeneic keratinocytes requires *in vitro* culture. The predominant keratinocyte culture technique uses murine feeder cells and bovine serum that have a disease risk, notably prion diseases and murine viruses that could infect human cells. To limit risk, murine fibroblasts are subjected to lethal gamma radiation and serum is collected from countries free from bovine spongiform encephalitis.[160] It is possible to expand keratinocytes in xenogeneic-free conditions but the proliferative lifespan of cells cultured without murine fibroblasts and bovine serum was shown to be significantly reduced.[161] In addition, allogeneic cells have reduced survival time (only a few weeks) within the skin substitute and will ultimately be rejected by the host, thereby limiting their beneficial effect. For future applications, there is a need to reduce the risks for patients receiving cultured cells as well as develop new materials for dermal substitutes that present fewer risks than bovine collagen or human dermis.

6.2. *Vascularization*

When a skin graft is applied to a wound, graft capillaries anastomose to the recipient's existing capillary network to provide nutrients and oxygen for graft survival. This process is referred as inosculation or "take". Failure of engineered skin substitutes to adequately vascularize leads to cell death and subsequent graft loss. In 1996, Young *et al.* have described the vascularization process of human STSG grafted on athymic mice. They showed that immediately after transplantation, the grafts survive by diffusion of nutrients and oxygen through the graft. Inosculation takes place between 2 and 4 days after grafting, whereas neovascularization occurs after 2 weeks.[162,163] Studies have shown that skin equivalents such as epidermal sheets have a low survival rate when transplanted onto a non-vascularized wound bed confirming that dermal vascularization before grafting is essential.[160,164] These observations have prompted bioengineers to develop new strategies to improve graft vascularization by promoting angiogenesis and the subsequent ingrowth of bed vessels into the graft, or/and by constructing microvascularized tissue-engineered skin substitutes that rapidly develop functional anastomoses with the host's blood vessels.

6.3. *Scarring*

The main goal of current skin substitutes is to restore barrier function and retain full mobility at the defect site. However, most available technologies neglect to address scarring problems at the graft margins, wound contracture, and abnormal pigmentation.[165] Scar tissue is of inferior functional and aesthetic quality, with a decreased cellularity and vascularity and lack of epithelial appendages such as hair follicles, sweat glands, and sebaceous glands. The absence of differentiated structures impairs skin functionality with a lack of thermoregulation, and temperature and pressure sensation. One essential development to improve skin constructs in the future will be to elaborate strategies to incorporate or induce differentiated structures into skin substitutes.

7. New Approaches for Next Generation Skin Substitutes

7.1. *Growth factor-based therapies*

Over the last several decades, the biology of wound healing has been extensively studied and growth factors have been identified as critical signaling molecules that control migration, proliferation, and cellular differentiation over the tissue regeneration process. Thus, growth factor-based therapies appeared as an attractive strategy to stimulate wound angiogenesis and re-epithelialization. Despite much enthusiasm and numerous clinical trials using growth factors to treat deep and severe wounds, the results have been largely disappointing. For example, it has been known for 20 years that topically applied EGF accelerates wound healing, but when used in a clinical setting, the impact on tissue repair was not sufficiently dramatic to justify the additional expense. At this point, the only topical growth factor that has received regulatory approval for wound healing is recombinant human PDGF (under trade name Regranex®).[166] In the past, most growth factors have been delivered to the wounds in saline solutions. Such administration requires high and repeated doses of factors and may have limited efficiency because of their rapid clearance and short half lives in the harsh wound environment. Moreover, supraphysiological and repeated doses of these growth factors might trigger life-threatening side effects including an increased cancer risk from systemic exposure of the administered growth factor. The FDA is currently conducting a safety review based on study data suggesting there may be an increased risk of death from cancer for diabetic patients who were given three or more prescriptions for treatment with Regranex®. Interestingly, the most clinically successful studies and trials all employ some sort of delivery vehicle, strongly suggesting that spatio-temporal control over the location and bioactivity of factors is crucial to obtain concrete therapeutic effects. To improve unsatisfactory outcomes in classical delivery of growth factors, matrices able to present and proteo-lytically shield growth factors could be efficient platforms as delivery substrates. These new designed skin substitutes will achieve local and sustainable delivery to the wound side, thus allowing the reduction of therapeutic doses and associated risks.

7.2 *Cell-based skin therapies*

Many approaches have been developed to incorporate cell-based therapies including stem cells in cutaneous wound healing.[13] Cells can be delivered within a skin substitute, a gel, as a sheet, by spray, or by direct application. Cells can be sourced locally for fibroblasts, keratinocytes, melanocytes or adipocytes, or derived from blood, bone marrow or stem cell niches such as hair follicles. Strategies focused on stem cells have used epidermal progenitor cells, mesenchymal stem cells (MSC), adipose tissue-derived stem cells (ASC), and iPS. Stem cells facilitate wound closure and healing by releasing growth factors that enhance local cell proliferation and migration, increase angiogenesis, and regulate ECM production.[167] Both MSCs and ASCs have shown encouraging results in promoting wound healing when delivered directly into the wound or incorporated into a matrix.[168,169] Both types of cells secrete molecules that decrease inflammation, stimulate angiogenesis, and promote faster wound closure. In addition, MSCs and ASCs could be used as allo- or autografts as they possess immunomodulatory potential that allows immune evasion and can be both harvested directly from the patient.[170] While this approach has not yet been translated to clinical application, the use of stem cells by themselves or in association with skin graft substitutes holds great promise.

7.3. *Reproducing skin complexity*

The ultimate goal of skin engineering would be to restore all the functional components including hair follicles, sweat glands, sebaceous glands, and nerves of the skin as well as skin pigmentation. Skin scaffolds do not contain melanin producing melanocytes, hence resulting in a very pale color. While generating normally pigmented human skin is possible in the laboratory, melanocytes embedded into skin constructs are rapidly lost after grafting in absence of the basement membrane necessary for their attachment and survival.[170] In addition, it was shown that fibroblasts play a critical role in determining pigmentation in reconstructed skin by regulating melanocyte behavior, suggesting that the ratio of melanocytes/fibroblasts would have to be finely controlled. In order to obtain pigmented engineered skin, further studies are required to fully understand the biology of melanocytes in a 3D environment. Generation of skin appendages in engineered skin substitutes remain limited by lack of trichogenic potency in cultured post-natal cells. Human skin appendages form completely during embryogenesis and cannot be restored in a postnatal individual after full-thickness skin loss. So far, success in hair regeneration depends upon the use of fresh fetal or newborn cells. By combining neonatal dermal and epidermal cells, normal hair morphogenesis was obtained.[171,172] Such approaches give new insights in appendages generation and is of great interest to develop the next generation of skin substitutes that will more completely mimic the anatomy and physiology of native skin.

8. Conclusions

Thirty years of tissue engineering and biomaterials research has significantly impacted wound care strategies as evidenced by the development of skin substitute technologies. The large number of products currently available has considerably enhanced the treatment of burns, chronic wounds, and congenital disorders. Continued research will focus on producing a construct that offers complete regeneration of functional skin, including all skin appendages and layers, and establishes a functional vascular and nerve network with scarless integration of the substitute. It has to be noted that costs of the future tissue engineered materials might be an issue to address in the current cost-conscious medical environment. The progress of our understanding of embryonic development, stem cell biology, and biomaterials engineering will be the key to achieve significant development in the field in the coming years.

References

1. Biedermann, T. Boettcher-Haberzeth, S. and Reichmann, E. Tissue engineering of skin for wound coverage. *Eur J Pediatr Surg* 23(5), 375–382 (2013).
2. Pereira, R. F. *et al.* Advanced biofabrication strategies for skin regeneration and repair. *Nanomedicine* 8(4), 603–621 (2013).
3. Fife, C. E. and Carter, M. J. Wound care outcomes and associated cost smong patients treated in US outpatient wound centers: Data from the us wound registry. *Wounds* 24(1), 10–17 (2012).
4. Greaves, N. S. *et al.* Current understanding of molecular and cellular mechanisms in fibroplasia and angiogenesis during acute wound healing. *J Dermatol Sci* 72(3), 206–217 (2013).
5. Kanitakis, J. Anatomy, histology and immunohistochemistry of normal human skin. *Eur J Dermatol* 12(4), 390–399; quiz 400–401 (2001).
6. Lechler, T. and Fuchs, E. Asymmetric cell divisions promote stratification and differentiation of mammalian skin. *Nature* 437(7056), 275–280 (2005).
7. Coward, K. and Wells, D. *Textbook of Clinical Embryology.* Cambridge University Press (2013).
8. Okano, J. *et al.* Cutaneous retinoic acid levels determine hair follicle development and downgrowth. *J Biol Chem* 287(47), 39304–39315 (2012).
9. Breitkreutz, D. *et al.* Skin basement membrane: the foundation of epidermal integrity — BM functions and diverse roles of bridging molecules nidogen and perlecan. *Biomed Res Int* 2013 (2013).
10. Paulsson, M. *Basement membrane proteins: Structure, assembly, and cellular interactions.* (2008).
11. Freinkel, R. K. and Woodley, D. T. *The Biology of the Skin.* CRC Press (2001).
12. Teller, P. and White, T. K. The physiology of wound healing: Injury through maturation. *Perioperative Nursing Clinics* 6(2), 159–170 (2011).
13. Sun, B. K. Siprashvili, Z. and Khavari, P. A. Advances in skin grafting and treatment of cutaneous wounds. *Science* 346(6212), 941–945 (2014).
14. Stadelmann, W. K. Digenis, A. G. and Tobin, G. R. Physiology and healing dynamics of chronic cutaneous wounds. *Am J Surg* 176(2), 26S–38S (1998).
15. Adams, R. L. and Bird, R. J. Review article: Coagulation cascade and therapeutics update: relevance to nephrology. Part 1: Overview of coagulation, thrombophilias and history of anticoagulants. *Nephrology* 14(5), 462–470 (2009).
16. Steffel, J. Lüscher, T. F. and Tanner, F. C. Tissue factor in cardiovascular diseases molecular mechanisms and clinical implications. *Circulation* 113(5), 722–731 (2006).
17. Enoch, S. and Leaper, D. J. Basic science of wound healing. *Surgery* 26(2), 31–37 (2008).
18. Eming, S. A. Krieg, T. and Davidson, J. M. Inflammation in wound repair: Molecular and cellular mechanisms. *J Invest Dermatol* 127(3), 514–525 (2007).

19. DiPietro, L. A. Angiogenesis and scar formation in healing wounds. *Curr Opin Rheumatol* **25**(1), 87–91 (2013).

20. Greenhalgh, D. G. The role of apoptosis in wound healing. *Int J Biochem Cell Biol* **30**(9), 1019–1030 (1998).

21. Pickwell, K. M. *et al.* Diabetic foot disease: Impact of ulcer location on ulcer healing. *Diabetes Metab Res Rev* **29**(5), 377–383 (2013).

22. Guo, S. and DiPietro, L. A. Factors affecting wound healing. *J Den Res* **89**(3), 219–229 (2010).

23. Mayet, N. *et al.* A comprehensive review of advanced biopolymeric wound healing systems. *J Pha Sci* **103**(8), 2211–2230 (2014).

24. Kearns, R. D. Holmes, J. H. T. and Cairns, B. A. Burn injury: What's in a name? Labels used for burn injury classification: A review of the data from 2000–2012. *Ann Burns Fire Disasters* **26**(3), 115–120 (2013).

25. Herndon, D. N. *et al.* A comparison of conservative versus early excision. Therapies in severely burned patients. *Ann Surg* **209**(5), 547–552; discussion 552–553 (1989).

26. Lazarus, G. S. *et al.* Definitions and guidelines for assessment of wounds and evaluation of healing. *Wound Repair Regen* **2**(3), 165–170 (1994).

27. Diegelmann, R. F. and Evans, M. C. Wound healing: An overview of acute, fibrotic and delayed healing. *Front Biosci* **9**(1), 283–289 (2004).

28. Menke, N. B. *et al.* Impaired wound healing. *Clin Dermatol* **25**(1), 19–25 (2007).

29. Nunan, R. Harding, K. G. and Martin, P. Clinical challenges of chronic wounds: Searching for an optimal animal model to recapitulate their complexity. *Dis Model Mech* **7**(11), 1205–1213 (2014).

30. Phillips, T. *et al.* A study of the impact of leg ulcers on quality of life: Financial, social, and psychologic implications. *J Am Acad Dermatol* **31**(1), 49–53 (1994).

31. Falanga, V. Chronic wounds: Pathophysiologic and experimental considerations. *J Invest Dermatol* **100**(5), 721–725 (1993).

32. Phillips, T .J. Chronic cutaneous ulcers: Etiology and epidemiology. *J Invest Dermatol* **102**(6), 38S–41S (1994).

33. Gauglitz, G. G. *et al.* Hypertrophic scarring and keloids: Pathomechanisms and current and emerging treatment strategies. *Mol Med* **17**(1–2), 113 (2011).

34. Viera, M. H. Vivas, A. C. and Berman, B. Treatment of Keloids and Scars. Ethnic Dermatology 159–172 (2013).

35. Huang, C. *et al.* Keloids and hypertrophic scars: Update and future directions. *Plast Reconstr Surg Glob Open* **1**(4) (2013).

36. Sen, C. K. *et al.* Human skin wounds: A major and snowballing threat to public health and the economy. *Wound Repair Regen* **17**(6), 763–771 (2009).

37. Nicolaides, A. N. *et al.* Management of chronic venous disorders of the lower limbs: Guidelines according to scientific evidence. *Int Angiol* **27**(1), 1–59 (2008).

38. Schreml, S. *et al.* Wound dressings in chronic wound therapy. *Phlebgie* **42**(4), 189–196 (2013).

39. Fonder, M. A. *et al.* Treating the chronic wound: A practical approach to the care of nonhealing wounds and wound care dressings. *J Am Acad Dermatol* 58(2), 185–206 (2008).

40. Park, K. H. A retrospective study using the pressure ulcer scale for healing (PUSH) tool to examine factors affecting stage II pressure ulcer healing in a Korean acute care hospital. *Ostomy Wound Manag* 60(9), 40–51 (2014).

41. Sørensen, L. T. Wound healing and infection in surgery: The pathophysiological impact of smoking, smoking cessation, and nicotine replacement therapy: A systematic review. *Ann Surg* 255(6), 1069–1079 (2012).

42. Khalil, H. *et al.* Elements affecting wound healing time: An evidence based analysis. *Wound Repair Regen* (2015).

43. Söderström, M. *et al.* Angiosome-targeted infrapopliteal endovascular revascularization for treatment of diabetic foot ulcers. *J Vasc Surg* 57(2), 427–435 (2013).

44. Varela, C. *et al.* The role of foot collateral vessels on ulcer healing and limb salvage after successful endovascular and surgical distal procedures, according to an angiosome model. *Vasc Endovasc Surg* 44(8), 654–660 (2010).

45. Adamskaya, N. *et al.* Light therapy by blue LED improves wound healing in an excision model in rats. *Injury* 42(9), 917–921 (2011).

46. Houreld, N. N. Shedding light on a new treatment for diabetic wound healing: A review on phototherapy. *Scientific World* 2014 (2014).

47. Whelan, H. T. *et al.* Effect of NASA light-emitting diode irradiation on molecular changes for wound healing in diabetic mice. *J Clin Laser Med Surg* 21(2), 67–74 (2003).

48. Berman, B. *et al.* Prevention and management of hypertrophic scars and keloids after burns in children. *J Craniofac Surg* 19(4), 989–1006 (2008).

49. Aarabi, S. Longaker, M. T. and Gurtner, G. C. Hypertrophic scar formation following burns and trauma: New approaches to treatment. *PLoS Med* 4(9), e234 (2007).

50. Andreassi, A. *et al.* Classification and pathophysiology of skin grafts. *Clin Dermatol* 23(4), 332–337 (2005).

51. Scobie, L. *et al.* Long-term IgG response to porcine Neu5Gc antigens without transmission of PERV in burn patients treated with porcine skin xenografts. *J Immunol* 191(6), 2907–2915 (2013).

52. Peck, M. D. Epidemiology of burns throughout the world. Part I: Distribution and risk factors. *Burns* 37(7), 1087–1100 (2011).

53. Sheridan, R. L. and Greenhalgh, D. Special problems in burns. *Surg Clin North Am* 94(4), 781–791 (2014).

54. Pashuck, E. T. and Stevens, M. M. Designing regenerative biomaterial therapies for the clinic. *Sci Trans Med* 4(160), 160sr4–160sr4 (2012).

55. Mogoşanu, G. D. and A. M. Grumezescu, Natural and synthetic polymers for wounds and burns dressing. *Int J Pharm* 463(2), 127–136 (2014).

56. Jayakumar, R. *et al.* Biomaterials based on chitin and chitosan in wound dressing applications. *Biotech Adv* **29**(3), 322–337 (2011).

57. Gil, E. S. *et al.* Functionalized silk biomaterials for wound healing. *Adv Healthc Mater* **2**(1), 206–217 (2013).

58. Gomes, S. *et al.* Natural and genetically engineered proteins for tissue engineering. *Prog Polym Sci* **37**(1), 1–17 (2012).

59. Zahedi, P. *et al.* A review on wound dressings with an emphasis on electrospun nanofibrous polymeric bandages. *Poly Adv Tech* **21**(2), 77–95 (2010).

60. Chattopadhyay, S. and Raines, R. T. Review collagen-based biomaterials for wound healing. *Biopolymers* **101**(8), 821–833 (2014).

61. Holmes, C. *et al.* Collagen-based wound dressings for the treatment of diabetes-related foot ulcers: A systematic review. *Diabetes, Metab Syndr Obes* **6**, 17 (2013).

62. Moura, L. I. *et al.* Recent advances on the development of wound dressings for diabetic foot ulcer treatment — A review. *Acta Biomate* **9**(7), 7093–7114 (2013).

63. Nyame, T. T. Chiang, H. A. and Orgill, D. P. Clinical Applications of Skin Substitutes. Surg *Clin North Am* **94**(4), 839–850 (2014).

64. Ortega-Zilic, N. *et al.* EpiDex® Swiss field trial 2004–2008. *Dermatology* **221**(4), 365–372 (2010).

65. Renner, R. Harth, W. and Simon, J. C. Transplantation of chronic wounds with epidermal sheets derived from autologous hair follicles — the Leipzig experience. *Int Wound J* **6**(3), 226–232 (2009).

66. Tausche, A. K. *et al.* An autologous epidermal equivalent tissue–engineered from follicular outer root sheath keratinocytes is as effective as split-thickness skin autograft in recalcitrant vascular leg ulcers. *Wound Repair Regene* **11**(4), 248–252 (2003).

67. Zweifel, C. *et al.* Initial experiences using non-cultured autologous keratinocyte suspension for burn wound closure. *J Plast Reconstr Aes* **61**(11), e1–e4 (2008).

68. Gravante, G. *et al.* A randomized trial comparing ReCell® system of epidermal cells delivery versus classic skin grafts for the treatment of deep partial thickness burns. *Burns* **33**(8), 966–972 (2007).

69. Jean, J. Garcia-Pérez, M. and Pouliot, R. Bioengineered skin: The self-assembly approach. *J Tissue Sci Eng S* **5**, 2 (2011).

70. Hu, D.-h. *et al.* A potential skin substitute constructed with hEGF gene modified HaCaT cells for treatment of burn wounds in a rat model. *Burns* **38**(5), 702–712 (2012).

71. Flasza, M. *et al. Development and manufacture of an investigational human living dermal equivalent (ICX-SKN)* (2007).

72. Marston, W. A. Dermagraft®, a bioengineered human dermal equivalent for the treatment of chronic nonhealing diabetic foot ulcer. *Expert Rev Med Devices* **1**(1), 21–31 (2004).

73. Marston, W. A. *et al.* The efficacy and safety of dermagraft in improving the healing of chronic diabetic foot ulcers results of a prospective randomized trial. *Diabetes Care* **26**(6), 1701–1705 (2003).

74. Noordenbos, J. Doré, C. and Hansbrough, J. F. Safety and efficacy of trans-cyte* for the treatment of partial-thickness burns. *J Burn Care Res* **20**(4), 275–281 (1999).

75. Vacher, D. *Cell therapy product (Dermagen) for treatment of hard-to-heal skin wounds.* in Annales Pharmaceutiques Francaises. (2005).

76. Hartwell, R. *et al.* An in-situ forming skin substitute improves healing outcome in a hypertrophic scar model. *Tissue Eng Part A* **21**(5–6), 1085–1094 (2015).

77. Lee, Y. *et al.* Enzyme-catalyzed in situ forming gelatin hydrogels as bioactive wound dressings: Effects of fibroblast delivery on wound healing efficacy. *J Mater Chem B* **2**(44), 7712–7718 (2014).

78. Zhou, B. *et al.* Rapidly in situ forming platelet-rich plasma gel enhances angiogenic responses and augments early wound healing after open abdomen. *Gastroenterol Res Prac* **2013** (2013).

79. Bello, Y. M. Falabella, A. F. and Eaglstein, W. H. Tissue-engineered skin. *Am J Clin Dermatol* **2**(5), 305–313 (2001).

80. Boyce, S. T. and Warden, G. D. Principles and practices for treatment of cutaneous wounds with cultured skin substitutes. *Am J Surg* **183**(4), 445–456 (2002).

81. Eaglstein, W. H. and V. Falanga, Tissue engineering and the development of Apligraf®, a human skin equivalent. *Clin Ther* **19**(5), 894–905 (1997).

82. MacNeil, S. Biomaterials for tissue engineering of skin. *Mater Today* **11**(5), 26–35 (2008).

83. Schurr, M. J. *et al.* Phase I/II clinical evaluation of StrataGraft: A consistent, pathogen-free human skin substitute. *J Trauma* **66**(3), 866 (2009).

84. Supp, D. M. *Skin substitutes for burn wound healing: current and future approaches.* (2011).

85. Waymack, P. *et al.* The effect of a tissue engineered bilayered living skin analog, over meshed split-thickness autografts on the healing of excised burn wounds. *Burns* **26**(7), 609–619 (2000).

86. Falanga, V. and Sabolinski, M. A bilayered living skin construct (APLIGRAF) accelerates complete closure of hard-to-heal venous ulcers. *Wound Repair Regen* **7**(4), 201–207 (1999).

87. Koch, L. *et al.* Skin tissue generation by laser cell printing. *Biotech Bioeng* **109**(7), 1855–1863 (2012).

88. Lee, W. *et al.* Multi-layered culture of human skin fibroblasts and keratinocytes through three-dimensional freeform fabrication. *Biomaterials* **30**(8), 1587–1595 (2009).

89. Michael, S. *et al.* Tissue engineered skin substitutes created by laser-assisted bioprinting form skin-like structures in the dorsal skin fold chamber in mice. *PloS One* **8**(3), e57741 (2013).

90. Sriwiriyanont, P. *et al.* Morphogenesis of chimeric hair follicles in engineered skin substitutes with human keratinocytes and murine dermal papilla cells. *Exp Dermatol* **21**(10), 783–785 (2012).

91. Burke, J. F. *et al.* Successful use of a physiologically acceptable artificial skin in the treatment of extensive burn injury. *Ann Surg* **194**(4), 413 (1981).

92. Dagalakis, N. *et al.* Design of an artificial skin. Part III. Control of pore structure. *J Biomed Mater Res* **14**(4), 511–528 (1980).

93. Yannas, I. *et al.* Design of an artificial skin. II. Control of chemical composition. *J Biomed Mater Res* **14**(2), 107–132 (1980).

94. Yannas, I. and Burke, J. F. Design of an artificial skin. I. Basic design principles. *J Biomed Mater Res* **14**(1), 65–81 (1980).

95. Yannas, I. *et al.* Synthesis and characterization of a model extracellular matrix that induces partial regeneration of adult mammalian skin. *Proc Nat Acad Sci USA* **86**(3), 933–937 (1989).

96. Heimbach, D. M. *et al.* Multicenter postapproval clinical trial of Integra dermal regeneration template for burn treatment. *J Burn Care Rehabil* **24**(1), 42–48 (2003).

97. Heimbach, D. *et al.* Artificial dermis for major burns. A multi-center randomized clinical trial. *Ann Surg* **208**(3), 313–320 (1988).

98. Branski, L. K. *et al.* Longitudinal assessment of Integra in primary burn management: A randomized pediatric clinical trial. *Crit Care Med* **35**(11), 2615–2623 (2007).

99. Lagus, H. *et al.* Prospective study on burns treated with Integra(R), a cellulose sponge and split thickness skin graft: Comparative clinical and histological study — randomized controlled trial. *Burns* **39**(8), 1577–1587 (2013).

100. Chou, T. D. *et al.* Reconstruction of burn scar of the upper extremities with artificial skin. *Plast Reconstr Surg* **108**(2), 378–384: discussion 385 (2001).

101. Moiemen, N. S. *et al.* Reconstructive surgery with a dermal regeneration template: Clinical and histologic study. *Plast Reconstr Surg* **108**(1), 93–103 (2001).

102. Stiefel, D. Schiestl, C. and Meuli, M. Integra Artificial Skin for burn scar revision in adolescents and children. *Burns* **36**(1), 114–120 (2010).

103. Moiemen, N. *et al.* Long-term clinical and histological analysis of Integra dermal regeneration template. *Plast Reconstr Surg* **127**(3), 1149–1154 (2011).

104. Nguyen, D. Q. Potokar, T. S. and Price, P. An objective long-term evaluation of Integra (a dermal skin substitute) and split thickness skin grafts, in acute burns and reconstructive surgery. *Burns* **36**(1), 23–28 (2010).

105. Loss, M. *et al.* Artificial skin, split-thickness autograft and cultured autologous keratinocytes combined to treat a severe burn injury of 93% of TBSA. *Burns* **26**(7), 644–652 (2000).

106. Kahn, S. A. Beers, R. J. and Lentz, C. W. Use of acellular dermal replacement in reconstruction of nonhealing lower extremity wounds. *J Burn Care Res* **32**(1), 124-128 (2011).

107. Shores, J. T. *et al.* Tendon coverage using an artificial skin substitute. *J Plast Reconstr Aesthet Surg* **65**(11), 1544–1550 (2012).

108. Weigert, R. H. Choughri, and Casoli, V. Management of severe hand wounds with Integra(R) dermal regeneration template. *J Hand Surg Eur Vol* **36**(3), 185–193 (2011).

109. Chalmers, R. L. Smock, E. and Geh, J. L. Experience of Integra((R)) in cancer reconstructive surgery. *J Plast Reconstr Aesthet Surg* **63**(12), 2081–2090 (2010).

110. Burd, A. and Wong, P. S. One-stage Integra reconstruction in head and neck defects. *J Plast Reconstr Aesthet Surg* **63**(3), 404–409 (2010).

111. Ou, L. F. *et al.* Use of Biobrane in pediatric scald burns — experience in 106 children. *Burns* **24**(1), 49–53 (1998).

112. Lang, E. M. *et al.* Biobrane in the treatment of burn and scald injuries in children. *Ann Plast Surg* **55**(5), 485–489 (2005).

113. Lal, S. *et al.* Biobrane improves wound healing in burned children without increased risk of infection. *Shock* **14**(3), 314–318; discussion 318–319 (2000).

114. Hubik, D. J. *et al.* Biobrane: A retrospective analysis of outcomes at a specialist adult burns centre. *Burns* **37**(4), 594–600 (2011).

115. Tan, H. *et al.* Effective use of Biobrane as a temporary wound dressing prior to definitive split-skin graft in the treatment of severe burn: A retrospective analysis. *Burns* **41**(5), 969–976 (2015).

116. Kumar, R. J. *et al.* Treatment of partial-thickness burns: A prospective, randomized trial using Transcyte. *ANZ J Surg* **74**(8), 622–626 (2004).

117. Barret, J. P. *et al.* Biobrane versus 1% silver sulfadiazine in second-degree pediatric burns. *Plast Reconstr Surg* **105**(1), 62–65 (2000).

118. Lesher, A. P. *et al.* Effectiveness of Biobrane for treatment of partial-thickness burns in children. *J Pediatr Surg* **46**(9), 1759–1763 (2011).

119. Rahmanian-Schwarz, A. *et al.* A clinical evaluation of Biobrane((R)) and Suprathel((R)) in acute burns and reconstructive surgery. *Burns* **37**(8), 1343–1348 (2011).

120. Austin, R. E. *et al.* A comparison of Biobrane and cadaveric allograft for temporizing the acute burn wound: Cost and procedural time. *Burns* **41**(4), 749–753 (2015).

121. Schulz, A. *et al.* A prospective clinical trial comparing Biobrane Dressilk and PolyMem dressings on partial-thickness skin graft donor sites. *Burns* (2015).

122. Cassidy, C. *et al.* Biobrane versus duoderm for the treatment of intermediate thickness burns in children: A prospective, randomized trial. *Burns* **31**(7), 890–893 (2005).

123. Hartmann-Fritsch, F. *et al.* A new model for preclinical testing of dermal substitutes for human skin reconstruction. *Pediatr Surg Int* **29**(5), 479–488 (2013).

124. Danin, A. *et al.* Assessment of burned hands reconstructed with Integra((R)) by ultrasonography and elastometry. *Burns* **38**(7), 998–1004 (2012).

125. Papa, G. *et al.* Compared to coverage by STSG grafts only reconstruction by the dermal substitute Integra(R) plus STSG increases TcPO2 values in diabetic feet at 3 and 6 months after reconstruction*. *G Chir* **35**(5–6), 141–145 (2014).

126. Sheridan, R. *et al.* Acellular allodermis in burns surgery: 1-year results of a pilot trial. *J Burn Care Rehabil* **19**(6), 528–530 (1998).

127. Wainwright, D. *et al.* Clinical evaluation of an acellular allograft dermal matrix in full-thickness burns. *J Burn Care Rehabil* **17**(2), 124–136 (1996).

128. Truong, A. T. *et al.* Comparison of dermal substitutes in wound healing utilizing a nude mouse model. *J Burns Wounds* **4**, e4 (2005).

129. de Moya, M. A. *et al.* Long-term outcome of acellular dermal matrix when used for large traumatic open abdomen. *J Trauma* **65**(2), 349–353 (2008).

130. Bochicchio, G. V. *et al.* Comparison study of acellular dermal matrices in complicated hernia surgery. *J Am Coll Surg* **217**(4), 606–613 (2013).

131. Alves, L. B. *et al.* Acellular dermal matrix graft with or without enamel matrix derivative for root coverage in smokers: A randomized clinical study. *J Clin Periodontol* **39**(4), 393–399 (2012).

132. Moslemi, N. *et al.* Acellular dermal matrix allograft versus subepithelial connective tissue graft in treatment of gingival recessions: A 5-year randomized clinical study. *J Clin Periodontol* **38**(12), 1122–1129 (2011).

133. Barker, T. S. *et al.* A comparative study of root coverage using two different acellular dermal matrix products. *J Periodontol* **81**(11), 1596–1603 (2010).

134. Butterfield, J. L. 440 Consecutive immediate, implant-based, single-surgeon breast reconstructions in 281 patients: A comparison of early outcomes and costs between SurgiMend fetal bovine and AlloDerm human cadaveric acellular dermal matrices. *Plast Reconstr Surg* **131**(5), 940–951 (2013).

135. Glasberg, S. B. and Light, D. AlloDerm and Strattice in breast reconstruction: A comparison and techniques for optimizing outcomes. *Plast Reconstr Surg* **129**(6), 1223–1233 (2012).

136. Antony, A. K. *et al.* Acellular human dermis implantation in 153 immediate two-stage tissue expander breast reconstructions: Determining the incidence and significant predictors of complications. *Plast Reconstr Surg* **125**(6), 1606–1614 (2010).

137. Breuing, K. H. and S. M. Warren, Immediate bilateral breast reconstruction with implants and inferolateral AlloDerm slings. *Ann Plast Surg* **55**(3), 232–239 (2005).

138. Chun, Y. S. *et al.* Implant-based breast reconstruction using acellular dermal matrix and the risk of postoperative complications. *Plast Reconstr Surg* **125**(2), 429–436 (2010).

139. Kim, J. Y. *et al.* A meta-analysis of human acellular dermis and submuscular tissue expander breast reconstruction. *Plast Reconstr Surg* **129**(1), 28–41 (2012).

140. Parks, J. W. *et al.* Human acellular dermis versus no acellular dermis in tissue expansion breast reconstruction. *Plast Reconstr Surg* **130**(4), 739–746 (2012).

141. Hill, J. L. *et al.* Infectious complications associated with the use of acellular dermal matrix in implant-based bilateral breast reconstruction. *Ann Plast Surg* **68**(5), 432–434 (2012).

142. Hoppe, I. C. *et al.* Complications following expander/implant breast reconstruction utilizing acellular dermal matrix: A systematic review and meta-analysis. *Eplasty* **11**, e40 (2011).

143. Ho, G. *et al.* A systematic review and meta-analysis of complications associated with acellular dermal matrix-assisted breast reconstruction. *Ann Plast Surg* **68**(4), 346–356 (2012).

144. Mendenhall, S. D. *et al.* The BREASTrial: Stage I. Outcomes from the time of tissue expander and acellular dermal matrix placement to definitive reconstruction. *Plast Reconstr Surg* **135**(1), 29e–42e (2015).

145. Reyzelman, A. *et al.* Clinical effectiveness of an acellular dermal regenerative tissue matrix compared to standard wound management in healing diabetic foot ulcers: A prospective, randomised, multicentre study. *Int Wound J* **6**(3), 196–208 (2009).

146. Brigido, S. A. The use of an acellular dermal regenerative tissue matrix in the treatment of lower extremity wounds: A prospective 16-week pilot study. *Int Wound J* **3**(3), 181–187 (2006).

147. Winters, C. L. *et al.* A multicenter study involving the use of a human acellular dermal regenerative tissue matrix for the treatment of diabetic lower extremity wounds. *Adv Skin Wound Care* **21**(8), 375–381 (2008).

148. Romanelli, M. V. Dini, and Bertone, M. S. Randomized comparison of OASIS wound matrix versus moist wound dressing in the treatment of difficult-to-heal wounds of mixed arterial/venous etiology. *Adv Skin Wound Care* **23**(1), 34–38 (2010).

149. Purdue, G. F. *et al.* A multicenter clinical trial of a biosynthetic skin replacement, Dermagraft-TC, compared with cryopreserved human cadaver skin for temporary coverage of excised burn wounds. *J Burn Care Rehabil* **18**(1 Pt 1), 52–57 (1997).

150. Gentzkow, G. D. *et al.* Use of dermagraft, a cultured human dermis, to treat diabetic foot ulcers. *Diabetes Care* **19**(4), 350–354 (1996).

151. Hanft, J. R. and M. S. Surprenant, Healing of chronic foot ulcers in diabetic patients treated with a human fibroblast-derived dermis. *J Foot Ankle Surg* **41**(5), 291–299 (2002).

152. Omar, A. A. *et al.* Treatment of venous leg ulcers with Dermagraft. *Eur J Vasc Endovasc Surg* **27**(6), 666–672 (2004).

153. Hansbrough, J. F. Dore, C. and Hansbrough, W. B. Clinical trials of a living dermal tissue replacement placed beneath meshed, split-thickness skin grafts on excised burn wounds. *J Burn Care Rehabil* **13**(5), 519–529 (1992).

154. Veves, A. *et al.* Graftskin, a human skin equivalent, is effective in the management of noninfected neuropathic diabetic foot ulcers: A prospective randomized multicenter clinical trial. *Diabetes Care* **24**(2), 290–295 (2001).

155. Edmonds, M. Apligraf in the treatment of neuropathic diabetic foot ulcers. *Int J Low Extrem Wounds* **8**(1), 11–18 (2009).
156. Kirsner, R. S. *et al.* Advanced biological therapies for diabetic foot ulcers. *Arch Dermatol* **146**(8), 857–862 (2010).
157. Ehrenreich, M. and Ruszczak, Z. Update on tissue-engineered biological dressings. *Tissue Eng* **12**(9), 2407–2424 (2006).
158. Eaglstein, W. H. *et al.* Acute excisional wounds treated with a tissue-engineered skin (Apligraf). *Dermatol Surg* **25**(3), 195–201 (1999).
159. Shevchenko, R. V. James, S. L. and James, S. E. A review of tissue-engineered skin bioconstructs available for skin reconstruction. *J R Soc Interface* **7**(43), 229–258 (2010).
160. MacNeil, S. Progress and opportunities for tissue-engineered skin. *Nature* **445** (7130), 874–880 (2007).
161. Papini, S. *et al.* Isolation and clonal analysis of human epidermal keratinocyte stem cells in long-term culture. *Stem Cells* **21**(4), 481–494 (2003).
162. Young, D. M. Greulich, K. M. and Weier, H. G. Species-specific in situ hybridization with fluorochrome-labeled DNA probes to study vascularization of human skin grafts on athymic mice. *J Burn Care Rehabil* **17**(4), 305–310 (1996).
163. Tremblay, P. L. *et al.* Inosculation of tissue-engineered capillaries with the host's vasculature in a reconstructed skin transplanted on mice. *Am J Transplant* **5**(5), 1002–1010 (2005).
164. Gibot, L. *et al.* A preexisting microvascular network benefits *in vivo* revascularization of a microvascularized tissue-engineered skin substitute. *Tissue Eng Part A* **16**(10), 3199–3206 (2010).
165. Soller, E. C. *et al.* Common features of optimal collagen scaffolds that disrupt wound contraction and enhance regeneration both in peripheral nerves and in skin. *Biomaterials* **33**(19), 4783–4791 (2012).
166. Murphy, P. S. and Evans, G. R. Advances in wound healing: A review of current wound healing products. *Plast Surg Int* **2012**, 190436 (2012).
167. Caplan, A. I. and Correa, D. The MSC: An injury drugstore. *Cell Stem Cell* **9**(1), 11–15 (2011).
168. Jackson, W. M. Nesti, L. J. and Tuan, R. S. Concise review: Clinical translation of wound healing therapies based on mesenchymal stem cells. *Stem Cells Transl Med* **1**(1), 44–50 (2012).
169. Mizuno, H. M. Tobita, and Uysal, A. C. Concise review: Adipose-derived stem cells as a novel tool for future regenerative medicine. *Stem Cells* **30**(5), 804–810 (2012).
170. Hedley, S. J. *et al.* Fibroblasts play a regulatory role in the control of pigmentation in reconstructed human skin from skin types I and II. *Pigment Cell Res* **15**(1), 49–56 (2002).
171. Zheng, Y. *et al.* Organogenesis from dissociated cells: Generation of mature cycling hair follicles from skin-derived cells. *J Invest Dermatol* **124**(5), 867–876 (2005).

172. Weinberg, W. C. *et al.* Reconstitution of hair follicle development *in vivo*: determination of follicle formation, hair growth, and hair quality by dermal cells. *J Invest Dermatol* **100**(3), 229–236 (1993).
173. Singh, R. and Singh, D. Radiation synthesis of PVP/alginate hydrogel containing nanosilver as wound dressing. *J Mater Sci Materials in Medicine* **23**(11), 2649–2658 (2012).
174. Shamshina, J. *et al.* Chitin–calcium alginate composite fibers for wound care dressings spun from ionic liquid solution. *J Mater Chem B* **2**(25), 3924–3936 (2014).
175. Lin, W.-C. *et al.* Bacterial cellulose and bacterial cellulose–chitosan membranes for wound dressing applications. *Carbohyd Polym* **94**(1), 603–611 (2013).
176. Liu, J. *et al.* Application of crystalline cellulose membrane (Veloderm®) on split-thickness skin graft donor sites in burn or reconstructive plastic surgery patients. *J Burn Care Res* **34**(3), e176–e182 (2013).
177. Fanti, P. A. *et al.* Repair of the donor areas defects after split-thickness skin grafts utilizing an advanced epithelialization dressing. *J Dermatol Treat* **25**(5), 434–437 (2014).
178. Madhumathi, K. *et al.* Development of novel chitin/nanosilver composite scaffolds for wound dressing applications. *J Mater Sci* **21**(2), 807–813 (2010).
179. Huang, Y. *et al.* Novel fibers fabricated directly from chitin solution and their application as wound dressing. *J Mater Chem B* **2**(22), 3427–3432 (2014).
180. Naseri, N. *et al.* Electrospun chitosan-based nanocomposite mats reinforced with chitin nanocrystals for wound dressing. *Carbohydr Polym* **109**, 7–15 (2014).
181. Archana, D. *et al. In vivo* evaluation of chitosan–PVP–titanium dioxide nanocomposite as wound dressing material. *Carbohydr Polym* **95**(1), 530–539 (2013).
182. Helary, C. *et al.* Evaluation of dense collagen matrices as medicated wound dressing for the treatment of cutaneous chronic wounds. *Biomater Sci* **3**(2), 373–382 (2015).
183. de Carvalho, V. F. *et al.* Clinical trial comparing 3 different wound dressings for the management of partial-thickness skin graft donor sites. *J Wound Ostomy Continence Nurs* **38**(6), 643–647 (2011).
184. Siritientong, T. *et al.* Clinical potential of a silk sericin-releasing bioactive wound dressing for the treatment of split-thickness skin graft donor sites. *Pharm Res* **31**(1), 104–116 (2014).
185. Mittermayr, R. *et al.* Fibrin biomatrix-conjugated platelet-derived growth factor AB accelerates wound healing in severe thermal injury. *J Tissue Eng Reg Med* (2013).
186. Xu, R. *et al.* Novel bilayer wound dressing composed of silicone rubber with particular micropores enhanced wound re-epithelialization and contraction. *Biomaterials* **40**, 1–11 (2015).

187. Pyun, D. G. *et al.* Evaluation of AgHAP-containing polyurethane foam dressing for wound healing: Synthesis, characterization, *in vitro* and *In vivo* studies. *J Mater Chem B* **3**(39), 7752–7763 (2015).

188. Nguyen, T. T. T. *et al.* Characteristics of curcumin-loaded poly (lactic acid) nanofibers for wound healing. *J Mater Sci* **48**(20), 7125–7133 (2013).

189. Wieman, T. J. Smiell, J. M. and Su, Y. Efficacy and Safely of a Topical Gel Formulation of Recombinant Human Platelet-Derived Growth Factor-BB (Becaplermin) in Patients With Chronic Neuropathic Diabetic Ulcers: A phase III randomized placebo-controlled double-blind study. *Diabetes Care* **21**(5), 822–827 (1998).

190. Niiyama, H. and Kuroyanagi, Y. Development of novel wound dressing composed of hyaluronic acid and collagen sponge containing epidermal growth factor and vitamin C derivative. *J Artif Organs* **17**(1), 81–87 (2014).

191. Koria, P. *et al.* Self-assembling elastin-like peptides growth factor chimeric nanoparticles for the treatment of chronic wounds. *Proc Nat Acad Sci USA* **108**(3), 1034–1039 (2011).

192. Losi, P. *et al.* Healing effect of a fibrin-based scaffold loaded with platelet lysate in full-thickness skin wounds. *J Bioact Compat Polym* **30**(2), 222–237 (2015).

193. Lan, Y. *et al.* Therapeutic efficacy of antibiotic-loaded gelatin microsphere/silk fibroin scaffolds in infected full-thickness burns. *Acta Biomater* **10**(7), 3167–3176 (2014).

194. Wainwright, D. Use of an acellular allograft dermal matrix (AlloDerm) in the management of full-thickness burns. *Burns* **21**(4), 243–248 (1995).

195. Romanelli, M. *et al.* OASIS wound matrix versus Hyaloskin in the treatment of difficult-to-heal wounds of mixed arterial/venous aetiology. *Int Wound J* **4**(1), 3–7 (2007).

196. Rommer, E. A. Peric, M. and Wong, A. Urinary bladder matrix for the treatment of recalcitrant nonhealing radiation wounds. *Adv Skin Wound Care* **26**(10), 450–455 (2013).

197. Kavros, S. J. *et al.* The use of PriMatrix, a fetal bovine acellular dermal matrix, in healing chronic diabetic foot ulcers: A prospective multicenter study. *Adv Skin Wound Care* **27**(8), 356–362 (2014).

198. Kanokpanont, S. *et al.* An innovative bi-layered wound dressing made of silk and gelatin for accelerated wound healing. *Int J Pharm* **436**(1), 141–153 (2012).

199. Huang, S. *et al.* Functional bilayered skin substitute constructed by tissue-engineered extracellular matrix and microsphere-incorporated gelatin hydrogel for wound repair. *Tissue Eng Part A* **15**(9), 2617–2624 (2009).

10. Biomaterial-Based Systems for Pharmacologic Treatment of Wound Repair

Mara A. Pop, Julia B. Sun and Benjamin D. Almquist

Department of Bioengineering
Imperial College London
South Kensington Campus
London SW7 2AZ, UK

Abstract

Cutaneous wound healing is a highly complex and dynamic process involving the transient expression and function of various cytokines, growth factors, and cell types. Any disruption in this tightly controlled process can in principle cause impaired healing, turning a healing wound into a non-healing one that does not follow the normal process of wound repair. Non-healing wounds encompass two very different types of wounds: at one end of the spectrum there are hypertrophic scars, characterised by fibrosis and excessive deposition of fibroblast-derived collagen and other extracellular matrix (ECM) proteins; at the opposite side of the spectrum are chronic ulcers, open wounds that fail to close due to lack of collagen deposition and re-epithelialization. Cells from chronic wounds exhibit decreased proliferation, poor response to growth factors, diminished migratory capacity, and are in a persistent inflammatory state. The wound bed exhibits little ECM production, increased concentration of proteases, and impaired angiogenesis. Moreover, chronic wounds are often accompanied by persistent pain and bacterial infection, leading in many cases to lower limb amputations and a significant permanent burden for patients. Finding effective long-term treatment options are therefore necessary to improve patient outcomes.

Pharmaceutical-based approaches for treating wounds are attractive as they have the potential to abrogate or at least mitigate the cellular-level deficiencies within the wound tissue. These deficiencies include improper gene expression, protein expression, or signal activation/propagation. To date, many different pharmacological-based strategies have been explored to correct these changes, although with very limited success. One reason for this may be that the method of delivery has not been optimized for wound healing applications. In this chapter, various therapeutic options are discussed including small molecule drugs and biologics that are of interest for wound healing experiments, with a focus on non-healing ulcers. This discussion is followed by a basic introduction to controlled drug delivery systems and an overview of various strategies that have or can be employed for therapeutic intervention to promote wound closure.

1. Pharmacologic Treatment of Wound Repair

Small molecule drugs (SMDs) represent the majority of therapeutic agents used in clinical medicine today. They are generally chemically synthesized low-weight molecules that interact with intracellular and extracellular targets, such as cell surface receptors and signaling proteins, in turn modulating their function. Because of their small size they can diffuse easily across cell membranes and can be resistant to degradation, making them viable for intravenous administration.[1] However, many times it is desirable to localize their delivery to the wound site to minimize systemic side effects, which makes their small size and relatively high diffusivity barriers to achieving localized, controlled delivery.

Biologics, on the other hand, tend to be larger substances derived or extracted from a biological source, and include protein and nucleic acid-based drugs (e.g. growth factors (GFs), antibodies, siRNAs), along with living cells. Biologics are generally more vulnerable to be broken down due to their susceptibility to endogenous mechanisms for degradation. Because of this they are often administered locally via injection or topical delivery to ensure high bioavailability,[1] and require high dosage levels to see therapeutic efficacy. However due to their potent biological activity, they are actively being evaluated as new therapeutics for many diseases including chronic wounds.

Neither SMDs nor biologics are necessarily advantageous over the other for applications in wound healing. Due to the high complexity and multifactorial pathogenesis of chronic wounds, many therapeutic options and strategies are needed that target a wide range of biological processes important for successful wound healing. These include promoting processes such as angiogenesis and re-epithelialization, while inhibiting those that are detrimental including excessive ECM destruction and prolonged inflammation. This complexity makes designing a therapeutic program quite challenging, since each pharmacologic possesses unique delivery requirements, both in terms of timings/dosage and physico-chemical properties. Here, we begin by discussing various pharmacologic options before delving into strategies that can be used to facilitate their delivery.

1.1. *Small molecule drugs*

1.1.1 *Prostaglandins*

Prostaglandins are endogenous molecules that primarily regulate inflammation, but also affect blood flow. One such molecule, prostaglandin E_1 (alprostadil) has been shown to reduce hematic viscosity and to contribute to vasodilation by relaxing vascular smooth muscle.[2] This has led to prostaglandin E_1 being evaluated clinically in two different studies for the treatment of ischemic and venous leg ulcers. In the first study patients with ischemic ulcers of the lower limbs were treated for 15 days with multiple daily subcutaneous prostaglandin injections (0.02 μg alprostadil/cm^2 of skin ulcer area) around the wound area, at a distance of 1 cm from the ulcer edge.[2] In the second study, patients with venous leg ulcers were treated with daily intravenous prostaglandin E_1 infusion (60 mg alprostadil/250 mL saline solution) for 20 days.[3] In both cases, the ulcers of patients treated with prostaglandin E_1, either subcutaneously or intravenously, healed at a faster rate and their size decreased by a statistically significant amount compared to the control groups receiving just saline solution.[2,3] In another recent study, patients with mixed arterial and venous ulcers of the lower limbs received prostaglandin E_1 intravenous infusion (60mg alprostadil/250 mL saline solution) for 21 days. In these patients the microcirculation to the limbs was improved and healing occurred in 80% of the limbs, compared to 52% of the limbs in the control group.[4] Taken together, these studies show that prostaglandin E_1 therapy may be a promising strategy for treating ulcers where the underlying cause of ulceration is due to poor blood circulation.

1.1.2 *Glucocorticoids inhibitors*

Glucocorticoids (e.g. cortisol) are inhibitors of wound healing and affect many of the essential steps of the normal wound repair process. Cortisol has been shown to inhibit cell proliferation, migration[5] and angiogenesis,[6] as well as prevent collagen synthesis by blocking the action of TGF-β in wounds.[7] Furthermore, it also participates in the resolution of inflammation by inhibiting the activity of pro-inflammatory cytokines, such as IL-1β upon tissue injury, thereby preventing an excessive pro-inflammatory response during acute wound healing.[8] Cortisol production in the skin is tightly regulated during acute wound healing, with synthesis increased during the inflammatory stage of wound healing followed by a decrease as the inflammatory stage declines.[8]

However, in chronic wound tissue the glucocorticoid synthesis and the glucocorticoid receptor pathways have been found to be erroneously activated, with the expression of the cortisol-producing enzyme CYP11B1 being 2.9-fold higher than in normal skin and the glucocorticoid receptor being constitutively active.[9] As such, inhibition of this enzyme has been suggested as a potential treatment to improve the outcome of wound healing. *Ex vivo* human acute wounds treated with the CYP11B1 inhibitor metyrapone (1 mM) exhibited enhanced epithelialization compared to untreated skin.[8] Similarly, topical treatment of *in vivo* partial thickness porcine wounds with metyrapone (500 μM) exhibited faster epithelialization and an improved rate of healing compared to the control vehicle (100% ethyl alcohol).[8] Despite there being no clinical evidence so far to report the efficacy of CYP11B1 inhibitors in the treatment of chronic wounds, these studies show that modulation of cortisol production may be a promising method to improve wound healing. However, care must be taken when modulating the cortisol synthesis pathway, as a continuous blockade of cortisol production might exacerbate the inflammatory response within the wound bed, which is detrimental to successful wound healing.

1.1.3. Wnt/β-catenin inhibitors

The abnormal increase of glucocorticoid synthesis in chronic wounds has also been shown to cause keratinocytes at the leading edge of non-healing diabetic and pressure ulcers to accumulate β-catenin, a Wnt signal transducer, in their nuclei.[10] This nuclear stabilisation of β-catenin activates the expression of genes including c-Myc, which can further contribute to the impairment of wound healing. Increased expression of c-Myc in primary human keratinocytes has been shown to block the effects of EGF and synergise with glucocorticoids to suppress the cytoskeletal proteins K6, K16, and β-tubulin, proteins required for keratinocyte migration.[10] In contrast with chronic wounds, nuclear accumulation of β-catenin has not been observed in the epidermis of normal skin or acute wounds,[10] suggesting that activation of β-catenin may contribute to the pathology of chronic wounds.

In order to address this aberrant expression of β-catenin, SMDs have been developed that efficiently inhibit the Wnt/β-catenin signaling pathway, one of which is astragaloside IV.[11] In an *in vitro* mouse wound healing model, murine keratinocytes were induced to over-express β-catenin by treatment with lithium chloride, which led to strong inhibition of cell proliferation and migration. Subsequent treatment with astragaloside IV (80 μg/mL for 72 h) significantly reduced the enhanced β-catenin expression, recovering keratinocyte migration and proliferation.[11] These early studies indicate that downregulation of β-catenin may represent a promising therapeutic target for promoting re-epithelialization of chronic wounds, although clinical studies are currently lacking. However, manipulation of this pathway must be very carefully approached as the Wnt signaling plays a crucial role in the maintenance of adult tissue homeostasis and stem cell pluripotency,[12] such that safe and precise targeting of this pathway is essential to prevent serious side effects.

1.1.4. *Curcumin*

Curcumin (diferuloylmethane) is a naturally occurring compound in the spice turmeric that has been shown to have positive effects in wound healing.[13] Curcumin 0.1% in polyethylene-glycol base applied topically onto skin biopsy punches in a genetically diabetic rat model has been found to enhance fibroblast proliferation and granulation tissue formation by increasing the expression of TGF-β,[14] accelerate epithelialization, as well as increase the vascular density compared to vehicle-control wounds.[14] Similarly, a recent study further investigated the healing effect of curcumin on burn wounds. Topical application of curcumin at a concentration of 100 mg/kg body weight for 12 days administered onto experimental 2nd degree burn injuries in albino rats resulted in significantly increased collagen deposition, angiogenesis, granulation tissue formation and re-epithelialization compared to untreated groups.[15]

However, despite its multiple benefits on modulating wound healing, curcumin-based therapy is still limited due to curcumin's increased cytotoxicity on cells at high concentration (>25 μM), which has been observed to induce fibroblast apoptosis in an *in vitro* cell culture model,[16] as well as curcumin's low bioavailability due to its low cellular uptake.[13] Therefore, delivery methods that increase its uptake and target its biodistribution may unlock its potential as a therapeutic for treating chronic wounds.

1.2. *Biologics*

1.2.1. *Peptide growth factors*

GFs released during wound healing have a myriad of roles including stimulating the formation of granulation tissue, driving epithelialization, and promoting angiogenesis.[17] In acute wounds GFs are transiently expressed to coordinate the process of tissue repair and wound closure. However, in chronic wounds GF levels are many times dysregulated,[17] which can be a result of many factors including aberrant gene expression and elevated levels of matrix metalloproteinases (MMPs).[18] This common alteration in the levels of GFs has driven an interest in using exogenous GFs as a therapeutic strategy to promote healing. Several recombinant GFs have been evaluated in the past and reached the preclinical or clinical stages, such as PDGF, VEGF, FGF-7, and FGF-10 (discussed below),[17] although most have achieved only limited success. As a result, only three GFs have been approved for the treatment of chronic wounds, namely recombinant human (rh) - PDGF, - EGF and, - FGF-2.[19]

Topical PDGF-BB (Regranex Gel, Smith and Nephew) is the only GF based gel approved by the FDA for the treatment of neuropathic diabetic foot ulcers with an area lower or equal to 5 cm^2.[20] Regranex (becaplermin) is formulated using a carboxymethylcellulose gel, a high viscosity and non-toxic compound that enables delivery of active proteins.[19] In a mouse wound healing model, topically administered Regranex (10 μg of PDGF-BB/wound) for 5 consecutive days improved granulation tissue formation and angiogenesis.[21] Several clinical studies evaluated topical administration of 100 μg becaplermin per 1 g of gel for 20 weeks compared to a placebo gel or standard practice of good ulcer care. The studies revealed controversial results, with some studies showing the incidence of complete healing in patients receiving becaplermin gel to be approximately twice that of patients treated with good ulcer care alone, while others showed no significant difference between groups.[22] Despite the good tolerability of the gel, prolonged use of Regranex (more than 3 tubes) is not recommended due to its associated increased risk of cancer metastasis,[23] resulting in a "Black Box" warning by the FDA. This is likely because of the high concentration of PDGF-BB required for physiological effect, which stems from inefficient delivery via the carboxymethylcellulose gel.

Topical FGF-2 (Fiblast Spray, Kaken Pharmaceutical) is approved in Japan and China for the treatment of pressure and burn ulcers.[24] The spray device is designed to spray 1 μg/cm^2 of FGF-2 to the wound surface. In clinical trials, patients with pressure ulcers treated with a daily topical application of 30 μg of FGF-2 for 8 weeks exhibited a rate of healing approximately twice as fast as that of the control group with increased formation of granulation tissue and faster epithelialization.[25] Topical application of FGF-2 has also been investigated for the healing of chronic diabetic neuropathic foot ulcers, in which patients received spray doses of FGF-2 (50 μL at a concentration of 5 μg/mL saline solution) daily for 6 weeks, then twice a week for 12 weeks.[26] The study found that the weekly reduction in ulcer perimeter area was the same in the treatment group as the placebo group, and the percent healed area at the end of the study did not differ significantly between the two groups, thus concluding that local application of topical FGF-2 is no more effective than the placebo to promote wound closure. The authors hypothesized that using a single GF might have been insufficient to accelerate wound closure in diabetic foot ulcers.[26]

Regarding EGF, three medications have been developed for its administration: **Heberprot-P, Regen-D, and Easyef.**[19]

Easyef (Daewoong Pharmaceuticals) has been approved for the treatment of diabetic foot ulcers in Korea and comes in spray form at 0.005%.[27] Clinical trials revealed that a daily topical treatment of diabetic ulcers with Easyef for 8 weeks resulted in reduction in the size of wounds by more than 80% in the eighth week, regardless of grade and size.[28] EGF stimulated the production of fibronectin, increased the number of fibroblasts in the wound, and enhanced epithelial proliferation and migration,[28] in turn promoting successful wound healing.

Similarly, clinical trials of **Heberprot-P** carried out in Cuba for the treatment of chronic diabetic foot ulcers as intralesional injections of 75 μg of Heberprot-P three times per week for 8 weeks, showed that 77% of the patients receiving treatment healed compared to only 56% of patients receiving placebo.[29] As a result, Heberprot-P has been approved since 2006 and is currently undergoing clinical trials in several additional countries.[29]

Regen-D 150 (Bharat Biotech) is an EGF-based gel approved in India for the treatment of diabetic foot ulcers, which contains 150 μg of EGF per 1 g of gel. Clinical trials demonstrated that daily topical application of Regen-D onto the wound area for 15 weeks caused the EGF-treated ulcers to heal faster than the control group (13 weeks versus 9 weeks) by promoting formation of granulation tissue and driving epithelialization.[30]

VEGF has been tested as a potential therapeutic modality for the treatment of chronic wounds due to its strong angiogenic/lymphangiogenic effects. Members of the VEGF family, such as VEGF-A, VEGF-C, and VEGF-165 have been investigated. One study has shown that a single intravenous administration of VEGF-A (1 mg VEGF/3mL of saline solution) in ischemic limbs of rabbits enhanced re-vascularization compared to the placebo group.[31] A second *in vivo* experiment in db/db diabetic mice showed that topical administration of VEGF-165 to dorsal skin wounds (20 μg/wound, every other day for 5 days) accelerated wound healing through increased angiogenesis, formation of granulation tissue, and faster epithelialization compared to the untreated control group.[32]

However, despite these promising results exogeneous adminstration of VEGF has been shown to promote formation of disorganised blood vessels and poorly functional lymphatic vessels,[33] thereby limiting its use in the clinical treatment of chronic wounds. With that being said, in a phase I clinical trial patients receiving topical application of VEGF-165 (telbermin) at 72 μg/cm^2 three times per week for up to 6 weeks showed a good tolerability and a higher incidence of complete wound healing compared to the control group (41.4% telbermin versus 26.9% placebo at day 43).[34] Additional studies are required to further characterise the clinical efficacy of telbermin.

FGF-7 (KGF-1) and FGF-10 (KGF-2) are keratinocyte-specific GFs in the context of dermal wound healing. They are key signaling molecules for promoting proliferation and migration of keratinocytes, in turn promoting epithelialization. A small phase II clinical trial investigated the topical application of KGF-2 (Repifermin) at 60 μg/cm^2 in venous ulcer patients with initial positive results.[35] However, follow up studies involving a much larger number of patients showed the efficacy of Repifermin to be low compared to the placebo, and further development as a potential therapeutic agent for chronic wounds has been dismissed.[36]

It is important to note that all the above experiments involved delivering a single GF at a time to the wound site. The efficacy of single-agent treatment is likely limited due to the process of wound healing relying on combinations of GFs to coordinate healing. Delivery of multiple GFs at strategic time points of wound healing (multiple-agent treatment) has been shown to be more effective than either GF alone in several studies.

A study carried out on guinea pigs with dorsal partial-thickness burns showed that topical application of an amnion derived cellular cytokine solution, consisting of PDGF, VEGF, TGF-β, and angiogenin at 0.007 mL/cm^2, significantly enhanced wound epithelialization compared to control wounds treated with saline.[37] This therapy entered Phase III clinical trials for treating partial thickness burns, but the study has been terminated early due to slow accrual of patients and no study results have been published so far.[38]

Similarly, in a partial thickness wound healing model in pigs, topical application of PDGF (500 ng) and IGF-I (500 ng) suspended in a gel was shown to cause a 2.2-fold increase in granulation tissue formation compared to either GF alone.[39] Another study conducted in a diabetic rat wound model investigated the sequential delivery onto the wound site of FGF-2 and EGF initially to accelerate epithelialization and granulation tissue formation, and VEGF and PDGF later to stimulate vasculature maturation.[40] The GFs were encapsulated in collagen and hyaluronic acid based nanofibers and released over 1 month by gradual degradation of the nanofibers. It was shown that the wound closure rate and angiogenesis were enhanced when all four GFs were administered compared to only FGF-2 and EGF being administered.[40]

Moreover, delivering more than one angiogenic factor to the wound site has been shown to promote a mature vascular network, thus overcoming the limitation of using VEGF as a monotherapeutic. One study showed that dual delivery of VEGF-165 and PDGF-BB, each with different kinetics, from a structural polymer scaffold, induced rapid formation of mature vasculature in a polymer scaffold.[41] This finding has been subsequently verified in the context of diabetic ulcer healing using db/db diabetic mice, thus illustrating the importance of multiple GF action in wound healing.[42]

Despite the encouraging results observed so far with the external application of GFs, the GF therapy possesses several critical limitations, such as the low *in vivo* stability of the GFs,[19] the impaired response of the cells to the administered GFs,[43] and regulatory hurdles for advancing combination therapies. The first two limitations have led to large concentrations of GFs being administered over long periods of time in order to have a therapeutic effect on wound healing. However, increased concentrations of GFs can lead to unwanted side effects such as increased risk of cancer, especially in cases where high systemic levels are reached.

1.2.2 *Nucleic acids*

An alternative to GFs is the use of nucleic acids to stimulate healing by modulating the expression of various proteins within the wound. Many times chronic wounds display a different profile of gene expression compared to acutely healing wounds. For example, biopsies from non-healing venous ulcers identified over 60 differentially regulated genes compared to healing ulcers, with significant differences being found for genes related to epidermal structural integrity, tissue repair, differentiation, proliferation, and transcription factors.[44] Exogenous modulation of final produced proteins can be accomplished using several strategies:

Exogeneous DNA can be inserted into wound cells to constitutively code for genes whose expression levels are abnormally low in the wound site, such as those of GFs. This stimulates the cells themselves to produce the GFs directly in the wound site, in turn accomplishing a similar goal to exogenous GF delivery.

Experimentally, hydrogels containing DNA coding for TGF-β gene have been shown to promote wound healing in a diabetic mouse wound model by accelerating re-epithelialization and enhancing collagen synthesis.[45] In humans, intramuscular transfer of naked plasmid DNA encoding for the VEGF-165 gene to patients with non-healing ischemic ulcers induced therapeutic angiogenesis, resulting in limb salvage of some patients recommended for below-knee amputation.[46] Similarly, in db/db mice overexpression of VEGF-C via an adenoviral vector was shown to increase the healing rate of full-thickness biopsy punches by enhancing angiogenesis and lymphangiogenesis in the wound compared to the control groups receiving β-galactosidase gene transfer.[47] Furthermore, adenoviral vectors coding for PDGF-BB have been administered topically on diabetic ulcers using a collagen gel (GAM501 gel).[48] The gel is currently in Phase II clinical trials, although no results have been published so far.[49]

Alternatively, the **RNA interference (RNAi) pathway** can be stimulated in wound cells via small interfering RNAs (siRNAs) or microRNAs (miRNAs). Both siRNAs and miRNAs are small non-coding RNAs (19–25 bp long) that bind to specific messenger RNAs (mRNAs) via complementary base pairing. This in turn causes degradation of the target mRNAs, silencing expression of the protein that the mRNA codes for.

siRNAs require complete complementarity to their target mRNA to cause degradation, thus yielding highly specific gene knockdown. This mechanism has been exploited in the treatment of chronic wounds, where topical application of exogenous siRNAs in a diabetic db/db mouse wound model resulted in knockdown of abnormally up-regulated genes such as p53, a master cell-cycle suppressor, which has been found to be up-regulated in cells from diabetic wounds.[50] The resulting local silencing of p53 expression within the wound bed resulted in faster wound closure than the control wounds due to enhanced angiogenesis,[50] illustrating the potency of siRNA-based therapeutic strategies in the context of wound healing.

Unlike siRNAs that generally target a specific mRNA, **miRNAs** are short RNA molecules that have been shown to act as regulators of wound healing by simultaneously targeting large numbers of mRNA molecules;[51] this difference arises due to miRNA only requiring partial complementarity to their targets with the main targeting region of the miRNA consisting of a "seed sequence" at the 5′ end. In the context of dermal wound repair, miRNA dysregulation has been linked with many impairments in wound repair. For example, biopsies from non-healing venous leg ulcers identified aberrantly over-expressed levels of miRNA-16, -20a, -21, -106a, -130a and -203, as compared to healthy tissue, which resulted in suppressed epithelialization and impaired formation of granulation tissue.[52] Restoring the levels of these dysregulated miRNAs represents a novel therapeutic approach for wound healing, that involves up-regulating the levels of potentially beneficial miRNAs via administration of miRNA mimics, and down-regulating the levels of potentially harmful miRNAs via miRNA inhibitors.

Several studies so far have investigated this approach in the context of wound healing. In a hypoxic ischaemic mouse wound model up-regulated miRNA-210 levels were identified and shown to impair wound closure by silencing the expression of E2F3, a key facilitator for keratinocyte proliferation at the wound edges.[53] Subsequent liposomal-based delivery of miR-210 inhibitors to hypoxic human immortalised HaCaT keratinocytes *in vitro* significantly increased E2F3 levels and cell proliferation,[53] suggesting that targeting miR-210 might be of therapeutic promise for improving healing of ischaemic wounds in the clinic.

Similarly, skin biopsies extracted from edges of non-healing chronic diabetic ulcers have been shown to exhibit persistent expression of miRNA-198, in contrast to the epidermis of acute cutaneous wounds whose levels were found to be low.[54] miRNA-198 directly targets and inhibits pro-migratory proteins in keratinocytes such as DIAPH1, PLAU, and LAMC2.[54] In an *in vitro* model of human immortalised N/TERT-1 keratinocytes over-expressing miRNA-198, functional inhibition of miRNA-198 with anti-miRNA-198 via liposomal-based delivery promoted cell migration.[54] These findings suggest that miRNA-198 could be a therapeutic target for improving diabetic ulcer healing by enhancing the rate of epithelialization.

1.2.3 *MMP and elastase inhibitors*

Chronic wounds commonly exhibit elevated levels of proteases, such as elastase and MMPs, as well as a low MMP to TIMP (tissue inhibitor of metalloproteinases) ratio.[55,56] Prolonged increased levels of MMPs and elastase in the wound bed result in increased degradation of ECM, GFs, and reduction in the ability to form granulation tissue.[55,57] Therefore, reducing the levels or impact of these destructive proteases through inhibition with peptides may be a potential therapeutic strategy. In the case of MMPs, this can be achieved by either blocking the MMPs directly, blocking the degradation of TIMPs, or using an artificial substrate for the MMPs.[58] The latter method involves using peptides with sequences related to the cleavage region of MMPs that act as a replacement substrate. The efficacy of such peptide-based MMP inhibitors has been investigated in a db/db diabetic mouse wound model, where wounds treated with 5 μL of a 20 μg/mL 19-mer peptide closed 5 days faster than wounds treated with saline which served as the control.[59] Similarly, several elastase inhibitors including alpha-1-antitrypsin and Elafin have been patented,[58] but no clinical studies on their action in wound healing has yet been carried out.

It is important to acknowledge that both MMPs and elastase play an important and beneficial role during wound healing, as they allow cell migration and remodeling of the provisional matrix. However, when these levels increase beyond the required ones they become detrimental to wound healing. Protease reduction via inhibitors has been shown to have beneficial effects in wound healing, but continuous and complete blockade of protease action in a chronic wound is undesirable.

1.2.4 *Connexin inhibitors*

Connexins are the channel-forming components of gap junctions that directly couple the cytoplasm between cells, permitting the exchange of small molecules to facilitate intercellular communication. It has been observed that signals of injury, inflammation, epithelialization, and wound closure spread through such gap junctions.[60] When the skin is injured, the expression of connexins in the wound area dramatically changes, in particular that of Cx43.[61] Its expression drops at the wound edges in the early phases of normal wound healing, but in chronic wounds Cx43 has been found to be permanently up-regulated.[61] This has led to the hypothesis that transient inhibition of Cx43 expression may be a potential therapeutic strategy to promote wound healing. Cx43 knockdown has been achieved using two classes of compounds, namely Cx mimetic peptides that directly interact with Cx43 binding partners, and Cx43 antisense oligodeoxynucleotides that prevent Cx43 mRNA translation.

Mimetic peptides such as Gap27 have been shown in a murine keratinocyte culture to significantly accelerate wound closure compared to the control.[62] Similarly, topical administration of ACT1, a small mimetic peptide of Cx43 (Granexin Gel, FirstString Research) to patients with chronic venous leg ulcers in a Phase II clinical trial significantly accelerated fibroblast proliferation and migration, as well as the rate of epithelialization compared to the control group.[63]

In addition, an *in vivo* rat wound model found that topically applied Cx43 antisense oligonucleotides resulted in wounds closing significantly faster than the untreated wounds causing reduced inflammation and an enhanced rate of re-epithelialization.[64] Furthermore, a Phase II clinical trial for the treatment of venous leg ulcers with Cx antisense oligonucleotides (Nexagon, CoDa Therapeutics) has been completed, but no study results have been reported at this time.[60] Taken together, these studies indicate that inhibition of Cx43 may be a promising method for accelerating wound healing.

1.3 Discussion

Currently there are a surprisingly small number of pharmacological-based options available for treating chronic wounds. Despite a great deal of understanding regarding the process of wound repair, the high complexity and multifactorial pathogenesis of chronic wounds have led to a dearth of therapies that have had promising results during clinical trials. Some of the reasons for this may include:

(1) The complex multifactorial pathogenesis of chronic wounds: Administration of a single agent to the wound site cannot restore all the impairments seen in chronic wounds. As such, different classes of agents should be administered simultaneously during a particular phase of wound healing to achieve a full therapeutic effect.

(2) The need of specific conditions in each phase of wound healing: This is important as administration of a certain therapeutic agent can be beneficial in one particular phase of wound healing, but detrimental for wound healing in the next phase. Examples of this include the inhibition of Cx43, MMPs, and elastase. Ideally, modulation of a therapeutic agent should be temporary and achieved in a controlled manner and at a specific time.

(3) Challenges of patient selection for clinical trials: Chronic wounds are a highly heterogeneous condition, and there is likely no "golden bullet" for driving successful closure. Correctly identifying sub-populations that will benefit from a given therapy will likely result in higher success rates, similar to genetic profiling of cancers for highly targeted therapies.

In all cases, there is a need for efficient, targeted, and controllable drug delivery. Getting the drug to its right destination at the right time can be as important as the development of the drug itself. Many therapeutic agents exploited for the treatment of chronic wounds mentioned above, such as nucleic acids, curcumin, and GFs have had low efficacy when delivered in a free, non-encapsulated manner to the wound site due to the low cellular uptake and hostile wound environment. By developing new drug delivery systems (e.g. nanoparticles, microspheres, hydrogels, nanofibers), it is likely to improve the translational success of pharmacologics to meet unmet medical needs for the treatment of wound healing disorders.

2. Biomaterial-Based Therapeutic Delivery for Wound Healing

The field of biomaterials has produced many methods for the delivery of SMDs and biologics for the promotion of wound healing. Due to the wide variety of materials and strategies available for drug delivery, it is important to understand how different types of materials interact with the biological environment. Once placed into a wound, the material undergoes continuous or cyclic stress, abrasion, and flexure.[65] The aqueous, ionic environment of the wound causes the plasticization, or softening, of polymers and the enhanced electrochemical activity of metals. Cell-secreted oxidizing agents and enzymes coordinate in a temporal biological response to the biomaterial. Combining all these factors can give rise to synergistic degradation of materials as swelling and water uptake increases the number of reaction sites, lowered pH from the degradation products furthers the reaction mechanism, and production of hydrophilic species from polymer hydrolysis increases polymer swelling and enhances the entry of enzymatic species into the bulk polymer.[65]

2.1 Polymeric scaffolds for pharmacologic treatment of wound healing

Polymers are macromolecular chains made up of multiple, repeated subunits joined by covalent bonds. These long-chain polymers can be categorized by the molecular organization of its monomer subunits. Homopolymers consist of repeated single monomer (AAAAAAAA), random copolymers are made up of at least two randomly ordered monomers (AAABBABAAABA), alternating copolymers have repeating monomers in alternate positions (ABABABAB), and block copolymers consist of a minimum of two types of repeating chains of the same monomer joined together (AAAABBBBB). Due to the wide-ranging mechanical and chemical properties of polymers, a variety of polymers are utilized for wound healing, including natural inert polymers, natural bioactive polymers, and synthetic polymers. The mechanical properties and bioresorbability of the polymer can be tailored by controlling factors such as molecular weight and chain length, cross-link density, degree of crystallinity, degree of freedom for bond rotation, temperature, and time of loading.

2.2. Constant versus exponential delivery

A polymer is biodegradable if its initial solid- or gel-phase material can be reduced to water soluble fragments that the body can metabolize or excrete under normal physiological conditions (37°C, pH = 7.4). Polymer degradation occurs when the polymer is placed in contact with bodily fluids and absorbs water, proteins, and lipids, in turn initiating the process of degradation. The mechanism of degradation depends on the functional groups of the subunits in the polymer. Esters, anhydrides, and carbonates primarily break down via hydrolysis, whereas the degradation of amides must also be facilitated by other enzymatic activities. In general the rate of hydrolysis is fastest in anhydrides, followed by esters, carbonates, and slowest in amides, which are usually stable *in vivo* without enzyme catalysis.

There are three different chemical mechanisms of solid polymer degradation including cross-link degradation, side chain degradation, and backbone degradation. Cross-link degradation generally occurs in water-soluble polymer chains where the cross-links between chains are cleaved, resulting in small, bioeliminable products. This is the main mechanism for hydrogel degradation and degradation of polyanhydride networks. During side chain degradation, water insoluble polymers are converted into soluble polymers through the cleavage or transformation of side chains into polar or charged groups. This occurs in polymers such as poly(methyl vinyl ether-co-maleic anhydride) and poly(alkyl cyanoacrylates). Finally, backbone degradation occurs when the backbone linkages between repeating subunits of a water insoluble polymer are cleaved to produce smaller, water soluble monomer units. This is the main degradation mechanism for solid polymers and polyphosphazenes.

The rate of degradation depends on the physical mechanisms of dissolution and can be explained by the Göpferich theory of polymer erosion.[66] If a polymer is initially water-insoluble and its main mechanism of degradation is hydrolysis, then two rates dominate erosion behavior: the rate of water diffusion into the matrix, t_{diff}, and the rate of chain cleavage by hydrolysis, t_h. The relative ratios of these two processes determine whether the polymer degrades primarily by surface erosion or bulk erosion, and in turn will influence the temporal release profile of encapsulated therapeutics (Figure 1).

In surface erosion (Figures 1A and B), the rate of water penetration into the polymer is much slower than the time for hydrolysis of the bonds in the same depth ($t_{\text{diff}} \ll t_h$). A constant loss of mass occurs at the surface until the entire polymer is degraded. Because there is a linear loss of mass with respect to time, degradation via surface erosion results in 0th order release kinetics of encapsulated therapeutic molecules (assuming a uniform distribution of therapeutic throughout the polymer). An example of surface erosion is the degradation of polyanhydride networks in which there is a constant decrease in molecular weight over time due to loss of mass at the surface.

In bulk erosion (Figures 1C and D) water penetrates into the polymer at a rate that is faster than the hydrolytic degradation of the polymer. This results in formation of micropores throughout the thickness that allows diffusion of therapeutic molecules out of the matrix ($t_{\text{diff}} \gg t_h$). Due to internal bond cleavage happening faster than the dissolution of the polymeric matrix,

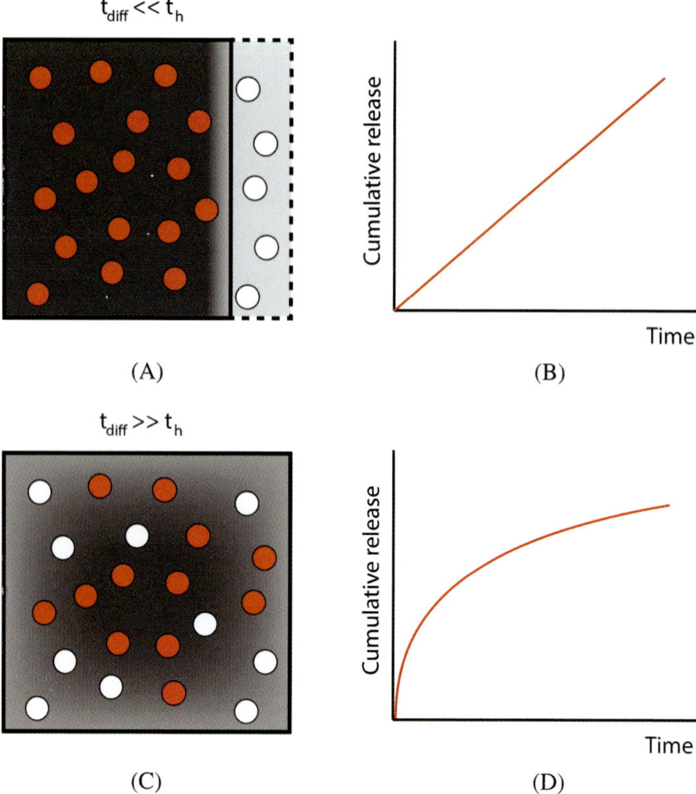

Figure 1. (A) Surface-based erosion occurs when the rate of polymer hydrolysis is faster than the rate of water diffusion into the polymer. This method of degradation results in decreasing sample dimensions. (B) Assuming a uniform therapeutic distribution and no appreciable burst release, surface-based erosion results in zero-order release kinetics. (C) Bulk erosion occurs when the rate of water diffusion into the polymer is faster than hydrolysis of the polymer. This method of degradation maintains sample dimensions until failure of the sample. (D) In this case, microvoids and channels form that increase diffusion of the therapeutic out of the polymer, resulting first-order release kinetics.

the dimensions of the matrix remain constant, while molecular weight of the polymer decays with respect to time, until total failure of the polymeric solid. The formation and growth of pores throughout the thickness of the material results in a release rate that varies with time, where the rate initially decreases due to diffusion-based degradation, and exponentially increases as matrix-degradation dominates. An example of bulk erosion is degradation of poly(lactic-co-glycolic acid) (PLGA).

Several aspects of the polymers can be manipulated to alter the kinetics of polymer degradation including bond stability, hydrophobicity, steric effects, production of autocatalytic fragments during degradation, and polymeric microstructure (crystallinity, porosity, and phase separation). Furthermore, the ratio and repeating structure of constituent polymer blocks (e.g. AAABBB, ABABAB) can influence the rate of degradation.

In the context of biomaterial-based drug delivery, it is necessary to couple the desired rate of degradation and subsequent therapeutic release with the physicochemical properties of the therapeutic of interest. For instance, encapsulating active GFs within a polymer matrix must not denature the protein and render it inactive. Furthermore, preventing phase separation and crystallization of the therapeutic from the polymer matrix, such as a hydrophobic SMD within a more hydrophilic polymer matrix will enable more well-controlled release kinetics and enhanced bioavailability of the SMD following delivery.

2.2.1. *Modeling degradation of polymeric systems using kinetics*

As an example of how to use kinetic models to design drug delivery systems, one approach was taken by Chen *et al.* (2011) to computationally model degradation of bulk-erosion polymeric devices for drug delivery.[67] Their hybrid model is based on a stochastic hydrolysis and diffusion-driven auto-catalysis representation of polymer degradation and erosion. Previous experimental data suggests that polymer matrix degradation follows pseudo first-order kinetics as given by,

$$M_a^t = M_a^0 e^{-\lambda t},$$

where M_a^0 is the initial average molecular weight of the polymer matrix, λ is the experimental degradation rate constant, and M_a^t is the average molecular weight at time t. Building upon previous models established by Göpferich and accounting for a hysteretic delay, t_{add}, in which the porous matrix is assumed to start from a solid state with initial porosity α and degrade over time to a porous matrix, the matrix degradation process can be described as a first order stochastic event given by the probability density function, $P(\lambda, t)$,

$$P(\lambda,t) = \frac{\lambda e^{-\lambda(t+t_{\mathrm{add}})}}{V(t)} = \frac{\lambda e^{-\lambda t}}{V_0 V(t)},$$

where V_0 is the initial volume fraction of the polymer matrix and $V(t)$ is the volume fraction at time t. The probability of hydrolysis of a single hydrolysable element increases with time due to a lower number of hydrolysable elements available as degradation proceeds.

Next, the model simulates the autocatalytic process, which accelerates local polymer hydrolysis as a multistep process that includes the release of hydrolysed monomers, diffusion of the acid catalysts, and an acid-accelerated hydrolytic reaction. Diffusion of the monomers following hydrolysis is modelled with Fick's second law,

$$\frac{\partial C_m}{\partial t} = \nabla\left(D_m \nabla C_m\right) + S(t),$$

where C_m is the time-dependent concentration of hydrolysed monomers, D_m is the diffusivity of degraded products through the polymer matrix, and $S(t)$ represents the source term of monomers generated during acid-catalysed hydrolysis. A constant β controls the contribution of the autocatalytic reaction to allow the model to be fit to known experimental data.

Summing the fundamental hydrolysis and autocatalysis probability density functions, P_F and P_C respectively, gives the accelerated probability density function, P_A,

$$P_A = P_F + P_C = \frac{\lambda_0 e^{-\lambda_0 t}\left(1 + \beta\left(e^{C_m} - 1\right)\right)}{V_0 V(t)}$$

which models the overall polymer degradation and erosion process. Polymeric materials of different composition, size, and configuration can be modeled using this equation by choosing appropriate autocatalysis-free degradation rate constants, λ_0, and varying the contribution of autocatalysis with the variable β.

The hybrid model proposed by Chen *et al.* (2011) was used to model degradation of square poly(DL-lactide) films over a range of matrix sizes and β, and applied for systems of varying geometries. In the case of drug-loading PLGA-based microparticles, the model was in good agreement with experimental data for a range of differently sized microparticles over a timeframe of 30 days. Both the computational model and experimental data showed that as the radius of the microparticles increased, the rate of average molecular weight loss also increased. By examining a number of matrix parameters, the degradation profiles predicted by this model suggests that fundamental hydrolysis rate, polymer architecture, and matrix size help determine how a controlled delivery system will degrade and erode.

As exemplified by the above hybrid model, kinetic modeling is a powerful tool in the design of drug-releasing polymeric systems. These models have the potential to increase understanding of mass transport mechanisms in biodegradable polymers, enable enhanced prediction of the effect of design parameters on drug release rates, and improve the design of controlled degradation systems for drug delivery.

2.3 *Delivery of small molecule drugs*

Though SMDs have the advantage of generally being more robust and stable than biologics, their smaller size and increased hydrophobicity raises challenges in controlling the rapid diffusion of drugs out of the wound, sustaining long-term release, and preventing phase separation and crystallization due to poor solubility. A number of different delivery platforms have been utilized in the delivery of SMDs including polymeric scaffolds, nanoparticles, liposomes, and microspheres.

2.3.1 *Controlling diffusion using polymeric scaffolds*

Wound dressings present a straightforward method for incorporating therapeutics into the clinical wound care setting. They can be easily substituted for standard gauze dressings and can minimize patient compliance problems that are associated with daily patient-applied creams. Dressings can take the form of coatings on traditional wound dressings (e.g. Tegaderm contact layers), modification of commercially available hydrogels (e.g. Intrasite, Nu-gel, Kikgel, Aqua-gel, and Aquaform), or specially designed dressings such as electrospun nanofiber mats, polymer wafers, and sponges.

These bioactive dressings generally allow for more controlled temporal delivery than with cream ointments and solutions, which are usually inefficient as a delivery system due to biological agents rapidly diffusing out of the wound bed upon application.[68] Wound dressings solve this issue by controlling the slow release of therapeutics into the local wound site. This direct delivery can also prevent exposing patients to high-systemic doses that may be detrimental elsewhere in the body if delivered orally or intravenously.[69]

Since SMDs have low molecular weights, they are easily susceptible to rapid release from swelled polymeric scaffolds, leading to large bolus releases. One technique to extend the release profile of small molecules is to conjugate the drug to the polymeric backbone. This can have a dramatic influence on the rate that small molecules are released. Recently, multi-month release of small molecules was demonstrated in layer-by-layer assembled films of biologically derived poly(L-glutamatic acid) (PGA) conjugated with the nonsteroidal anti-inflammatory drug diclofenac.[70] While multi-month release is not required for applications in dermal wound healing, the strategy of conjugation to larger polymer chains via hydrolytically degradable links provides a tuneable and customizable strategy for achieving steady release of SMDs.

2.3.2 Enhancing bioavailability with nanoparticles

Solid nanoparticles can be synthesized from polymers, gold, and silver for drug delivery and injected via fine-gauge needles into tissue to manipulate healing within the wound.[71] The release profiles of these particles can be finely tuned to allow for complex profiles that cannot be achieved by wound dressing and scaffold systems due to injection at specified timepoints.[72,73] It should be noted, though, that this does require more time from clinical staff than coordinated release from a wound dressing.

Small molecular drugs that have poor aqueous solubility, limited biodistribution, or quick degradation and clearance can be encapsulated within nano and microparticles for delivery. The particles can be designed to enable intracellular access and reduce toxicity by facilitating the slow and sustained release of encapsulated drugs within the ideal therapeutic window.

PLGA is one of the most commonly used polymers for the synthesis of drug delivery particles due to its degradation products being natural compounds processed during cell metabolism. Encapsulation of curcumin in PLGA nanoparticles facilitated sustained released over the course of 8 days, resulting in increased angiogenesis, re-epithelialization, and granulation tissue formation in excisional mouse wounds.[74] Nanoparticles made out of biodegradable poly(β-amino esters) (PBAE) have intradermally delivered sonic hedgehog plasmid to promote the upregulation of multiple GFs in mice.[75] Nitric oxide has also been previously delivered using nanoparticles of silica, polyethylene glycol (PEG), and chitosan composites to promote angiogenesis and collagen synthesis.[76]

2.3.3 Co-delivery of hydrophobic and hydrophilic molecules using liposomes

Liposomes are vesicular structures formed from hydrated lipid bilayers and have the unique advantage of delivering high concentrations of both hydrophobic and hydrophilic drugs from a single particle.[77,78] The lipid envelope of the liposome contains the hydrophobic drug, while the aqueous interior encloses the hydrophilic drug, allowing for the sequential staggered release of the hydrophobic drug, followed by the hydrophilic drug. Liposomes can also improve the membrane permeability and stability of large, polar chemicals for delivery. As a proof of concept, liposomes have been employed for the dual delivery of the hydrophobic EGFR inhibitor erlotinib and hydrophilic doxorubicin and exhibited precise time-staggered *in vivo* drug release in xenograft tumor mouse models.[79] While not specifically designed for wound healing, this strategy of combinatorial delivery of two SMDs over different timescales is potentially attractive for wound healing applications.

2.3.4 *Delivery of multiple drugs using microspheres and microemulsions*

Microspheres are synthesized through the application of thin coatings to small solid particles or liquid dispersion droplets. These 1–1000 μm particles can be biocompatible, bioadhesive, adaptable to a large range of dosage forms, and can incorporate multiple drugs for simultaneous delivery.[80] Microspheres can also have an occlusive effect in skin to prevent wound dehydration.[81] Microemulsions have been frequently utilized to deliver plant-based formulations. These small droplet mixtures have large surface areas, increased solubility, good stability, and have an easy formulation process.[80] For instance, a number of plant derived essential oils, such as cinnamon oil[82] and eucalyptus oil[83] have been delivered in the form of microemulsions to prevent sepsis in wistar rats.

2.4 *Delivery of biologics*

Due to their larger size and increased susceptibility to degradation, the delivery of biologics for wound healing poses different challenges than the delivery of SMDs. Though their natural role in the process of wound repair can yield high specificity and potency, their complex structure and susceptibility to specific degradation mechanisms leads to challenges maintaining the biologic's activity during delivery. Changes in the environment such as pH, moisture content, or temperature during both storage and delivery in the body can alter the formulation of the biologic. Their higher molecular weight substantially reduces the biologics' ability to freely permeate across biological barriers such as skin, mucosal membranes, and cell membranes.[84] This inability to permeate across cell membranes makes it especially challenging in the delivery of therapeutics like nucleic acids, which have intracellular targets. Furthermore, delivery of nucleic acids to the extracellular space can simulate viral and bacterial infection, further stimulating an immune response that is many times overactive in chronic wound environments.

2.4.1 *Modifications in injection formulation for short-term delivery*

The stability of the biologics during short-term injectable delivery can be enhanced with changes to the formulation. The tendency of proteins to aggregate can be reduced through the introduction of small sugars such as trehalose and dextrans, pluronics, and non-ionic surfactants like polysorbates.[84] Reduction of viscosity is important when delivering proteins in highly concentrated formulations. The viscosity of these concentrated solutions is controlled using the addition of hydrophobic salts, inorganic salts, and amino acids including lysine or arginine.[84] Lastly, co-injection with recombinant human hyaluronidase, which aids in the degradation of hyaluronic acid in tissues, has been shown to manipulate the diffusion rate of drugs at the injection site.[84]

2.4.2 *Microparticles and nanoparticles for longer-term delivery*

Longer-term delivery of biologics can be achieved by encapsulating peptides or proteins into microparticles to prevent early degradation or elimination from the body. As in the case of the delivery of small molecular drugs, solid microparticles are synthesized from various polymers and delivered using needle injection. These microparticles are often mechanically robust, hydrophobic, biocompatible, and degrade into products that can be safely eliminated from the body.[84] In this case, the degradation rate of the polymer microparticle or diffusion rate of the protein from the polymer controls the release rate of the protein. These rates are controlled by a number of factors such as porosity, size, polydispersity, surface properties, and shape.[84]

Increased porosity of the microparticle allows enhanced water penetration into the polymer matrix. This in turn increases matrix degradation and drug release. Microsphere particle size also affects the duration of drug release. In general, the larger the particle, the smaller the surface area to volume ratio, and the longer the release profile. This means that increased polydispersity of microparticle size will increase the variability of release rates, so biologics requiring tight delivery windows are best served by microparticle systems with a small polydispersity. Modification of the microparticle surface with polymers like PEG can influence interaction and clearance by immune cells, while the shape of the particle affects orientation-dependent interaction with macrophages.[85]

Similar to microparticles, nanoparticles can also be used as carriers of biologics. However, due to their size approaching the size of the biologics to be delivered, there are fewer examples of nanoparticles used to sustain delivery of proteins and peptides than SMDs. GFs such as EGF have been successfully encapsulated within PLGA nanoparticles and released within 24 h, resulting in a higher level of fibroblast proliferation and faster healing rate than free EGF delivery,[86] although longer timescales of release generally require multiple doses or embedding in a host matrix.

Though drug release from nanoparticles generally follow first-order kinetics due to the increase in surface area as the particles degrade,[72] the release kinetics can be extended or multiplexed by incorporating particles into other delivery systems. Biphasic release of PDGF for up to 21 days has been accomplished using a polyurethane (PU) scaffold embedded with gelatin-coated PLGA microparticles.[87] Also, chitosan and poly(ethylene oxide) electrospun nanofibers loaded with VEGF and PDGF-encapsulated PLGA nanoparticles have been used to facilitate dual release of VEGF and PDGF in rat wound models.[88] Again using electrospinning to create bioactive fiber mats, GF-encapsulated gelatin nanoparticles embedded in collagen and hyaluronic acid fibers sustained release of four different GFs (VEGF, PDGF, FGF-2, and EGF).[40]

Another class of nanoparticles are solid lipid nanoparticles, which have been utilized for delivering therapeutics to heavily damaged or inflamed skin. Though they are currently limited to short-term release of encapsulated proteins, lipid nanoparticles have the advantages of being non-irritant, non-toxic, and can protect proteins from rapid degradation.[80] Both solid lipid nanoparticles and nanostructured lipid carriers have been created for the delivery of EGF. This has resulted in significantly accelerated wound closure, reduction in inflammation, and increase in the rate of epithelization in a diabetic mouse model, however this strategy required four applications of the nanoparticles due to the limited ability of lipid nanoparticles to sustain long-term release.[89]

Similar to the case with SMDs, microparticles and nanoparticles commonly have the limitation of burst release of encapsulated biologics and a relatively short timescale of release. While the short timescales can be beneficial for some therapeutics, they may require frequent dosing and can lead to local toxicity and inefficient therapies if the burst release is undesired.

2.4.3 *Polymeric scaffolds and wound dressings for sustained, controlled release*

One way to reduce burst release and extend the release profile of biologics is to incorporate them into polymeric scaffolds and wound dressings. Similar to bioactive wound dressings for the delivery of SMDs, these dressings control the release of therapeutics into the local wound site by immobilizing or covalently attaching the biologics into or onto a polymeric scaffold. There are many potential polymers that are available for use when designing bioactive dressings for wound care; these include both synthetic and natural polymers. Each class of polymer has its own unique benefits and drawbacks that must be considered when selecting the therapeutic/s of interest.

(1) Synthetic polymers have the benefit of being chemically synthesized, which allows for tight control over the chemical composition, polymer

properties (e.g. molecular weight), and purity. This can lead to better control and consistency over the degradation rate, along with consistent tuning of properties such as moisture absorption, water vapour transmission rate, and mechanical strength.[90] However, synthetic polymers many times lack specific bioactivity that natural polymers can provide, and can break down into unnatural constituents that must be excreted from the body. Synthetic polymers commonly used in wound dressings include nylon, polyvinylalcohol (PVA), PEG, polycaprolactone (PCL), and PU.

(2) Natural polymers are derived from plants, bacteria, fungi, and animals, and have a wide variety of biological activity and physical properties. This allows for incorporation of a range of bioactivity that works in combination with biologics. Many natural polymers such as bacterial cellulose and silk have the advantage of biocompatibility and biodegradability when incorporated into a wound dressing without significant biological activity. However, the various properties and batch-to-batch variability of naturally derived polymers are much harder to control than with synthetic polymers, which can lead to difficulties in scaling up production while maintaining consistent bioactivity and degradability. Futhermore, purification of bacteria-derived polymers of contaminants such as lipopolysaccharides (LPS) can present significant challenges to minimizing potential stimulation of the immune system.

2.4.4 *Sustained delivery of biologics using polymeric dressings*

There are many potential avenues to pursue for controlling the temporal release of biologics into a wound. For instance, the temporal and sequential release of biologics such as VEGF and PDGF-BB has been achieved using layer-by-layer assembly of hydrolytically degradable PBAE, poly(acrylic acid) (PAA), and heparan sulfate into therapeutic dressings.[42] In this work, cysteine conjugated to PAA created 2D diffusion barriers that enabled biphasic release of the GFs over the course of 2 weeks, promoting significant improvements in angiogenesis and formation of granulation tissue in a db/db diabetic mouse wound model.

In another study, block copolymers of biodegradable PCL and PEG were electrospun to create biocompatible nanofibers that can be subsequently functionalized with EGF. The release of conjugated EGF from these nanofibers prevented burst release traditionally observed in encapsulated EGF delivery systems and promoted proliferation and migration of keratinocytes in diabetic mouse models.[91] Also, coaxial incorporation of FGF-2 within the PCL/PEG nanofibers allowed for the release of FGF-2 followed by EGF, which greatly enhanced wound healing rates in mice.[92]

Finally, electrospun poly(ethylene glycol)-poly(D,L-lactide) (PELA) nanofibers mats have been used to enable the release of FGF-2 to promote angiogenesis in diabetic rat models.[93] Since the scaffold was fabricated using electrospinning, its high surface area and porosity allowed for rapid hydrolytic degradation of the polymer nanofibers, and the scaffold had the advantage of being fully degraded within 4 weeks of application. Looking at all the examples together, it is apparent that methods to control the diffusivity of the biologics in order to prevent rapid burst release are key to sustaining the release of therapeutics within the local wound bed.

2.4.5 *Enhanced delivery of biologics via functional conjugation*

Another strategy to promote enhanced delivery of GFs is the conjugation to glycosaminoglycans that have specific affinity for the desired GFs and cytokines. This not only slows down the release due to increasing residence time via the addition of binding sites, but can also be used to protect GFs from proteolytic degradation, in turn increasing their *in vivo* half-lives. Sustained delivery of GFs like PRP, VEGF, PDGF, and FGF-2 with heparin-conjugated fibrin has been shown to enhance the epithelialization and angiogenesis in a mouse wound model.[94] Additionally, by mimicking native ECM function, cross-linking chondroitin sulfate and heparan sulfate to enable the delivery of FGF-2 enhanced wound closure in a diabetic mouse model.[95]

Another innovative solution that has been recently shown is the fusion of the ECM binding domain from placenta growth factor-2 (PLGF-2), which binds strongly and exclusively to ECM proteins, to GFs such as PDGF-BB and BMP-2. These new fusion proteins displayed greatly improved therapeutic efficacy over wild-type counterparts in diabetic mouse models.[96] Similarly, collagen-binding domains and VEGF fusion proteins loaded into a collagen scaffold enhanced vascularization and delivery of a higher level of VEGF to granulation tissue as compared to native VEGF in a diabetic rat model.[97]

2.4.6 *Delivery of nucleic acids*

Unlike proteins which have extracellular targets, the delivery of nucleic acid is complicated by the challenge of facilitating delivery into cells, while protecting against rapid degradation by native nucleases. However, it is a highly promising therapeutic strategy since it can transiently limit endogenous protein production at the cell source within the wound bed. To achieve sustained knockdown of protein targets within diabetic wounds, a layer-by-layer coated wound dressing has been developed that enables sustained release and knockdown of overexpressed proteins by siRNA.[98] In this study, siRNA targeting MMP-9 was directly incorporated into nanoscale films, which degraded over the course of approximately 2 weeks and resulted in significantly lower levels of MMP-9 protein levels and activity. This had the subsequent effect of significantly improving the rate of wound healing compared to control therapies in a db/db diabetic wound model. New developments such as this are now unlocking new potential to actively manipulate protein expression within wound environments. If coupled with the delivery of other therapeutics, these strategies may enable comprehensive transient manipulation of the signaling that occurs during wound repair.

3. Conclusion

Pharmaceutical-based treatment of dermal wounds is a promising area for addressing abnormal or deficient wound repair. Biomaterials provide an attractive avenue to enable efficient and effective delivery of both SMDs and biologics to a variety of wounds. By designing new delivery systems such as bioactive wound dressings, it is possible to incorporate sustained and temporally coordinated delivery of therapeutics in a context that is easily amenable to clinical settings. This is especially true in situations such as burns and chronic wounds, where dressings are routinely used to reduce the risk of infection. However, in order to design optimal delivery systems it is necessary to understand both the function of the biological targets one is aiming to treat, along with the physicochemical properties of the therapeutics. As biological systems are dynamic, the design of delivery systems must appropriately match the temporal changes in biology. This combined knowledge can then be used to rationally guide the development of delivery systems that optimize the temporal dosing, and in the case of multi-therapeutics their transient coordination, in order to maximize therapeutic benefit and efficacy.

References

1. Sekhon, B. S. and Saluja, V. Biosimilars: an overview. *Biosimilars* **1**, 1–11 (2011).
2. Tondi, P. *et al.* Treatment of ischemic ulcers of the lower limbs with alprostadil (prostaglandin E1). *Dermatol Surg* **30**, 1113–1117 (2004).
3. Milio, G. Minà, C. Cospite, V. Almasio, P. L. and Novo, S. Efficacy of the treatment with prostaglandin E-1 in venous ulcers of the lower limbs. *J Vasc Surg* **42**, 304–308 (2005).

4. De Caridi, G. *et al.* Effectiveness of prostaglandin E1 in patients with mixed arterial and venous ulcers of the lower limbs. *Int Wound J* (2014).

5. Stojadinovic, O. Gordon, K. A. Lebrun, E. and Tomic-Canic, M. Stress-induced hormones cortisol and epinephrine impair wound epithelization. *Adv Wound Care* **1**, 29–35 (2012).

6. Shikatani, E. A. *et al.* Inhibition of proliferation, migration and proteolysis contribute to corticosterone-mediated inhibition of angiogenesis. *PLoS One* **7**, e46625 (2012).

7. Wang, A. S. Armstrong, E. J. and Armstrong, A. W. Corticosteroids and wound healing: Clinical considerations in the perioperative period. *Am J Surg* **206**, 410–417 (2013).

8. Vukelic, S. *et al.* Cortisol synthesis in epidermis is induced by IL-1 and tissue injury. *J Biol Chem* **286**, 10265–10275 (2011).

9. *De novo* synthesis of glucocorticoids in the epidermis and its uses and applications. US 20080274079 A1; 2008.

10. Stojadinovic, O. *et al.* Molecular pathogenesis of chronic wounds: The role of beta-catenin and c-myc in the inhibition of epithelialization and wound healing. *Am J Pathol* **167**, 59–69 (2005).

11. Li, F. L. *et al.* Astragaloside IV downregulates β-Catenin in rat keratinocytes to counter LiCl-induced inhibition of proliferation and migration. *Evid Based Complement Alternat Med* **2012**, (2012).

12. Kahn, M. Can we safely target the WNT pathway? *Nat Rev Drug Discov* **13**, 513–532 (2014).

13. Akbik, D. Ghadiri, M. Chrzanowski, W. and Rohanizadeh, R. Curcumin as a wound healing agent. *Life Sci* **116**, 1–7 (2014).

14. Sidhu, G. S. *et al.* Curcumin enhances wound healing in streptozotocin induced diabetic rats and genetically diabetic mice. *Wound Repair Regen* **7**, 362–74 (1999).

15. Kulac, M. *et al.* The effects of topical treatment with curcumin on burn wound healing in rats. *J Mol Histol* **44**, 83–90 (2013).

16. Scharstuhl, a. *et al.* Curcumin-induced fibroblast apoptosis and *in vitro* wound contraction are regulated by antioxidants and heme oxygenase: Implications for scar formation. *J Cell Mol Med* **13**, 712–725 (2009).

17. Barrientos, S. Stojadinovic, O. Golinko, M. S. Brem, H. and Tomic-Canic, M. Perspective article: Growth factors and cytokines in wound healing. *Wound Repair Regen* **16**, 585–601 (2008).

18. Menke, N. B. Ward, K. R. Witten, T. M. Bonchev, D. G. and Diegelmann, R. F. Impaired wound healing. *Clin Dermatol* **25**, 19–25 (2007).

19. Gainza, G. Villullas, S. Pedraz, J. L. Hernandez, R. M. and Igartua, M. Advances in drug delivery systems (DDSs) to release growth factors for wound healing and skin regeneration. *Nanomedicine* **11**, 1551–1573 (2015).

20. Goldman, R. Growth factors and chronic wound healing: Past, present, and future. *Adv Skin Wound Care* **17**, 24–35 (2004).

21. LeGrand, E. K. Preclinical promise of becaplermin (rhPDGF-BB) in wound healing. *Am J Surg* **176**, 48S–54S (1998).

22. Wieman, T. J. Clinical efficacy of becaplermin (rhPDGF-BB) gel. Becaplermin gel studies group. *Am J Surg* **176**, 74S–79S (1998).

23. FDA, U.S. Food and Drug Administration. Update of Safety Review: Follow-up to the March 27, 2008, Communication about the Ongoing Safety Review of Regranex (becaplermin). [ONLINE] Available at: http://www.fda.gov/Drugs/DrugSafety/PostmarketDrugSafetyInformationforPatientsand Providers/DrugSafetyInformationforHeathcareProfessionals/ucm072148.htm [Accessed 9 August 2016].

24. Kaken Pharmaceutical Co., LTD. KAKEN's Innovation Product for Regeneration Fiblast® (Recombinant human basic fibroblast growth factor, rh bFGF). [ONLINE] Available at: http://www.kaken.co.jp/english/business/rd_pipeline.html [Accessed 9 August 2016].

25. Ohura, T. *et al.* Clinical efficacy of basic fibroblast growth factor on pressure ulcers: case-control pairing study using a new evaluation method. *Wound Repair Regen* **19**, 542–551 (2011).

26. Richard, J.-L. *et al.* Effect of topical basic fibroblast growth factor on the healing of chronic diabetic neuropathic ulcer of the foot: A pilot, randomized, double-blind, placebo-controlled study. *Diabetes Care* **18**, 64–69 (1995).

27. Easyef Hikma pharmaceuticals. [ONLINE] Available at: http://www.meppo.com/pdf/drugs/592-EASYEF-1444208053.pdf [Accessed 9 August 2016].

28. Tuyet, H. Le *et al.* The efficacy and safety of epidermal growth factor in treatment of diabetic foot ulcers: The preliminary results. *Int Wound J* **6**, 159–166 (2009).

29. Berlanga, J. *et al.* Heberprot-P: A novel product for treating advanced diabetic foot ulcer. *Medicc Rev* **15**, 11–15 (2013).

30. Viswanathan, V. and Pendsey, S. A phase III study to evaluate the safety and efficacy of recombinant human epidermal growth factor (REGEN-D^{TM} 150) in healing diabetic foot ulcers. *Wounds* **18**, 186–196 (2006).

31. Bauters, C. *et al.* Site-specific therapeutic angiogenesis after systemic administration of vascular endothelial growth factor. *J Vasc Surg* **21**, 314–324; discussion 324–325 (1995).

32. Galiano, R. D. *et al.* Topical vascular endothelial growth factor accelerates diabetic wound healing through increased angiogenesis and by mobilizing and recruiting bone marrow-derived cells. *Am J Pathol* **164**, 1935–1947 (2004).

33. Nagy, J. A. *et al.* Vascular permeability factor/vascular endothelial growth factor induces lymphangiogenesis as well as angiogenesis. *J Exp Med* **196**, 1497–1506 (2002).

34. Hanft, J. R. *et al.* Phase I trial on the safety of topical rhVEGF on chronic neuropathic diabetic foot ulcers. *J Wound Care* **17**, 30–32, 34–37 (2008).

35. Robson, M. C. *et al.* Randomized trial of topically applied repifermin (recombinant human keratinocyte growth factor-2) to accelerate wound healing in venous ulcers. *Wound Repair Regen* **9**, 347–352 (2001).

36. Human Genome Sciences Reports Results of Clinical Trial of Repifermin in Patients with Chronic Venous Ulcers. [ONLINE] Available at: http://www.prnewswire.com/news-releases/human-genome-sciences-reports-results-of-clinical-trial-of-repifermin-in-patients-with-chronic-venous-ulcers-71171297.html [Accessed 9 August 2016].

37. Payne, W. G. *et al.* Effect of amnion-derived cellular cytokine solution on healing of experimental partial-thickness burns. *World J Surg* **34**, 1663–1668 (2010).

38. ACCS (Amnion-derived Cellular Cytokine Solution) Versus Standard Care In Treating Partial Thickness Burns — Full Text View — ClinicalTrials.gov.

39. Lynch, S. E. Nixon, J. C. Colvin, R. B. and Antoniades, H. N. Role of platelet-derived growth factor in wound healing: Synergistic effects with other growth factors. *Proc Natl Acad Sci USA* **84**, 7696–7700 (1987).

40. Lai, H.-J. *et al.* Tailored design of electrospun composite nanofibers with staged release of multiple angiogenic growth factors for chronic wound healing. *Acta Biomater* **10**, 4156–4166 (2014).

41. Richardson, T. P. Peters, M. C. Ennett, a B. and Mooney, D. J. Polymeric system for dual growth factor delivery. *Nat Biotechnol* **19**, 1029–1034 (2001).

42. Almquist, B. D. Castleberry, S. A. Sun, J. B. Lu, A. Y. and Hammond, P. T. Combination growth factor therapy via electrostatically assembled wound dressings improves diabetic ulcer healing *in vivo. Adv Healthc Mater* **4**, 2090–2099 (2015).

43. Pastar, I. *et al.* Attenuation of the transforming growth factor beta-signaling pathway in chronic venous ulcers. *Mol Med* **16**, 92–101 (2010).

44. Charles, C. a *et al.* A gene signature of nonhealing venous ulcers: Potential diagnostic markers. *J Am Acad Dermatol* **59**, 758–771 (2008).

45. Lee, P.-Y. Li, Z. and Huang, L. Thermosensitive hydrogel as a Tgf-beta1 gene delivery vehicle enhances diabetic wound healing. *Pharm Res* **20**, 1995–2000 (2003).

46. Baumgartner, I. *et al.* Constitutive expression of phVEGF165 after intramuscular gene transfer promotes collateral vessel development in patients with critical limb ischemia. *Circulation* **97**, 1114–1123 (1998).

47. Saaristo, A. *et al.* Vascular endothelial growth factor-C accelerates diabetic wound healing. *Am J Pathol* **169**, 1080–1087 (2006).

48. Mulder, G. *et al.* Treatment of nonhealing diabetic foot ulcers with a platelet-derived growth factor gene-activated matrix (GAM501): Results of a Phase 1/2 trial. *Wound Repair Regen* **17**, 772–779 (2009).

49. Phase 2b Study of GAM501 in the Treatment of Diabetic Ulcers of the Lower Extremities — Full Text View — ClinicalTrials.gov.

50. Nguyen, P. D. *et al.* Improved diabetic wound healing through topical silencing of p53 is associated with augmented vasculogenic mediators. *Wound Repair Regen* **18**, 553–559 (2010).

51. Pastar, I. *et al.* Micro-RNAs: New regulators of wound healing. *Surg Technol Int* **XXI**, 51–60 (2011).

52. Pastar, I. *et al.* Induction of specific microRNAs inhibits cutaneous wound healing. *J Biol Chem* **287**, 29324–29335 (2012).

53. Biswas, S. *et al.* Hypoxia inducible microRNA 210 attenuates keratinocyte proliferation and impairs closure in a murine model of ischemic wounds. *Proc Natl Acad Sci USA* **107**, 6976–6981 (2010).

54. Sundaram, G. M. *et al.* 'See-saw' expression of microRNA-198 and FSTL1 from a single transcript in wound healing. *Nature* **495**, 103–106 (2013).

55. McCarty, S. M. and Percival, S. L. Proteases and delayed wound healing. *Adv Wound Care* **2**, 438–447 (2013).

56. Liu, Y. *et al.* Increased matrix metalloproteinase-9 predicts poor wound healing in diabetic foot ulcers. *Diabetes Care* **32**, 117–119 (2009).

57. Chen, S. M. Ward, S. I. Olutoye, O. O. Diegelmann, R. F. and Kelman Cohen, I. Ability of chronic wound fluids to degrade peptide growth factors is associated with increased levels of elastase activity and diminished levels of proteinase inhibitors. *Wound Repair Regen* **5**, 23–32 (1997).

58. van Koppen, C. J. and Hartmann, R. W. Advances in the treatment of chronic wounds: A patent review. *Expert Opin Ther Pat* **3776**, 1–7 (2015).

59. Kimberly–Clark Worldwide, Inc. Metalloproteinase inhibitors for wound healing. US7186693 B2; 2007.

60. Becker, D. L. Thrasivoulou, C. and Phillips, A. R. J. Connexins in wound healing; perspectives in diabetic patients. *Biochi Biophys Acta-Biomembr* **1818**, 2068–2075 (2012).

61. Churko, J. M. and Laird, D. W. Gap junction remodeling in skin repair following wounding and disease. *Physiology* **28**, 190–198 (2013).

62. Kandyba, E. E. Hodgins, M. B. and Martin, P. E. A murine living skin equivalent amenable to live-cell imaging: Analysis of the roles of connexins in the epidermis. *J Invest Dermatol* **128**, 1039–1049 (2008).

63. Ghatnekar, G. Grek, C. Armstrong, D. G. Desai, S. C. and Gourdie, R. The Effect of a Connexin43-based peptide on the healing of chronic venous leg ulcers: A multicenter, randomized trial. *J Invest Dermatol* **135**, 289–298 (2014).

64. Qiu, C. *et al.* Targeting connexin43 expression accelerates the rate of wound repair. *Curr Biol* **13**, 1697–1703 (2003).

65. Ratner, B. Hoffman, A. Schoen F. and Lemons, J. Eds. *Biomaterials Science*, 1st edition. Elsevier (1996).

66. Göpferich, A. Mechanisms of polymer degradation and erosion. *Biomaterials* **17**, 103–114 (1996).

67. Chen, Y. Zhou, S. and Li, Q. Mathematical modeling of degradation for bulk-erosive polymers: Applications in tissue engineering scaffolds and drug delivery systems. *Acta Biomater* **7**, 1140–1149 (2011).

68. Boateng, J. S. Matthews, K. H. Stevens, H. N. E. and Eccleston, G. M. Wound healing dressings and drug delivery systems: A review. *J Pharm Sci* **97**, 2892–2923 (2008).

69. Langer, R. Polymeric delivery systems for controlled drug release. *Chem Eng Commun* (2007). Available at: http://www.tandfonline.com/doi/abs/10.1080/00986448008912519.

70. Hsu, B. B. Park, M.-H. Hagerman, S. R. and Hammond, P. T. Multimonth controlled small molecule release from biodegradable thin films. *Proc Natl Acad Sci USA* **111**, 12175–12180 (2014).

71. Johnson, N. R. and Wang, Y. Drug delivery systems for wound healing. *Curr Pharm Biotechnol* **16**, 621–629 (2015).

72. Soppimath, K. S. Aminabhavi, T. M. Kulkarni, A. R. and Rudzinski, W. E. Biodegradable polymeric nanoparticles as drug delivery devices. *J Control Release* **70**, 1–20 (2001).

73. Kumari, A. Yadav, S. K. and Yadav, S. C. Biodegradable polymeric nanoparticles based drug delivery systems. *Colloids Surf B Biointerfaces* **75**, 1–18 (2010).

74. Chereddy, K. K. *et al.* Combined effect of PLGA and curcumin on wound healing activity. *J Control Release* **171**, 208–215 (2013).

75. Park, H.-J. *et al.* Sonic hedgehog intradermal gene therapy using a biodegradable poly(β-amino esters) nanoparticle to enhance wound healing. *Biomaterials* **33**, 9148–9156 (2012).

76. Han, G. *et al.* Nitric oxide-releasing nanoparticles accelerate wound healing by promoting fibroblast migration and collagen deposition. *Am J Pathol* **180**, 1465–1473 (2012).

77. Dai, W. *et al.* Spatiotemporally controlled co-delivery of anti-vasculature agent and cytotoxic drug by octreotide-modified stealth liposomes. *Pharm Res* **29**, 2902–2911 (2012).

78. Sengupta, S. *et al.* Temporal targeting of tumour cells and neovasculature with a nanoscale delivery system. *Nature* **436**, 568–572 (2005).

79. Morton, S. W. *et al.* A nanoparticle-based combination chemotherapy delivery system for enhanced tumor killing by dynamic rewiring of signaling pathways. *Sci Signal* **7**, ra44 (2014).

80. Pachuau, L. Recent developments in novel drug delivery systems for wound healing. *Expert Opin Drug Deliv* **12**, 1895–1909 (2015).

81. Değim, Z. Use of microparticulate systems to accelerate skin wound healing. *J Drug Target* **16**, 437–448 (2008).

82. Ghosh, V. Saranya, S. Mukherjee, A. and Chandrasekaran, N. Antibacterial microemulsion prevents sepsis and triggers healing of wound in wistar rats. *Colloids Surf B Biointerfaces* **105**, 152–157 (2013).

83. Sugumar, S. Ghosh, V. Nirmala, M. J. Mukherjee, A. and Chandrasekaran, N. Ultrasonic emulsification of eucalyptus oil nanoemulsion: Antibacterial activity against Staphylococcus aureus and wound healing activity in Wistar rats. *Ultrason Sonochem* **21**, 1044–1049 (2014).

84. Mitragotri, S. Burke, P. A. and Langer, R. Overcoming the challenges in administering biopharmaceuticals: Formulation and delivery strategies. *Nat Rev Drug Discov* **13**, 655–672 (2014).

85. Sharma, G. *et al.* Polymer particle shape independently influences binding and internalization by macrophages. *J Control Release* **147**, 408–412 (2010).

86. Chu, Y. *et al.* Nanotechnology promotes the full-thickness diabetic wound healing effect of recombinant human epidermal growth factor in diabetic rats. *Wound Repair Regen* **18**, 499–505.

87. Li, B. Davidson, J. M. and Guelcher, S. A. The effect of the local delivery of platelet-derived growth factor from reactive two-component polyurethane scaffolds on the healing in rat skin excisional wounds. *Biomaterials* **30**, 3486–3494 (2009).

88. Xie, Z. *et al.* Dual growth factor releasing multi-functional nanofibers for wound healing. *Acta Biomater* **9**, 9351–9359 (2013).

89. Gainza, G. *et al.* A novel strategy for the treatment of chronic wounds based on the topical administration of rhEGF-loaded lipid nanoparticles: *In vitro* bioactivity and *in vivo* effectiveness in healing-impaired db/db mice. *J Control Release* **185**, 51–61 (2014).

90. Boateng, J. and Catanzano, O. Advanced therapeutic dressings for effective wound healing-a review. *J Pharm Sci* **104**, 3653–3680 (2015).

91. Choi, J. S. Leong, K. W. and Yoo, H. S. *In vivo* wound healing of diabetic ulcers using electrospun nanofibers immobilized with human epidermal growth factor (EGF). *Biomaterials* **29**, 587–596 (2008).

92. Choi, J. S. Choi, S. H. and Yoo, H. S. Coaxial electrospun nanofibers for treatment of diabetic ulcers with binary release of multiple growth factors. *J Mater Chem* **21**, 5258 (2011).

93. Yang, Y. *et al.* Promotion of skin regeneration in diabetic rats by electrospun core-sheath fibers loaded with basic fibroblast growth factor. *Biomaterials* **32**, 4243–4254 (2011).

94. Yang, H. S. *et al.* Enhanced skin wound healing by a sustained release of growth factors contained in platelet-rich plasma. *Exp Mol Med* **43**, 622–629 (2011).

95. Liu, Y. Cai, S. Shu, X. Z. Shelby, J. and Prestwich, G. D. Release of basic fibroblast growth factor from a crosslinked glycosaminoglycan hydrogel promotes wound healing. *Wound Repair Regen* **15**, 245–251.

96. Martino, M. M. *et al.* Growth factors engineered for super-affinity to the extracellular matrix enhance tissue healing. *Science* **343**, 885–888 (2014).

97. Tan, Q. *et al.* Promotion of diabetic wound healing by collagen scaffold with collagen-binding vascular endothelial growth factor in a diabetic rat model. *J Tissue Eng Regen Med* **8**, 195–201 (2014).

98. Castleberry, S. A. *et al.* Self-assembled wound dressings silence mmp-9 and improve diabetic wound healing *in vivo*. *Adv Mat* (2015).

11. Laser Tissue Welding in Wound Healing and Surgical Repair

Russell Urie, Tanner Flake and Kaushal Rege

Chemical Engineering
Arizona State University
Tempe, AZ, 85287, USA

Abstract

Laser tissue welding (LTW) is a sutureless technique for sealing incised or wounded tissue and is an attractive alternative or supplement in various surgeries. In LTW, chromophores convert laser light to heat which ultimately results in tissue sealing. While LTW has had success without exogenous absorbers, introducing chromophores that absorb near-infrared light creates differential laser absorption and allows for a laser wavelength that minimizes tissue damage. Additionally, chromophore-embedded thin biopolymer films have been used to create robust tissue bonding in larger tissue geometries. Advances in LTW procedures and materials (chromophores and composites) can reduce treatment complexity, facilitate welding of larger tissue, and result in higher tensile strengths compared to suturing in many cases. However, the elevated temperatures reached due to laser light absorption and concomitant heat generation can cause significant tissue damage in some cases, and procedural and material improvements to LTW are underway to optimize the trade-off between tissue damage and tissue seal strength.

1. Introduction

Sutures have been used for thousands of years to close wounds and surgical incisions, with documented use occurring as early as ancient Egypt and

India. Despite the nearly ubiquitous use of sutures and recent developments in absorbable sutures, suturing still holds a number of shortcomings. Laparoscopic suturing is very technically demanding and requires long operating times,[1,2] the tissue is handled extensively during suturing, and fluid and bacterial leakage can occur across the sutured tissue during the initial period of healing.[3,4] Surgical staples, clamps, and glues have been investigated for overcoming some of the limitations of suturing. Yet, sutures, clips, and staples cannot be used safely for repairing or functionally rehabilitating a number of tissue types,[5] including the gastrointestinal (GI) tract,[6] nerves,[7] dura repair in spinal surgery,[8] urethroplasties,[9] laparoscopic mesh fixation in hernia repair,[10] lung,[11] vasculature,[12] cataract surgery,[13] traumatic injuries,[14] lens capsule bag,[15,16] and others.[17]

In 1979, a report described the joining of small blood vessels together by laser light to avoid the use of sutures, a technique called laser tissue welding (LTW).[18] LTW involves the rapid bonding of tissues or tissue edges through the conversion of laser light to heat, and can use either endogenous or exogenous chromophores. A chromophore is a molecule or part of a molecule or particle that can absorb light. In the context of LTW, a chromophore is an agent that absorbs laser energy and converts this energy to heat. This generation of heat results in an increase in the local temperature, which results in tissue sealing by inducing local changes in the tissue.

Following the introduction of the LTW technology, lasers began to be used in the early 1980s[19] to improve and hasten scar healing.[20] Since this time, LTW has been researched in various tissue types,[21,22] including hypospadias repair,[23] urinary tract reconstruction,[24] blood vessels,[25,26] GI tract,[27,28] cartilage,[29] skin,[30,31] bladder,[32] urethra,[33] cornea,[34] liver,[35] nerve,[36,37] aorta,[38] and small intestine[39–41]. LTW has potential benefits that include shortened procedure times compared to suturing,[5,42] scar reduction and prevention compared to sutures,[5,37,43,44] greater procedural ease and minimal tissue handling,[45,46] improved wound strength in the early postoperative period,[47,48] immediate fluid-tight seal,[40,41] and reduced healing times.[49] Due to these potential benefits as well as the possibility of wide application, LTW is an exciting approach with the potential to improve patient outcomes in surgery.

Endogenous LTW relies on naturally occurring chromophores within the tissue and is advantageous[13,50,51] because no foreign material is required, which can reduce scar formation and inflammatory response.[48] However, endogenous LTW is often limited to very thin, transparent, and homogenous tissue (such as the ocular lens), employing laser wavelengths in the mid and far infrared regions and fluences which can cause significant thermal damage to tissue.[52] LTW has been investigated extensively for eye surgeries because the tissue has little vasculature, pigment, or pathological changes, while thicker, less homogeneous tissues have been largely avoided. To address this shortcoming, many studies have relied on water as an endogenous chromophore because it is the major component in tissue, using wavelengths with high absorption by water including CO_2 lasers at 10.6 μm, diode lasers at 1900 nm, and erbium: YAG lasers at 2900 nm. Another common natural chromophore is hemoglobin with a maximum absorbance at 488–514 nm.[1] Due to limited penetration of laser light at these wavelengths, the outer layer of tissue, exposed to the laser, suffers from heat damage, while the lower layers are only weakly heated, indicating possibilities for robustly welding only thin tissues.[52] Skin has also been repaired in endogenous LTW studies because it is easily accessible and more robust than many internal tissues. Optical properties are vastly different in different tissue, and this variation results in different temperature fields and thermal damage.[53] In addition, the composition of endogenous chromophores varies depending on tissue type. Thus, tissues such as vasculature, nerves, and intestine, which either possess low content of endogenous chromophores or are significantly heterogeneous in their composition may not be ideal candidates for endogenous LTW.

Exogenous LTW, on the other hand, relies on introducing a chromophore to induce differential laser absorption from the tissue and is effective because these chromophores can convert laser light to heat with much greater efficiency than endogenous chromophores such as water, melanin, hemoglobin, and others.[45] Because of this, exogenous LTW can minimize thermal damage, limit the area of exposure, and generate heat more deeply into tissue, but may result in a foreign body response due to the introduced chromophores.

This chapter discusses recent research progress in exogenous LTW as well as other noteworthy progress towards advancing LTW to the clinic. A brief summary of the theory and mechanism underlying LTW and advancements in chromophores are discussed, with emphasis on the use of nanoparticle-based materials. Additionally, a brief comparison is made with other suture-less wound healing techniques.

2. Exogenous Laser Tissue Welding: Mechanisms

LTW utilizes a wavelength-specific chromophore to generate sufficient heat, which restructures the extracellular matrix (ECM) and disorganizes proteins at tissue edges. Upon cooling, this reorganization of native tissue proteins results in collagen interdigitation and noncovalent bonding of laminin and entactin.[54] LTW is an appropriate term for this technique because there are many parallels to conventional welding in the proposed mechanism of LTW. The most agreed upon mechanism is composed of two parts; first, thermal denaturation of the tissue constituents and, second, bond formation as the tissue cools. The exact bonding mechanism is not precisely known due to a number of factors including variation in laser wavelength and power density (power per unit volume or cross-sectional area), length of treatment, spot size and number of spots, variation in tissue-specific factors such as laser absorption, and differences in chromophore and solder types, compositions, and dimensions.

In conventional metal welding, a solder is a low-melting alloy that is capable of fusing metals that are less fusible. Similarly, in the context of tissue approximation and repair, a solder is an exogenous material that facilitates the bonding of tissues or tissue edges. Two distinct approaches, direct and indirect laser welding, are employed in exogenous LTW. In direct laser welding, an exogenous chromophore is used to stain the wound edges to be bonded. The tissue is irradiated directly. Direct LTW is sometimes referred to as stained LTW. Indirect LTW, sometimes referred to as laser tissue soldering (LTS), employs an exogenous chromophore within a liquid or solid "solder". The solder is placed or applied over the wound edges and irradiated to generate heat and seal the underlying tissue. This method is advantageous because the tissue is not directly irradiated; the wavelength of light can be tuned to minimize absorption by the natural chromophores in the tissue, and the exogenous chromophores can greatly increase photothermal conversion compared to endogenous chromophores. The tissue "therapeutic optical window" is the region from 700 nm to 1,300 nm in which little to no laser radiation is directly absorbed.[55]

The theory and mechanism of action of LTW have mainly been derived from the heating of rat tails[56] and other highly collagenous tissues such as cornea[13] and arteries.[55,57] Laser light converted to heat denatures and fuses tissue proteins.[58,59] Denaturation of collagenous tissue occurs in three phases, revealed by second-harmonic generation microscopy.[60,61] First, proteoglycan bridges between fibrils are broken around 45°C. Second, at 60°C collagen starts to collapse due to hydrolysis of intramolecular H-bonds. Third, complete collagen denaturation and tissue homogenization occurs above 80–85°C due to hydrolysis of covalent peptide bonds.[62] However, irreversible tissue thermal damage begins at temperatures as low as 40°C. Because of this large difference, it is important to find the minimum temperature necessary to create robust welds and also localize the heat and area of exposure to a feasible extent. Temperature at the wound bed surface in direct exogenous LTW and the solder-tissue interface in LTS is an indirect indication of the status of the wound during laser treatment.[2] As detailed by Fourier's law, the heat transferred to the wound bed by conduction is proportional to the temperature gradient of the solder-tissue interface.

An optimal temperature range to produce robust welds of pure Type I collagen may occur at approximately 60–66°C. Mouse skin was shown to coagulate *in vitro* above 60°C.[63] In another study, bovine aortic tissue was welded with dye-albumin solders. Below 66°C at the solder-tissue interface, the tissue was not sealed, but at 66°C the tissue was sealed with maximum tensile strength. Above this temperature, the tensile strength decreased with increasing temperature.[64] This suggests that complete collagen denaturation is not required for patent tissue welds. Because other ECM components have transition temperatures below 60°C, other tissue types may have an even lower optimal LTW temperature range.

In another study, porcine eyes were stained with ICG and corneal wounds were successfully welded with temperatures in the 55–65°C range. Atomic force microscopy and transmission electron microscopy of the welded corneal wounds revealed that normal fibrillar structure was largely disorganized, but the collagen was not denatured (Figure 1).[65] This work suggests that even mild changes in the stroma ultrastructure can facilitate new ocular tissue bonding.

Cezo et al. demonstrated that water bound to ECM proteins is driven off by laser irradiation, opening sites for bonding.[56] Elastin,[56,66] decorin, cells, and other ECM components have yet to be analyzed in detail following laser welding to better understand the effects of heat and laser irradiation. By analyzing these components in greater detail, more in-depth spatio-temporal modeling could be used to predict the optimal temperature range for a wider variety of tissue types.

A number of parameters can be used to assess the effectiveness of LTW treatment. Bond tensile strength is measured by pulling the tissue apart at a fixed rate and can provide a rapid preliminary assessment of the robustness of the welded closure. For LTS, the mode of failure — cohesive or adhesive[67] — is an important additional observation from tensile strength testing. Often more clinically relevant, the burst pressure is the most widely used mechanical parameter for determining the patency of welded tissue. The welded tissue is filled with fluid until leakage or bursting, as shown in Figure 2. Other less commonly reported but valuable treatment outcomes are water loss, cell viability, and biocompatibility in the case of laser tissue solders, scarring, bacterial leakage across the weld, and histological analysis. Additionally, the surface temperature of the tissue or solder can be monitored directly to provide valuable information regarding the correct

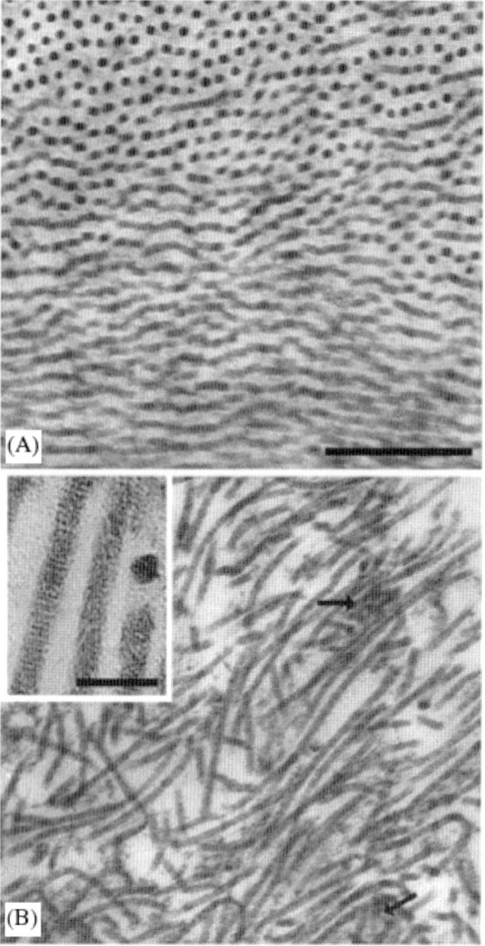

Figure 1. Transmission electron microscopy images of corneal stroma (top) without laser irradiation and (bottom) at the weld site. Scale bar is 400 nm and the insert scale bar is 100 nm. Reproduced with permission from Ref. 65.

laser parameters (power density and exposure time) to maintain the optimal temperature range.[41,52]

A number of factors can be cause for concern with regards to LTW, and these factors can hinder further growth and clinical translation. These factors include low weld strength in some cases,[68] thermal damage,[68] overcoming normative suture use, current application for only very thin transparent tissue,[16] training required,[1] subjectivity and ambiguity regarding endpoint of welding[1,64] and inconsistency in results.[5,13] These concerns are being improved through technical advances in chromophores and solders[15,22,40,41] as discussed in the subsequent sections. Full automation of laser dosimetry

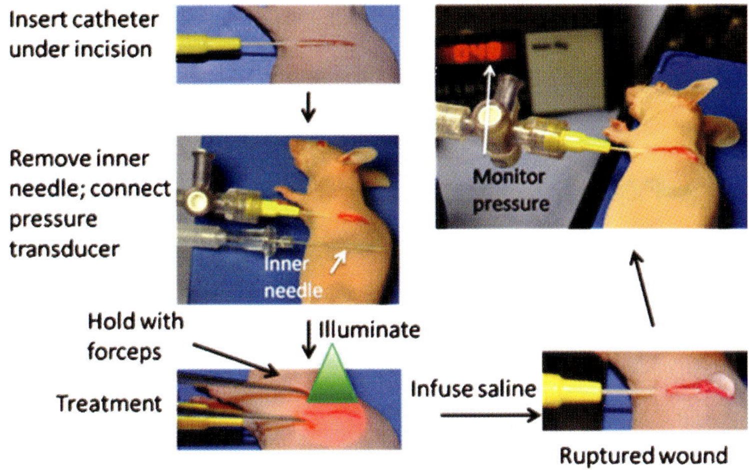

Figure 2. Procedure for measurement of weld strength in skin closures of mice. Saline solution is infused under the weld site, and the burst pressure is recorded. Reproduced with permission from Ref. 31.

may be necessary for more general application,[2] and surgical experience and surgeon motor reflexes also contribute significantly to LTW success rates.[2] Meticulous hemostasis is likely necessary prior to tissue welding,[1] and potential problems for high stress regions are possible.[69] Laser equipment can be costly, however, portable, handheld diode lasers with continuous or pulsed wave NIR light are becoming available, some at relatively low cost. The equipment used for LTW ranges in size and price. Some groups have incorporated large robots to automate welding,[13] while others simply use portable handheld laser fiber optic cables that are quite inexpensive.[22] Additionally, the treatment length compared to that of conventional suturing depends largely on the type of LTW. For endogenous LTW, the powers and wavelengths are so high as to take from 5 s to 15 s per spot. This is much shorter than suturing times. For thicker tissue and with exogenous LTW, especially with indirect LTW, the times are longer due to the necessary heat transfer through the exogenous solder and the thick tissue. In these cases, the time is on the order of minutes and is generally not spot treated. The time required may be equal to or even greater than suturing, depending on the laser power used. Additional innovation will be necessary to shorten the length of treatment in these cases. Also, treatment planning varies from study to study, where some spot weld the incision with a varying number of spots and others perform a continuous weld by moving the laser along the incision at varying speeds and with a varying number of passes. Additional work within the field is needed to assess the large parameter space involved in welding and optimize techniques to minimize laser exposure, treatment time, and peripheral thermal damage.

Sutured closures have been reported to elicit a stronger inflammatory response than laser soldered closures, and the inflammatory response has a longer duration with sutured wounds as well.[69,70] Additionally, laser exposure results in local hemostasis which prevents scabbing, excessive blood flow, and bacterial transfer.[71] The perfusion rates of blood vary greatly by tissue type and body location. At high blood perfusion rates, the convective heat loss will be stronger and will result in less thermal damage. A dual-phase lag model has been developed to evaluate thermal damage of laser irradiation of tissue[72,73] based on an assumption that thermal damage can be treated as a chemical reaction with a rate process for three steps: (1) laser deposition in tissue, (2) heat transfer, and (3) protein denaturation.[74,75] Nonlinear finite-element analysis to simulate the dynamic evolution of coagulation in laser-heated tissue using a prostate model showed that photo-coagulation results in three unique regions of blood flow.[76] First, a superficial region of coagulation where blood flow is limited or negated due to thermal damage which is typically quite small. Second, a deeper thin ring of increased blood flow around the coagulated region known as the hyperemic region. Third, a deeper region where blood perfusion remains normal, and the tissue remains at 40°C. The hyperemic region acts as a heat sink, preventing an expanded region of coagulation and thermal damage.

3. Advances in Laser Tissue Welding: Exogenous Chromophores and Sealants

Exogenous chromophores for LTW can be selected or tuned based on a peak absorption wavelength that does not coincide with the natural chromophores in tissue. By using laser irradiation in the near infrared (NIR) therapeutic optical window,[55] heat is localized and peripheral tissue damage is limited.[1] The NIR region is the region of light with wavelengths longer than visible light, beginning roughly at the red end of the visible spectrum at 700 nm and extending to 1300 nm.

In the 1980s, solders were introduced to improve the tensile strength of laser-sealed tissue.[1] The first exogenous chromophores introduced for use in LTW were FDA-approved organic dyes for ocular surgeries. The most widely researched of this group is the FDA-approved dye indocyanine green (ICG) coupled with an 810 nm wavelength NIR laser. ICG has been widely researched due to high biocompatibility compared to other organic dyes.

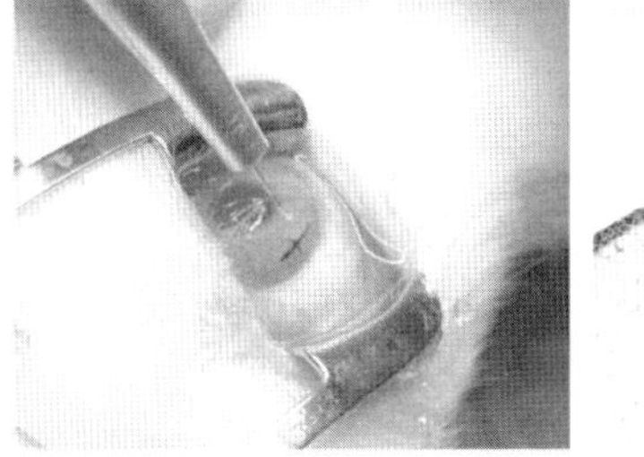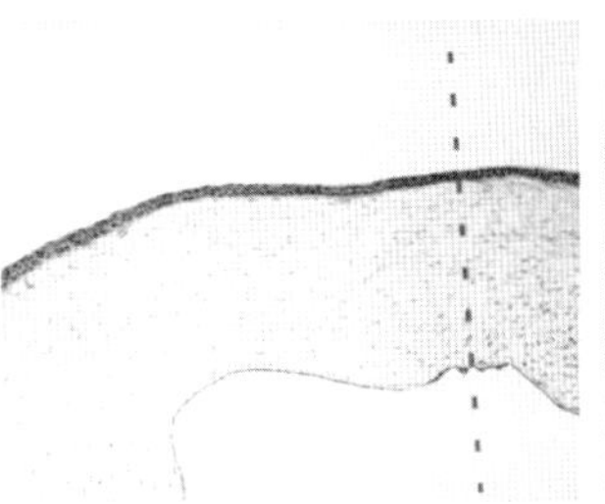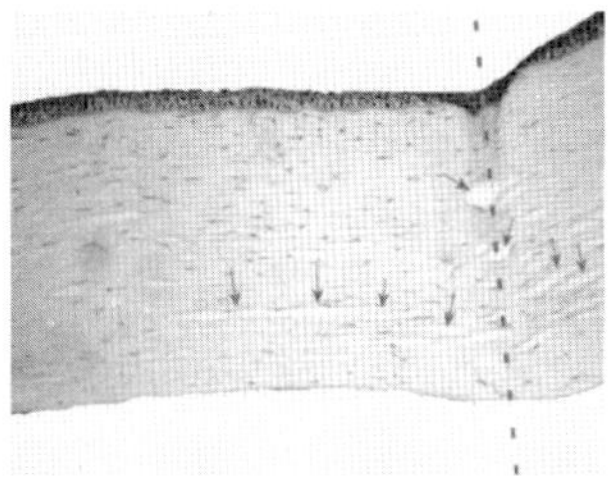

Figure 3. (Left) Welding a corneal cut held by a single stitch at the center. Histological section of (center) a laser-welded cornea and (Right) a sutured cornea at 30 days post operation. Large lacunae (arrows) are present in the sutured cornea but not seen in the welded cornea. Reproduced with permission from Ref. 52.

However, ICG and other organic dyes such as methylene blue and fluorescein suffer from a number of serious limitations. The dyes alone bleed into the tissue periphery, which may lead to increased thermal damage. These dyes have been shown to photobleach relatively quickly and are not stable over long periods of time.[64] Despite these concerns, direct LTW has shown promising results as discussed below.

Tests that simulated penetrating keratoplasty in four eyes of rabbits showed that ocular repair occurred more quickly and tissue restoration was more effective with ICG-stained tissue and laser treatment than with the suture control (Figure 3).[52] Tabakoğlu and Gülsoy compared closure techniques of rat dorsal skin.[47] Suturing, 809-nm laser with albumin-ICG solder, 980-nm laser without solder, and 1070-nm laser without solder were all compared over a 21-day recovery period. All welding techniques were successful at immediate skin closure (Figure 4), while suturing was not, and showed mechanically stronger closure than suturing at the end of the study.

Organic dyes do have a number of inherent advantages over metal nanoparticles: small molecular size and synthetic flexibility, and some research is exploring alternative small molecule chromophores to ICG that are not limited by low stability and rapid photothermal bleaching. For example, Spence *et al.* showed that croconaine dyes respond to NIR laser with unusually great stability and efficient heating, making them suitable candidates for LTW.[77]

Rose Bengal (RB) dye is unique in that it elicits a photochemical response rather than a photothermal response following laser irradiation, generating no significant heat in the process (Figure 5). While not following exactly the same mechanism as LTW, this "photochemical tissue bonding" (PTB) has shown promising results. Incisions in rat median nerves were repaired *in vivo* using a chitosan adhesive film mixed with RB and a 532 nm green laser. While the sutured group resulted in greater tensile strength than

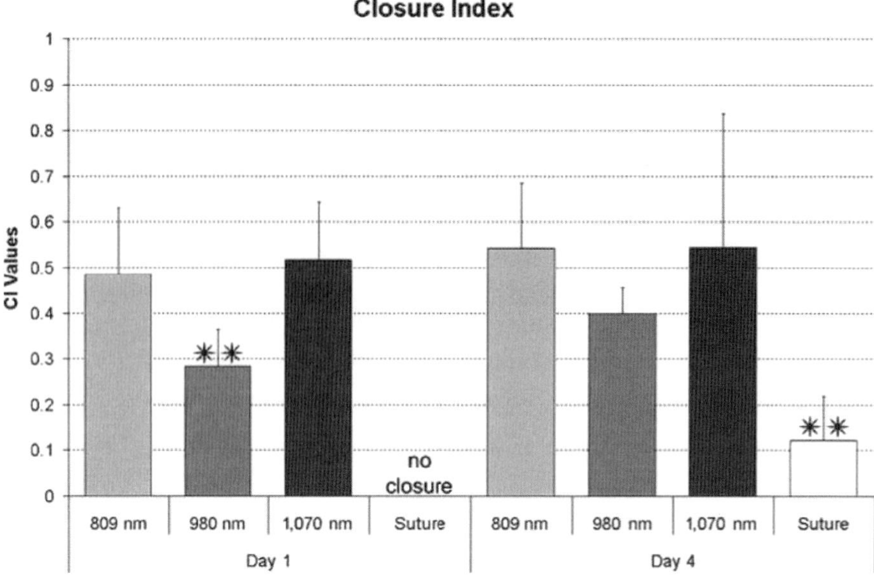

Figure 4. Closure index values of closing skin incisions using sutures or near infrared laser at 809 nm with solder, 980 nm, or 1070 nm. All laser modalities showed immediate closure at day 1. The 980-nm closure was statistically lower than 808-nm and 1070-nm (* *) ($p < 0.05$). On day 4 postrepair, the sutured group closure index was statistically lower than the laser-irradiated groups (* *) ($p < 0.05$). Reproduced with permission from Ref. 47.

RB–chitosan treatment after one week, this photochemical process resulted in better nerve realignment and organization compared to the disorganized suture repair.[78] Additionally, an *in vivo* study on the sciatic nerve of adult rats was performed using PTB and RB with green laser at 532 nm. Several combinations were tested with PTB, sutures, a human amniotic membrane wrap, and an autologous vein wrap. The study showed that PTB alone and PTB with the amnion wrap showed significant functional improvement over the suture alone.[79] In another study, *ex vivo* incised calf intestine was treated with a 532-nm laser and an adhesive albumin-RB film. The laser application significantly increased the adhesion strength and did not increase the average temperature of the adhesive. This is an example of PTB that causes crosslinking of fibers without producing heat.[80]

Inorganic nanoparticles including gold nanorods (GNRs) (Figure 6) are finding greater use as chromophores for LTW. While gold nanoparticles have been studied extensively, a fundamental examination of nanoparticle heating demonstrated that carbon black, single-walled carbon nanotubes, silver nanoparticles, and copper nanoparticles can all be used as NIR chromophores.[81] GNRs are being extensively researched in many medical applications, including laser welding.[22] GNRs are stable over long periods of time, do not leach easily

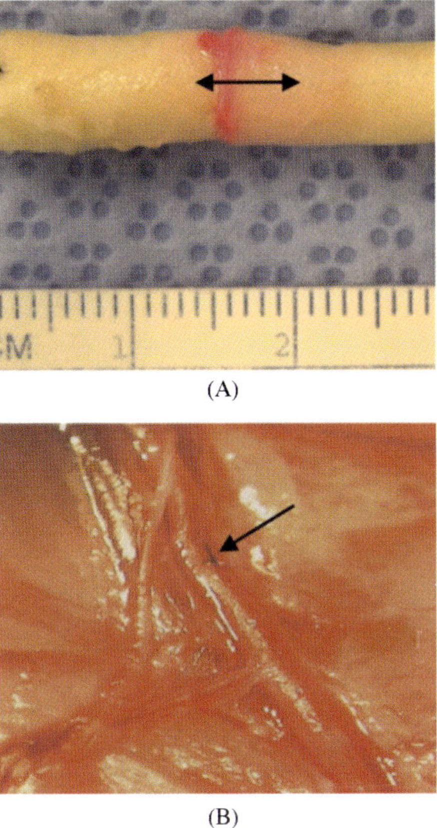

(A)

(B)

Figure 5. (A) *Ex vivo* porcine carotid artery closure by PTB. Arrow indicates are of artery overlap. (B) *In vivo* rat femoral artery 6 h following repair. Arrow shows the repair site with a marking stay suture. Reproduced with permission from Ref. 25.

due to their larger size, facilitate efficient photothermal conversion, and do not undergo photobleaching.[64] GNRs are widely employed due to tunable shape, functionalization,[82] and localized heat.[83,84] Light absorption by GNRs can be tuned to a maximum absorbance wavelength by adjusting the size ratio of the nanoparticles. Additionally, the surface of these nanoparticles can easily be modified to shield the nanoparticles from the immune system, alter their properties, and/or to conjugate bioactive molecules.

GNRs convert light to heat through collective electron oscillations.[16] Various groups have calculated that roughly five orders of magnitude fewer GNRs are required to produce the same amount of heat as ICG, making GNRs a million-fold more effective NIR light absorbers. Colloidal GNRs were used to directly stain porcine eye lens capsules *ex vivo* at the interface between a patch from a donor eye and the treated eye. Laser pulses (40 msec) locally denatured the ocular collagen, reaching temperatures above 50°C

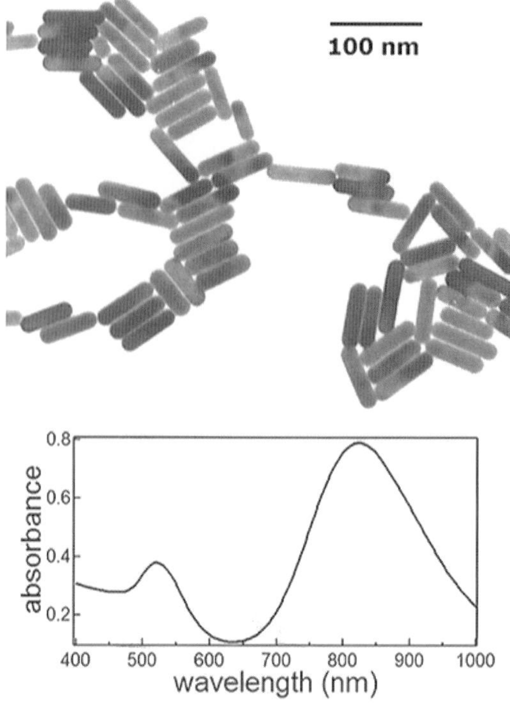

Figure 6. (Top) Transmission electron microscopy image of GNRs and (Bottom) gold nanorod absorption spectra. Reproduced with permission from Ref. 16.

confined to within 70 μm from the area of exposure (Figure 7),[16] showing the extent to which heat can be localized. In another study, the hetero-transplantation of *ex vivo* porcine lens capsules that were soaked in a collagen–GNR mixture and welded with an 810 nm diode laser was performed. Results showed successful welding at a heat flux of less than half of the tissue collateral damage threshold, a roughly two-fold improvement over ICG.[85]

Indirect LTW or LTS employs an exogenous chromophore within a liquid, semi-solid, or solid solder. Laser light induces conformational changes in collagen in many different types of tissue. In the cochlea basil membrane for example, delayed effects of laser radiation on collagen result in increased tissue stiffness.[86] LTS can potentially avoid these effects of tissue exposure to laser. These solders are typically polymers with high biocompatibility. This avoids direct staining and irradiation of the tissue and employs wavelengths in the therapeutic optical window.

Liquid solders with organic dyes are reported to be difficult to handle[87] and can lead to inconsistent results in weld strength. Albumin is among the most commonly studied solder material and is often coupled with ICG and

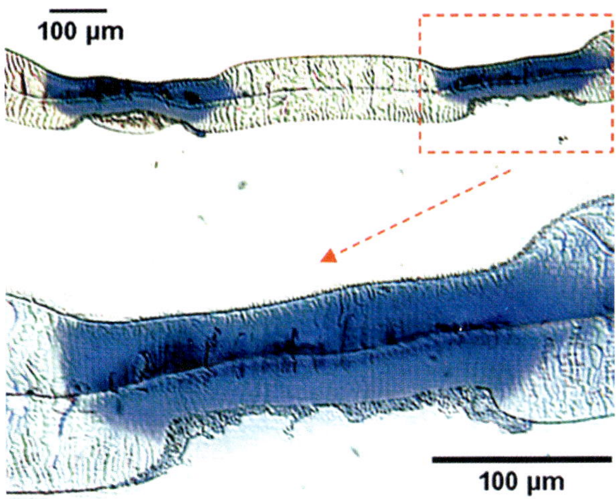

Figure 7. Histological section of spot laser welding of sandwiched anterior lens capsules with toluidine blue stain. The results from each spot are reproducible. Reproduced with permission from Ref. 16.

NIR laser at 800–810 nm. Higher concentrations of albumin (semi-solid solders) correlate with increased bond strength and higher burst threshold pressures.[54] On the other hand, solid thin films can be inflexible, and are not readily able to incorporate into or adapt to the tissue geometry.[15]

An *ex vivo* study on porcine aortas utilized laser-assisted vascular welding. Poly(ε-caprolactone) (PCL) and poly(lactic-co-glycolic acid) (PLGA) scaffolds were created with semi-solid albumin solder and albumin solder containing hydroxypropylmethylcellulose (HPMC) to improve bonding. A 670 nm diode laser was used, and the breaking strength, resilience under hydration, and thermal damage were assessed. PLGA welds were stronger in terms of breaking strength, but did not hold up well with hydration. The addition of HPMC increased the breaking strength of PLGA welds after hydration. Pulsed, as opposed to continuous wave, laser inflicted less tissue damage.[88] This work highlights the possibility of using multi-phase polymer composites to combine the ease and reproducibility of a solid solder with the wound integration of a viscous solder.

Chitosan is another solder matrix material being explored in LTW.[55,80] Chitosan is biocompatible, inexpensive, and can be processed into very thin transparent films. In one study, chitosan films embedded with GNRs were placed on rabbit tendons and irradiated with pulsed-wave 810-nm diode laser. Strong adhesion was measured using tensile strength tests.[49] Matteini *et al.* sealed rabbit carotid artery *in vivo* with an 810 nm diode laser and a

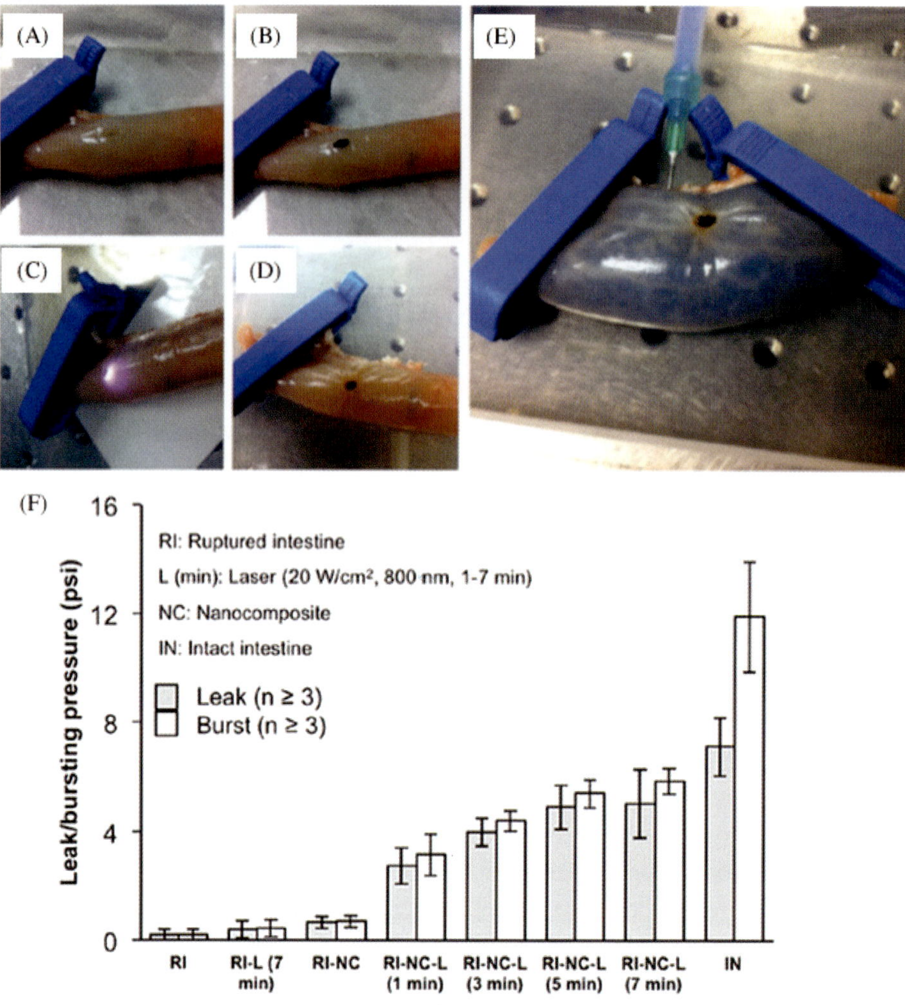

Figure 8. Leakage and burst pressures of welded intestine. (A) 5 mm incision made in the intestine. (B) Solder placed over incision, and (C) irradiated with a laser to (D) create a seal. (E) The leaking and bursting pressures were measured and reported. (F) Bursting and leakage pressures of tissues before and after LTW using nanocomposite solders. Reproduced with permission from Ref. 40.

hyaluronan-GNR composite. Immediately after the procedure, there was no bleeding and artery patency was confirmed. Follow-up showed patent arteries and no adverse host reaction to the soldering material. Hyaluronan provided stability to the solder.[89]

We employed thin elastin-like polypeptide (ELP) matrices cross-linked with GNRs to weld ruptured porcine intestine with 800-nm laser irradiation. Fluid-tight seals in the intestine were created (Figure 8), and bacterial leakage across the weld did not occur, indicating the application of these solders in preventing anastomotic leakage.[40] Collagen is vital in wound healing and mimics the ECM,[90,91] with low immunogenicity.[92] Previously, collagen applied by laser improved wound healing and reduced scarring in skin.[93] Our group additionally explored collagen-GNR thin hydrogel solders *ex vivo* in porcine intestine with 800-nm laser. The burst pressure of welded intestine reached up to 64% of that of native intestine, while greatly minimizing laser power density compared to other studies. Pulsed wave laser was able to create robust welds, while maintaining lower temperatures at the solder-tissue interface than continuous wave laser.[41]

The success of these improvements in chromophores and solders highlights the need for further development of next generation solders for LTW. Multi-wavelength chromophore systems, improved composite adhesiveness, drug loading, photochemical bonding, and plasmonic nanoparticles all have the potential to drive enhancement of LTW results.

4. Competing Sutureless Techniques

Different types of sutureless tissue fusion methods for wound healing include surgical sealants or adhesives and direct heat tissue compression devices. While a detailed review of these techniques cannot be given here, a brief and qualitative comparison to LTW is described here.

Adhesives, including cyanoacrylate or fibrin glues, often suffer from toxicity, brittleness, low wound strength (especially in wet wound settings), long polymerization times, difficult application, wound reopening, and poor uniformity. Some thin film adhesives are able to create an instant seal and prevent bacterial leakage but may not be suitable for load-bearing tissue, vasculature, nerves, or muscle. Alternatively, direct heat tissue fusion devices require a more available wound site and lead to greater and more widespread thermal damage. The tissue is sealed by clamping wound edges together and directly applying heat to the clamp.[94]

LTW was compared to fibrin glue in the closure of *ex vivo* lung and tracheal incisions on rabbits. An 808 nm diode laser was used with a solder composed of bovine serum albumin, ICG, and either hyaluronic acid or chitosan (gelling agents). LTW provided burst strengths that were twice as high as fibrin glue. Solders composed of chitosan required much less laser time than hyaluronic acid solders.[95]

5. Outlook and Future Progress

The chromophores, solders, and procedures employed for LTW have changed very little since the 1990s. Despite this, the results from LTW show that this technique has great potential for clinical translation and application. The more recent work highlighted in this chapter shows that technical advances in exogenous chromophores, tissue solders, and heat management will continue to improve LTW results.

Acknowledgements

The authors are grateful to the National Institute of Biomedical Imaging and Bioengineering, NIH (Grant R01 EB020690-01) and the ASU Fulton Undergraduate Research Initiative (FURI) award (to T.F.) for funding.

References

1. Schalow, E. L. and Kirsch, A. J. Laser tissue soldering: Applications in the genitourinary system. *Curr Urol Rep* **4**, 56–59 (2003).
2. Cilesiz, I. Controlled temperature photothermal tissue welding. *J Biomed Opt* **4**, 327–336 (1999).
3. Asencio-Arana, F. Effects of a low-power He–Ne laser on the healing of experimental colon anastomoses: Our experience. *Opt Eng* **31**, 1452 (1992).
4. Lauto, A. *et al.* Albumin-genipin solder for laser tissue repair. *Lasers Surg Med* **35**, 140–145 (2004).
5. Lauto, A. Mawad, D. and Foster, L. J. R. Adhesive biomaterials for tissue reconstruction. *J. Chem. Technol Biotechnol* **83**, 464–472 (2008).
6. Pecha, R. E. *et al.* Gastrointestinal hemorrhage consequent to foreign body reaction to silk sutures: Case series and review. *Gastrointest Endosc* **48**, 299–301 (1998).
7. Millesi, H. Peripheral nerve injuries. Nerve sutures and nerve grafting. *Scand. J Plast Reconstr* (1981). Available at: http://europepmc.org/abstract/med/6750775.
8. Hida, K. *et al.* Nonsuture dural repair using polyglycolic acid mesh and fibrin glue: Clinical application to spinal surgery. *Surg Neurol* **65**, 136–142 (2006).
9. Cimador, M. Castagnetti, M. and Milazzo, M. Suture materials: Do they affect fistula and stricture rates in flap urethroplasties? *Urol Int* **73**(4), 320–324 (2004). Available at: http://www.karger.com/Article/Abstract/81592.
10. Chelala, E. *et al.* The suturing concept for laparoscopic mesh fixation in ventral and incisional hernia repair: Mid-term analysis of 400 cases. *Surg Endosc* **21**, 391–395 (2007).
11. Murray, K. and Ho, C. The influence of pulmonary staple line reinforcement on air leaks. *Chest* **122**(6), 2146–2149 (2002). Available at: http://journal.publications.chestnet.org/article.aspx?articleid=1081108.
12. Lumsden, A. B. and Heyman, E. R. Prospective randomized study evaluating an absorbable cyanoacrylate for use in vascular reconstructions. *J Vasc Surg* **44**, 1002–1009 (2006).

13. Garcia, P. *et al.* Robotic laser tissue welding of sclera using chitosan films. *Lasers Surg Med* **41**, 59–67 (2009).

14. Sheffy, N. Mintz, Y. Rivkind, A. and Shapira, S. Terror-related injuries: A comparison of gunshot wounds versus secondary-fragments — induced injuries from explosives. *J Am Coll Surg* **203**, 297–303 (2006).

15. Rossi, F. *et al.* Experimental study on laser assisted vascular repair and anastomosis with ICG-infused chitosan films. in *IEEE–Biophotonics* 1–3 (2011). Available at: http://ieeexplore.ieee.org/xpls/abs_all.jsp?arnumber=5954804.

16. Ratto, F. *et al.* Photothermal effects in connective tissues mediated by laser-activated gold nanorods. *Nanomedicine* **5**, 143–151 (2009).

17. Thomsen, S. Pathologic analysis of photothermal and photomechanical effects of laser–tissue interactions. *Photochem Photobiol* **53**(6), 825–835 (1991). Available at: http: //onlinelibrary.wiley.com/doi/10.1111/j.1751-1097.1991.tb09897.x/full.

18. Jain, K. and Gorisch, W. Repair of small blood vessels with the Neodymium-YAG laser: A preliminary report. *Surgery* **85**(6), 684–688 (1979). Available at: http: //www.surgjournal.com/article/0039-6060(79)90157-0/abstract.

19. Abergel, R. and Meeker, C. Nonthermal effects of Nd: YAG laser on biological functions of human skin fibroblasts in culture. *Lasers Surg* **3**(4), 279–284 (1984). Available at: http: //onlinelibrary.wiley.com/doi/10.1002/lsm.1900030403/full.

20. Rabi, Y. and Katzir, A. Temporal heating profile influence on the immediate bond strength following laser tissue soldering. *Lasers Surg Med* **42**, 425–432 (2010).

21. Wolf -de Jonge, I. C. D. Y. M. Beek, J. F. and Balm, R. 25 Years of laser assisted vascular anastomosis (LAVA): What have we learned? *Eur J Vasc Endovasc Surg* **27**, 466–476 (2004).

22. Gobin, A. M. *et al.* Near infrared laser-tissue welding using nanoshells as an exogenous absorber. *Lasers Surg Med* **37**, 123–129 (2005).

23. Kirsch, A. Duckett, J. and Snyder, H. Skin flap closure by dermal laser soldering: A wound healing model for sutureless hypospadias repair. *Urology* **50**(2), 263–272 (1997). Available at: http: //www.sciencedirect.com/science/article/pii/S0090429597002781.

24. Kirsch, A. J. *et al.* Laser tissue soldering in urinary tract reconstruction: First human experience. *Urology* **46**, 261–266 (1995).

25. O'Neill, A. C. *et al.* Microvascular anastomosis using a photochemical tissue bonding technique. *Lasers Surg Med* **39**, 716–722 (2007).

26. Xie, H. Bendre, S. C. Burke, A. P. Gregory, K. W. and Furnary, A. P. Laser-assisted vascular end to end anastomosis of elastin heterograft to carotid artery with an albumin stent: A preliminary *in vivo* study. *Lasers Surg Med* **35**, 201–205 (2004).

27. Libutti, S. Oz, M. Forde, K. and Auteri, J. Canine colonic anastomoses reinforced with dye-enhanced fibrinogen and a diode laser. *Surg* **4**(2), 97–99 (1990). Available at: http: //link.springer.com/article/10.1007/BF00591269.

28. Spector, D. *et al. In vitro* large diameter bowel anastomosis using a temperature controlled laser tissue soldering system and albumin stent. *Lasers Surg Med* **41**, 504–508 (2009).

29. Züger, B. J. *et al.* Laser solder welding of articular cartilage: Tensile strength and chondrocyte viability. *Lasers Surg Med* **28**, 427–434 (2001).

30. Simhon, D. *et al. In vivo* laser soldering of incisions in juvenile pig skins using GaAs or CO lasers and a temperature control system. *Proc. SPIE-5-312, Lasers in Surgery: Advanced Characterization, Therapeutics, and Systems* XIV, pp. 162–175 (2004).

31. Yang, P. Yao, M. DeMartelaere, S. L. Redmond, R. W. and Kochevar, I. E. Light-activated sutureless closure of wounds in thin skin. *Lasers Surg Med* **44**, 163–167 (2012).

32. Lobik, L. Ravid, A. and Nissenkorn, I. Bladder welding in rats using controlled temperature CO 2 laser system. *J Urology* **161**(5), 1662–1665 (1999). Available at: http://www.sciencedirect.com/science/article/pii/S0022534705690003.

33. Shumalinsky, D. and Lobik, L. Laparoscopic laser soldering for repair of uretero-pelvic junction obstruction in the porcine model. *J Endoural* **18**(2), 177–181 (2004). Available at: http://online.liebertpub.com/doi/abs/10.1089/08927790 4322959833.

34. Matteini, P. Rossi, F. Menabuoni, L. and Pini, R. Microscopic characterization of collagen modifications induced by low-temperature diode-laser welding of corneal tissue. *Lasers Surg Med* **39**, 597–604 (2007).

35. Wadia, Y. Xie, H. and Kajitani, M. Liver repair and hemorrhage control by using laser soldering of liquid albumin in a porcine model. *Lasers Surg Med* **27**, 319–328 (2000).

36. Solhpour, S. Weldon, E. Foster, T. and Anderson, R. Mechanism of thermal tissue welding (Part 1). *Lasers Surg Med Suppl* **6**(295) 56 (1994). Available at: https://scholar.google.com/scholar?q=Mechanism+of+thermal+tissue+welding+%28Part+1%29.+&btnG=&hl=en&as_sdt=1%2C3#0.

37. Ngeow, W. C. Scar less: A review of methods of scar reduction at sites of peripheral nerve repair. *Oral Surg Oral Med Oral Pathol Oral Radiol Endodontology* **109**, 357–366 (2010).

38. Gayen, T. and Katz, A. Near-infrared laser welding of aortic and skin tissues and microscopic investigation of welding efficacy. *Biomed* **182** (2003). Available at: http://proceedings.spiedigitallibrary.org/proceeding.aspx?articleid=1313847.

39. Mercer, C. Minich, P. and Pauli, B. Sutureless bowel anastomosis using Nd: YAG laser. *Lasers Surg Med* **7**(6), 503–506 (1987). Available at: http://onlineli brary.wiley.com/doi/10.1002/lsm.1900070612/abstract.

40. Huang, H.-C. Walker, C. R. Nanda, A. and Rege, K. Laser Welding of Ruptured Intestinal Tissue Using Plasmonic Polypeptide Nanocomposite Solders. *ACS Nano* **7**, 2988–2998 (2013).

41. Urie, R. Quraishi, S. Jaffe, M. and Rege, K. Gold Nanorod–Collagen Nanocomposites as photothermal nanosolders for laser welding of ruptured porcine intestines. *ACS Biomater Sci Eng* **1**, 805–815 (2015).

42. Capon, A. *et al.* Laser assisted skin closure (LASC) by using a 815-nm diode-laser system accelerates and improves wound healing. *Lasers Surg Med* **28**, 168–175 (2001).

43. Atalay, M. *et al.* Heat shock proteins in diabetes and wound healing. *Curr Protein Pept Sci* **10**, 85 (2009).

44. Alster, T. and Zaulyanov-Scanlon, L. Laser scar revision: A review. *Dermatol Surg* **33**, 131–140 (2007).

45. Matteini, P. Ratto, F. Rossi, F. and Pini, R. Emerging concepts of laser-activated nanoparticles for tissue bonding. *J Biomed Opt* **17**, 0107011–0107019 (2012).

46. Bilici, T. Modulated and continuous-wave operations of low-power thulium (Tm: YAP) laser in tissue welding. *J Biomed Opt* **15**, 038001 (2010).

47. Tabakoğlu, H. Ö. and Gülsoy, M. *In vivo* comparison of near infrared lasers for skin welding. *Lasers Med Sci* **25**, 411–421 (2010).

48. Hu, L. *et al.* Closure of skin incisions by laser-welding with a combination of two near-infrared diode lasers: Preliminary study for determination of optimal parameters. *J Biomed Opt* **16**, 038001–038001 (2011).

49. Matteini, P. Ratto, F. Rossi, F. and Pini, R. Biofilms of chitosan-gold nanorods as a novel composite for the laser welding of biological tissue. *Proc. SPIE 7574, Nanoscale Imaging, Sensing and Actuation of Biomedical Applications* VII, 7574, p.4 (2010).

50. Sriramoju, V. and Alfano, R. R. Laser tissue welding analyzed using fluorescence, Stokes shift spectroscopy, and Huang-Rhys parameter. *J Biophotonics* **5**, 185–193 (2012).

51. Savage, H. E. *et al.* NIR laser tissue welding of *in vitro* porcine cornea and sclera tissue. *Lasers Surg Med* **35**, 293–303 (2004).

52. Rossi, F. and Pini, R. Experimental study on the healing process following laser welding of the cornea. *J Biomed Opt* **10**(2) (2005). Available at: http: //biomedicaloptics.spiedigitallibrary.org/article.aspx?articleid=1101911.

53. Li, C. Protsenko, D. E. Zemek, A. Chae, Y.-S. and Wong, B. Analysis of Nd: YAG laser-mediated thermal damage in rabbit nasal septal cartilage. *Lasers Surg Med* **39**, 451–457 (2007).

54. Bleier, B. Palmer, J. Sparano, A. and Cohen, N. Laser-assisted cerebrospinal fluid leak repair: An animal model to test feasibility. *Otolaryngo Head Neck Surg* **137**, 810–814 (2007).

55. Esposito, G. *et al. In vivo* laser assisted microvascular repair and end-to-end anastomosis by means of indocyanine green-infused chitosan patches: A pilot study: Laser Assisted Vascular Repair and Anastomosis. *Lasers Surg Med* **45**, 318–325 (2013).

56. Cezo, J. Kramer, E. Taylor, K. Ferguson, V. and Rentschler, M. Tissue fusion bursting pressure and the role of tissue water content. *Proc. SPIE 8584, Energy based Treatment of Tissue and Assessment* VII, 8584, p. 9 (2013).

57. Pabittei, D. R. *et al.* End-to-end scaffold-enhanced laser-assisted vascular anastomosis: *Ex vivo* proof-of-concept in porcine arteries. *J Vasec Surg* **62**(1) 200–209 (2013). Available at: http://dare.uva.nl/document/468488.

58. Constantinescu, M. A. *et al.* Effect of laser soldering irradiation on covalent bonds of pure collagen. *Lasers Med Sci* **22**, 10–14 (2007).

59. Murray, L. W. Su, L. Kopchok, G. E. and White, R. A. Crosslinking of extracellular matrix proteins: A preliminary report on a possible mechanism of argon laser welding. *Lasers Surg Med* **9**, 490–496 (1989).

60. Matteini, P. *et al.* Photothermally-induced disordered patterns of corneal collagen revealed by SHG imaging. *Opt Express* **17**, 4868–4878 (2009).

61. Matteini, P. *et al.* Thermal transitions of fibrillar collagen unveiled by second-harmonic generation microscopy of corneal stroma. *Biophys J* **103**, 1179–1187 (2012).

62. Small IV, W. Da Silva, L. B. and Matthews, D. L. Spectroscopic study of the effect of laser heating on collagen stability: Implications for tissue welding. *International Society for Optics and Photonics* 276–281 (1998). Available at: http://proceedings.spiedigitallibrary.org/proceeding.aspx?articleid=933496.

63. Jacques, S. and Prahl, S. Modeling optical and thermal distributions in tissue during laser irradiation. *Lasers Surg Med* **6**(6), 494–503 (1987). Available at: http://onlinelibrary.wiley.com/doi/10.1002/lsm.1900060604/full.

64. Hoffman, G. T. Byrd, B. D. Soller, E. C. Heintzelman, D. L. and McNally-Heintzelman, K. M. Alternative chromophores for use in light-activated surgical adhesives: Optimization of parameters for tensile strength and thermal damage profile. *International Society for Optics and Photonics* 174–181 (2003). Available at: http://proceedings.spiedigitallibrary.org/proceeding.aspx?articleid=1313840.

65. Matteini, P. Sbrana, F. Tiribilli, B. and Pini, R. Atomic force microscopy and transmission electron microscopy analyses of low-temperature laser welding of the cornea. *Lasers Med Sci* **24**(4) 667–671 (2009). Available at: http://link.springer.com/article/10.1007/s10103-008-0617-4.

66. Norman, G. Rabi, Y. Assia, E. and Katzir, A. *In vitro* conjunctival incision repair by temperature-controlled laser soldering. *J Biomed Opt* **14**, 064016 (2009).

67. Pabittei, D. R. *et al.* Optimization of suture-free laser-assisted vessel repair by solder-doped electrospun poly(ε-caprolactone) scaffold. *Ann Biomed Eng* **39**, 223–234 (2011).

68. Pabittei, D. R. *et al.* Laser-assisted vessel welding: A review on the current status and future outlook. (2013). Available at: http://dare.uva.nl/document/468489.

69. Sorg, B. S. and Welch, A. J. Preliminary biocompatibility experiment of polymer films for laser-assisted tissue welding. *Lasers Surg Med* **32**, 215–223 (2003).

70. Bass, L. S. and Treat, M. R. Laser tissue welding: A comprehensive review of current and future clinical applications. *Lasers Surg Med* **17**, 315–349 (1995).

71. D'Arcangelo, C. *et al.* A preliminary study of healing of diode laser versus scalpel incisions in rat oral tissue: A comparison of clinical, histological, and immunohistochemical results. *Oral Surgery, Oral Med Oral Pathol Oral Radiol Endodontology* **103**, 764–773 (2007).

72. Zhou, J. Chen, J. K. and Zhang, Y. Dual-phase lag effects on thermal damage to biological tissues caused by laser irradiation. *Comput Biol Med* **39**, 286–293 (2009).

73. Liu, K.-C. and Wang, J.-C. Analysis of thermal damage to laser irradiated tissue based on the dual-phase-lag model. *Int J Heat Mass Transf* **70**, 621–628 (2014).

74. Zhou, J. Zhang, Y. and Chen, J. K. Non-fourier heat conduction effect on laser-induced thermal damage in biological tissues. *Numer Heat Transf Part A Appl* **54**, 1–19 (2008).

75. Welch, A. J. The thermal response of laser irradiated tissue. *IEEE J Quantum Electron* **20**, 1471–1481 (1984).

76. Kim, B.-M. Jacques, S. L. Rastegar, S. Thomsen, S. and Motamedi, M. Nonlinear finite-element analysis of the role of dynamic changes in blood perfusion and optical properties in laser coagulation of tissue. *Sel Top Quantum Electron IEEE J* **2**, 922–933 (1996).

77. Spence, G. T. Hartland, G. V. and Smith, B. D. Activated photothermal heating using croconaine dyes. *Chem Sci* **4**, 4240 (2013).

78. Barton, M. *et al.* Laser-activated adhesive films for sutureless median nerve anastomosis: Laser-activated adhesive films. *J Biophotonics* **6**, 938–949 (2013).

79. O'Neill, A. C. *et al.* Photochemical sealing improves outcome following peripheral neurorrhaphy. *J Surg Res* **151**, 33–39 (2009).

80. Lauto, A. *et al.* Photochemical tissue bonding with chitosan adhesive films. *Biomed Eng Online* **9**, 47 (2010).

81. Schrand, A. M. Stacy, B. M. Payne, S. Dosser, L. and Hussain, S. M. Fundamental examination of nanoparticle heating kinetics upon near infrared (NIR) irradiation. *ACS Appl Mater Interfaces* **3**, 3971–3980 (2011).

82. Brust, M. Fink, J. and Bethell, D. Synthesis and reactions of functionalised gold nanoparticles. *J Chem* **16**, 1655–1656 (1995). Available at: http://pubs.rsc.org/en/content/articlehtml/1995/c3/c39950001655.

83. Huang, H.-C. Rege, K. and Heys, J. J. Spatiotemporal temperature distribution and cancer cell death in response to extracellular hyperthermia induced by gold nanorods. *ACS Nano* **4**, 2892–2900 (2010).

84. Daniel, M. and Astruc, D. Gold nanoparticles: Assembly, supramolecular chemistry, quantum-size-related properties, and applications toward biology, catalysis, and nanotechnology. *Chem Rev* **104**(1), 293–346 (2004). Available at: http://pubs.acs.org/doi/abs/10.1021/cr030698+.

85. Ratto, F. *et al.* Gold nanorods as exogenous chromophores in the welding of ocular tissues in (eds. Manns, F. Söderberg, P. G. Ho, A. Stuck, B. E. & Belkin, M.) 68441V–68441V–6 (2008).

86. Wenzel, G. I. Anvari, B. Mazhar, A. Pikkula, B. and Oghalai, J. S. Laser-induced collagen remodeling and deposition within the basilar membrane of the mouse cochlea. *J Biomed Opt* **12**, 021007 (2007).

87. Bleustein, C. and Walker, C. Semi-solid albumin solder improved mechanical properties for laser tissue welding. *Lasers Surg* **27**(2), 140–146 (2000). Available at:

http://onlinelibrary.wiley.com/doi/10.1002/1096-9101(2000)27: 2%3C140: AID-LSM5%3E3.0.CO;2-C/full.

88. Pabittei, D. R. *et al.* *In vitro* laser-assisted vascular welding: optimization of acute and post-hydration welding strength. *Sci Technol Adv Mater* (2013). Available at: http://dare.uva.nl/document/468487.

89. Matteini, P. *et al.* *In vivo* carotid artery closure by laser activation of hyaluronan-embedded gold nanorods. *J Biomed Opt* **15**, 041508 (2010).

90. Shoulders, M. D. and Raines, R. T. Collagen Structure and Stability. *Annu Rev Biochem* **78**, 929–958 (2009).

91. Fang, M. Yuan, J. Peng, C. and Li, Y. Collagen as a double-edged sword in tumor progression. *Tumor Biol* **35**, 2871–2882 (2014).

92. Alencar, H. Funovics, M. and Figueiredo, J. Colonic adenocarcinomas: near-infrared microcatheter imaging of smart probes for early detection — study in mice 1. *Radiology* **224**(1) (2007). Available at: http://pubs.rsna.org/doi/abs/10.1148/radiol.2441052114.

93. Steinstraesser, L. *et al.* Laser-mediated fixation of collagen-based scaffolds to dermal wounds. *Lasers Surg Med* **42**, 141–149 (2010).

94. Cezo, J. D. Passernig, A. C. Ferguson, V. L. Taylor, K. D. and Rentschler, M. E. Evaluating temperature and duration in arterial tissue fusion to maximize bond strength. *J Mech Behav Biomed Mater* **30**, 41–49 (2014).

95. Bleier, B. S. Laser tissue welding in lung and tracheobronchial repair: An animal model. *Chest J* **138**, 345 (2010).

12. Bioprinting for Wound Healing Applications

Aleksander Skardal, Sean Murphy, Anthony Atala and Shay Soker

Wake Forest Institute for Regenerative Medicine,
Wake Forest School of Medicine,
Medical Center Boulevard, Winston-Salem, NC 27157

Abstract

Bioprinting technology has emerged in recent years as an attractive method for biofabrication of 3D tissues and organs. The widespread utility of this promising technology has resulted in its exploration and implementation in a variety of regenerative medicine applications, including the production of skin or treatment of wounds for patients with skin wounds or burns. Bioprinting can take the shape of several printing modalities, including inkjet printing, extrusion-based printing, and laser-induced forward transfer printing, each of which has their own pros and cons, and lend themselves to different approaches and applications. Using these technologies, researchers can leverage engineered biomaterials and potent biological components such as stem cells or cytokines to produce effective therapies to treat patients. This chapter discusses skin regeneration through bioprinting, and the multiple technologies employed for skin regeneration, including the development bioprinter device and their general requirements, biological components that are printed, and the biomaterials and matrices that are integral in facilitating integration between the hardware, biologicals, and the patient.

1. Introduction and Current State of Care for Wound Healing

1.1 *Current wound healing treatment practices*

Extensive burns and full thickness skin wounds can be devastating to many patients, even when treated quickly in the clinic. Unfortunately, there are an estimated 500,000 burns that are treated in the US every year.[1,2] The overall mortality rate for burn injury was 4.9% between 1998–2007 and medical costs associated with burn treatments approach $2 billion per year.[3] Globally, there are over 11 million burn injuries per year, creating a significant demand for improved therapies.[4] In addition to burn wounds, full-thickness chronic skin wounds constitute a substantial patient base, and despite scientific advancement of therapies, less than 50% of these wounds are treated successfully.[5] These non-healing wounds are estimated to effect over 7 million people per year in the US alone, with annual expenses approaching $25 billion.[6] Patients who suffer from these types of skin injuries respond best when rapid treatments are available that result in closure and protection of the wounds as fast as possible. Patients who receive delayed treatments, or under-performing treatments, often are subject to extensive scarring that can result in long-term physiological defects such as disfigurements and loss of range of motion.

The gold standard therapy employed in the clinic is the autologous split-thickness skin graft. This requires removing a piece of skin from a secondary surgical site on that very patient, stretching the skin under sterile conditions, and re-applying the graft at a later time. Obviously, this approach has limitations when dealing with large or severe wounds, where secondary surgical sites may be limited, resulting in autografts being unusable in cases that require prompt, aggressive, and large-scale treatment measures to maintain the lives of wounded patients. Allografts, or skin grafts from other patients, are an additional option, but in many cases can suffer from the need of immunosuppressive drugs to prevent immune rejection of the allograft. As with any transplant surgery using allogenic tissues, immunosuppressive drugs can have a variety of side effects on a patient, some of which can be negative. These limitations have led to the development of non-cellular dermal substitutes, which are generally a polymer scaffold-based membrane. Examples include INTEGRA® Dermal Regeneration Template and Biobrane®, and while such materials often result in improved healing over untreated controls,[7,8] they are costly to produce and still result in relatively poor cosmetic outcomes compared to autografts. Furthermore, patient survival is inversely proportional to time required to cover and stabilize a wound in the clinic. Patients with burns greater than 15–20% of total body surface area are likely to go into shock without rapid treatment, making rapid procurement of an optimal graft type or application of a therapeutic material quickly absolutely integral to successful treatment and patient state. Without sufficient and rapid fluid resuscitation, patient conditions deteriorate quickly and mortality rates rapidly increase.[9] Inadequate treatment regimes will result in long-term complications for patients, including open wounds, pain, problems with temperature sensation and touch, prominent scarring, and itching.[10]

1.2 *Dressings and hydrogels*

Development of more modern, functional approaches that permit immediate burn wound stabilization and support functional skin regeneration are needed. Current commercially available hydrogel-based dressings have advantages in that they are often immunologically inert, can aid with regulation of fluid and gas exchange from the wound surface, and can be purchased as gels, sheets, and even impregnated in ordinary cotton gauze pads.[8,11] For relatively regular, shallow, and flat wounds, hydrogel sheets are quite easily employed as primary dressings and will remain in place for 4–7 days before fresh dressings are applied. For irregular and/or deep wounds, amorphous hydrogels can be used to fill the wound, but must be held in place with a secondary dressing or bandage that is usually changed once daily. While these materials have advantages over some traditional wound care regimes, the potential for hydrogels to promote wound stabilization and regenerative healing is not limited to inert wound dressings. Biofunctionalized or naturally derived hydrogel biomaterials and scaffolds can better interact with the biology of the wound environment, integrating and influencing cellular behavior and thus improving skin regeneration. Many of these new materials are being developed in tandem with technologies to improve delivery to the wound, such as cell sprayers and bioprinting devices.

1.3 *Bioprinting — A new technology for wound care*

Bioprinting has emerged as an exceedingly flexible tool in the broad field of regenerative medicine with potential in a wide variety of applications. Bioprinting has been described as a robotic additive biofabrication approach that has the potential to build organs or viable tissues.[12] In general, bioprinting uses a computer controlled 3D printing device to accurately deposit cells and biomaterials into precise and sometimes intricate geometries in order to fabricate anatomically correct tissue-like structures. These printing devices have the capability to print droplets of cells, cell aggregates, or cells encapsulated in hydrogels, — the "bio-ink", as well as cell-free polymers that provide structure.[13,14] 3D computer-assisted designs, or blueprints, can be used to guide the placement of specific types of cells and polymers into the appropriate geometries that mimic actual tissue construction, which can then be matured into a tissue or organ construct.[15,16] To date, complete organs are still difficult to create, but creation of a whole functional human organ remains the primary long-term goal of bioprinting. A number of bioprinting approaches have been recently explored, many dealing the production of vascular structures, which are necessary if larger organs are to be fabricated. Cell aggregates and cell rods have been printed layer-by-layer into tubular formations, showing the feasibility of one method for printing cellularized vascular structures.[17–20] After printing, the property of tissue liquidity allows these aggregates and rods to fuse into singular seamless structures.[21,22] Other approaches have relied on the development of hydrogels with specific mechanical properties and cross-linking chemistries to facilitate extrusion from printing devices to build other cellularized tubular structures.[23,24] An area that as recently been explored is the application of bioprinting as a technique for treating open wounds and burns. Our team is currently utilizing novel bio-printing technology[25] to aid in the healing of dermal wounds and full thickness burns by applying a layered gel and cells to the wound to promote appropriate cell regeneration and reduce inflammation and scarring.[26]

1.4 *New wound healing materials*

Rather than passive and inert materials that have minimal contribution to the regenerative process, we can bioprint hydrogels that actively promote wound healing by either acting as a substrate for endogenous cell migration and proliferation, or as a supportive delivery vehicle of potent cells and cytokines. In the bioprinting applications described below, the printer dispenses hydrogel bio-ink to fill the wound volume. Importantly, successful bioprinting requires a hydrogel with specific properties, including a fast gelling time (< 2 min) with a cross-linking mechanism that supports *in situ* cross-linking and stabilization at the wound site. The hydrogel must not only be non-immunogenic, but also able to facilitate desired cellular activities such as migration and secretion of trophic factors. We have explored promotion of endogenous cell migration and proliferation, as well as inclusion of autologous or allogeneic cell sources or cell or tissue-derived biologicals, employing the hydrogel bio-ink as a delivery vehicle as well. In the end, a hydrogel bioink needs to have suitable mechanical properties and swelling characteristics, shortterm stability, be biodegradable over the long-term, promote cell migration, proliferation, and function, and facilitate robust engraftment in the endogenous wound site. An appropriate hydrogel bio-ink must also eventually meet appropriate regulatory, commercial, and financial considerations to provide affordable products that can be FDA approved, and be employed easily enough to be accepted by clinicians as usable and effective therapy in the clinic.

2. Bioprinting

2.1 *Hardware for complex control in 3D space*

Bioprinting is tied to the hardware capabilities of the devices. In particular, the movement capabilities in 3D and deposition methodology, often referred to as the type of printing modality (Figure 1).

Movement hardware: Parameters for the movement system in a bioprinting platform include movement range, spatial precision, spatial accuracy, and movement speed. Movement range is the total distance the movement system can travel, spatial accuracy is the smallest distance that the movement system can travel, and lastly, spatial precision describes the error in spatial accuracy, which can be influenced by a variety of factors, including hardware-related, software-related, or environmental. Some sources of spatial

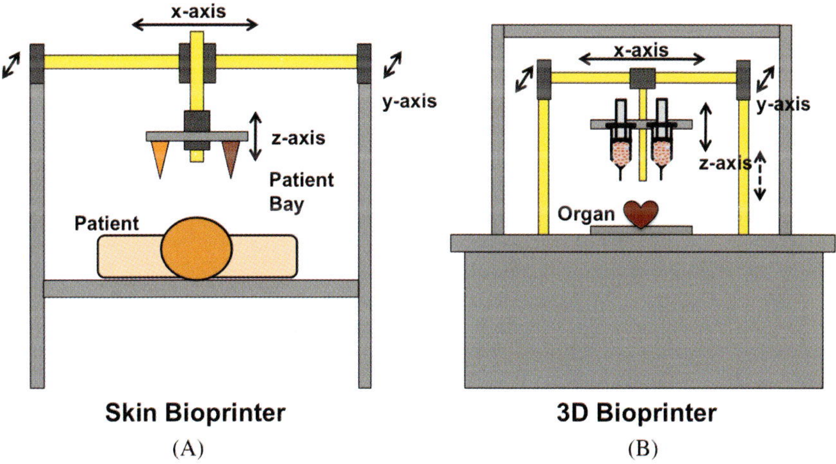

Figure 1. Typical bioprinter hardware. (A) A skin bioprinter is a large device, with room for a patient to lay down below the printing mechanisms in order to support *in situ* skin bioprinting. Movement is facilitated by building rails into the device along which other rails or the print head array can travel. Alternatively, (B) a more traditional 3D bioprinter used for biofabricating organs or tissue constructs is often a desktop device with a smaller footprint since it does not need to have room for an actual patient. Instead, a stage exists, on which the construct is printed. The movement system can be similar to that of the skin bioprinter, employing rails or tracks to support *X–Y–Z* motion in 3D space.

errors include vibrations, drop distance changes, and weight applied to various parts of the system (e.g. patient or target printing area on a moveable stage, size of printing bioink reservoir, etc). Vibrations caused by the movement of hardware during printing can create small errors in the final bioprinted object. Often, these vibrations can be ignored if the resolution does not need to be intricate, but smaller constructs or intricate regions of constructs, requiring micron precision can negatively be effected by vibration-causing imperfections. Drop distance error refers to errors spawned from the printhead being held at a different distance than intended above its target. Any discrepencies in the travel path of a printed droplet will be magnified as the distance between the stage and printhead increases. Weight applied to the system can affect the distance or speed the movement system travels due to changes in resistance or motor capabilities.

Modifying existing desktop printer hardware: Initial work in bioprinting was explored using modified desktop inkjet printers.[27,28] However, there are several limitations in employing a desktop inkjet printer as the hardware platform for a skin bioprinter. First, the movement range of an inkjet printer is severely limited due to the printhead being fixed on two axes. Thus, the size of a printable skin construct or size of wound that can be covered is limited to the lengths of travel of the printhead because this length is fixed. Spatial accuracy and spatial precision is limited based on the type of inkjet printer cartridge associated with that printer. As such, researchers must consider whether modification of the cartridges can overcome these limitations, or if other hardware platforms are more appropriate. Movement speed is excellent along one axis because this type of system leverages the pre-existing hardware of the inkjet printer. Given these limitations, we have found that modified inkjet printers are most useful for proof-of-concept testing, but more customized and built from the ground platforms are more useful for creating *in vitro* or *in vivo* skin constructs, as well as other 3D tissue constructs.

Transitioning from 2D to 3D bioprinting: Desktop inkjet bioprinters modifications can create 2D skin constructs with relatively few modifications, besides cartridge modifications. However, creating a 3D construct requires the addition of a Z-axis.[27,29–31] The addition of a Z-axis allows the printer to create multiple stacked 2D layers in succession, building up a 3D construct one layer at a time. Printing successive layers requires more significant modification, which amounts to building a Z-axis supported stage or printhead. This task is not trivial, and renders the ease of modification of a desktop inkjet printer much more complicated. Therefore, many researchers and engineers began building entirely customized printing platforms with 3-axis movement capability as a basic function.

Movable stage bioprinting: One approach to building a bioprinting system with 3-axis control, is rendering the printeheads stationary and allowing the target region to move beneath the printheads. In these types of system, the X- and Y-axes operate a stage that contains the bioprinting area and the Z-axis may be either part of the stage or part of the printhead apparatus. The system is calibrated to the printhead in relation to the stage and moves the stage along the X- and Y-axes until it reaches the desired location. The printheads then activate and print at this location. The system then transports the printing area to the next desired location beneath the printheads and continues to print. Our team's initial prototype operated in this way, and used stepper motors to move the stage in 1.57 μm increments which could provide single cell precision if necessary. Spatial accuracy is excellent with this type of system because most sources of error can be easily controlled. However, spatial precision of the printed material often varied as an effect of the bio-ink itself. Fortunately, vibrations could be minimized because the printheads were fixed. Drop distance error could be adjusted using the Z-axis and could be minimized as necessary. Weight applied to the system did not affect our constructs because we attached encoders to the stepper motors to verify that the motors reached their intended destination. Our system was designed to minimize micrometer errors and thus movement speed was decreased.[25,28] In addition, the entire construct needs to move to a new location for each cell drop, which increased build times compared to other prototypes. When applied to skin bioprinting, we found this system most useful for testing printhead prototypes because swapping of printheads could be easily performed without needing to design an entirely new movement system for each prototype. However, eventually we chose to focus on printheads mounted to a movement system, due to our need support the capability to bioprint over the area of large skin wounds.

XYZ printhead movement: To facilitate printing over large *in vivo* skin wounds, bioprinting benefits from printhead systems mounted to a movement system that is capable of moving in all 3Ds. The primary purpose of our first bioprinter was proof of concept testing of *in vivo* bioprinted fibroblasts and keratinocytes. This device was designed for wounds created in a murine wound model and did not require micron precision for deposition of the cells, as these cell types could self-arrange in constructs on their own. We used a pressure-based extrusion printhead array mounted to a movement system comprised of three stepper motors connected to threaded rods. The movement range of this type of system is limited only to the length of each axis, which can be changed according to the primary purpose. Spatial precision is dependent on the types of motors used, which in our case was 1.57 μm per step. Spatial accuracy in this type of system is dependent on the primary purpose. Movement speed was relatively high, because the system was designed to move to specific locations in small wounds quickly rather than achieving micrometer accuracy. This example illustrates how the end application of the bioprinter influences the design parameters of the movement system. This skin bioprinter prototype performed admirably for treating small skin wounds (4–9 cm^2), but would not be optimal for larger wounds in human patients or large animal models. The next bioprinter platform was thus designed with the primary application being rapid coverage and treatment of large skin wounds in a clinical environment. We developed a larger device framework using belt driven stepper motors instead of threaded rods. This type of system allowed faster coverage of larger areas.[32] Movement range and movement speed were the more important parameters in this system. A mobile lightweight frame where dimensions of the system are large enough to cover the torso of an average patient but small enough to easily pass through most doorframes. The movement system consisted of belt driven stepper motors capable of 100 μm movements. The X- and Y-axes were attached to the frame and the Z-axis was attached to the X- and Y-axes. This independent Z-axis allowed the printheads to track with the curvature of a patient's body. However, spatial precision and accuracy were sacrificed in this to support much faster printing over a larger area. We were able to bioprint 100 cm^2 skin constructs consisting of one layer of fibroblasts and one layer of keratinocytes in 45 min, including gelation time.[33] However, it should be noted that we were able to adapt this large scale printing system to the small wound murine model described above as well, demonstrating its versatility.[26]

2.2 *Bioprinting modalities*

In general, there are three main types of delivery systems that can be used for bioprinting. These consist of inkjet bioprinting, which is also sometimes referred to as drop-by-drop bioprinting, laser-induced forward transfer LIFT bioprinting, and laser-assisted bioprinting, and extrusion-based bioprinting.

Inkjet bioprinting: The first printing modality is inkjet bioprinting, of which there are two general categories: thermal and piezoelectric. These systems have a similar overall structure in which a cartridge is filled with bio-ink that is then forced through a chamber to an output aperture. Thermal inkjet bioprinters are typically derived from commercially available desktop inkjet printers, as we have described above. In these applications, commercially available ink cartridges often act as the printhead. The ink from these cartridges can be removed and replaced with a bio-ink.[29,34] Thermal inkjets then use a heating element that creates a bubble in the bio-ink, generating pressure that forces the bio-ink through the output aperture in the printhead. The parameters of these printheads are highly dependent on the type of cartridge used. Spatial resolution, in particular, is highly dependent on the type of inkjet printer and the size of the output aperture. Spatial precision can be manipulated by altering the composition and properties of the bio-ink in relation to the average drop volume of the cartridge.

Throughput in modified commercial inkjet printers can be highly variable, and these printers, with small output apertures have a tendency to rapidly clog. Some groups have experimented with adding chemical agents to bio-ink formulations in an effort to alleviate clogging.[35] Low throughput is another concern that prevents thermal inkjet printing from being used for skin bioprinting. Small droplet volume results in increased printing and treatment times. Cell viability is also a concern due to the thermodynamics of thermal inkjet printing. The thermal element in the cartridge briefly heats the ink to 300°C, which raises the temperature of the bio-ink by approximately 4–10°C.[36] Cell survival has been found to be 70–90%, although some groups have determined that cells require a recovery period after bioprinting to restore membrane integrity.[27,36–38] Thermal inkjet printers are useful for their low cost and ease of customization; however, the issues with spatial resolution and throughput prevent their use as skin bioprinters.

In contrast to thermal inkjet printing, piezoelectric inkjet printers do not alter the bio-ink temperature to create pressure inside the cartridge. Piezoelectric printers use either acoustic waves or a material that changes shape or size in order to generate pressure inside the cartridge that forces bio-ink through the output aperture.[39,40] Acoustic waves in particular are controllable through the manipulation of the wave parameters including duration, amplitude, and frequency. Therefore, while spatial resolution can be variable, it is far more controllable than in modified thermal inkjet printers. Single cell precision is excellent due to the ability to control cell droplet sizes. Throughput is also improved — These systems can be modified to remove the output nozzle, thus avoiding the issue of clogging that is prevalent with thermal inkjet printers that head and can cause evaporation or precipitation of bio-ink components. Throughput is also improved compared to thermal inkjet printers through the use of parallel cartridges, which can deliver multiple cell types simultaneously. Cell viability has been 80–90% in studies of encapsulated cells delivered through acoustic bioprinting.[41]

Even though piezoelectric inkjet bioprinting has several advantages over thermal inkjets for bioprinting, both types of inkjet printing suffer from limitations with respect to bio-inks.[42,43] The greatest limitation is the viscosity of the delivery matrix. In order to form a small droplet, the bio-ink must be liquid inside the cartridge and become solid once it has exited the printhead. This transformation is achieved by inducing gelation of the bio-ink chemically, enzymatically, or in some other manner. Practically, this limitation means that bio-inks for inkjet bioprinting are constrained to a maximum viscosity value.[42] The gelation requirment to form a solid or semi-solid gel also places a limitation on materials that can be employed. Gelation agents typically form cross-links between molecules of the matrix and can range from chemical reactions to ultraviolet radiation to pH changes. However, some gelation agents are toxic to cells, which further limits the available bioinks.[44]

LIFT bioprinting: LIFT-based bioprinting is an approach to bioprinting that has been adopted from other fields by researchers pursuing bioprinting.[45,46] LIFT technology was developed for high resolution deposition and patterning of metals and other non-biological materials for use in areas such as computer chip fabrication. More recently it has been employed to create micro- and nano-patterned peptide, DNA, and cell arrays. LIFT technology is comprised of a laser that is pulsed at desired time periods and a donor material "ribbon" comprised of the printable material. This is supported on a transport layer such as gold or titanium that absorbs the energy of the laser and transfers it to the ribbon. When the laser energy pulse reaches the ribbon, the focused energy pulse generates a small, but high pressure bubble that propels a droplet of the printable donor material onto a substrate. Patterning is achieved by moving the stage.[47–49] For LIFT-based bioprinting, ribbons have been developed that are comprised of a biopolymer or protein, sometimes similar to the materials used in other printing modalities. In this new scenario the laser pulse-driven ribbon droplets contain cells, which are then deposited in a pattern on the substrate on the movable stage to create cellular structures and patterns. The lack of a nozzle in LIFT is a departure from other printing modalities, and generally does away with the need to prevent clogging issues. This results in increased flexibility in the printing materials, as long as they can be sufficiently transferred by the laser energy pulse. Studies have shown few negative effects on cell viability[50–52] and the ability to print nearly at a single cell per droplet resolution,[53] positioning LIFT as a bioprinting technology with immense potential in the future.

The high resolution of LIFT is directed by a number of variables, such as the laser itself, the material properties of the printable ribbon material, the relative hydrophilic/hydrophobic nature of the substrate material to the printable material, cell density, and the path distance between the ribbon and the substrate that the droplet travels.[54] Thus, there are also some challenges that need to be overcome. The high resolution and subsequently small printing volume per pulse requires fast gelation kinetics of the printable material and a fast moving stage for fabrication. In current LIFT methods, preparation times of the ribbon, especially when containing cells and thus cell-friendly biomaterials, can be time consuming. Furthermore, to create structures of size, multiple ribbons are often employed, requiring multiple reloadings during the printing process. Nevertheless, with the potential to reach single cell positioning accuracy, LIFT bioprinting holds great promise for creating intricate structures.

Extrusion bioprinting: Extrusion-based deposition, generally from syringe-like pieces of equipment housed on XYZ-mobile carriages, is an additional approach for 3D bioprinting that primarily relies on the mechanical and temporal properties of the polymer materials being printed. In this modality, the properties of the printed polymer or hydrogel facilitate extrusion through a syringe tip, commonly driven by pneumatic pressure or mechanical pistons controlled by the computer. The reliance on the material properties for printing, means that the material must be soft or nearly fluidlike enough to facilitate extrusion through the small diameter tip or nozzle, but must also support itself mechanically after deposition. The rheological properties of the material can be tailored in such a manner that the materials have a high enough elastic modulus, or the solid component of its mechanical properties, such that extrusions maintain their shape. Simultaneously, the material may maintain a sufficient loss modulus, or the fluid-like component, such that extrusion is possible.[23,24] Regardless, this method of printing very much relies on the mechanical properties of the bioink. One common approach is to employ melt-curable polymers such as polycaprolactone, which when heated can be deformed and printed at relatively high resolutions, but can cool down to a solid material. These materials can be used to build rather large and intricate structures, capable of mimicking physiological structures. However, melt-cure printing cannot be done with cells at this time. Cells must be seeded at a later time, or other cell-supportive materials can be incorporated, such as hydrogels, where the cells are encapsulated in the printing process. Printing with hydrogels via extrusion techniques can be difficult when working with materials that rely on time for gelation to occur. Mistiming the deposition process can result in either a structure that collapses because crosslinks have not formed quickly enough, or conversely, clogging of the bioprinting device as a result of polymerization that was too fast. However, numerous studies have implemented novel cross-linking chemistries, photopolymerization techniques, and methods to facilitate spatial and temporal control over material properties that can contain encapsulated cells for printing 3D structures.[23,24,55]

Another approach to extrusion-based printing is what has been coined by some as "scaffold-free" bioprinting, which is based on the principles of tissue liquidity and tissue fusion of multi-cellular components.[19] In this approach, aggregates, rods, or tissue fragments comprised primarily of cells that are bound to one another are printed in geometric patterns or shaped and allowed to fuse over time to form larger constructs.[56] Multiple layers of aggregates or rods can be printed, and after fusing together during a post-print maturation period, singular 3D structures remain. In this work, an approach was used to build branched vascular structures,[19] and more recently nerve grafts.[57] Despite the name, it should be noted that this type of bioprinting often does rely on biomaterials. In most cases, the cell aggregates or cell rods are either printed into a cell-free biopolymer substrate or additively stacked using space-holding biomaterials to preserve the appropriate structures during the tissue fusion and maturation processes. These space-holders are generally removed when the construct is sufficiently fused and possesses sufficient mechanical properties to exist without the supporting spaceholders. The strength of this method lies within its high cell density, which allows for rapid fusion between discretely printed pieces. This method has been explored extensively and is the basis for the current commercial entity Organovo, one of the first bioprinting-based companies. Additionally, our laboratory previously explored a hydrogel-based approach that mimicked this technique, by printing hydrogel and cell-hydrogel rods that fused over time to create tubular constructs *in vitro*,[20] and Bertassoni *et al.* used this approach to print HepG2 liver cells within methacrylated gelatin rods.[58]

Each type of delivery system has its own set of advantages and disadvantages. Bioprinting can be performed in many different ways, and as such, the choice of bioprinting modality influences the capabilities of the end system and which applications are suitable. The hardware, in turn, is heavily influenced by the biomaterial and biological components chosen for the end bioprinting application as well.

3. Bioprinting of Skin

3.1 *Generation of layered tissue-like skin "equivalents"*

Recent advances in tissue engineering and material science have lead to more complex biological skin equivalents that may yield more suitable wound treatment options for patients in comparison to the inert dressings that see widespread use in the clinic. Some examples of the skin equivalents include cellularized graft-like products, such as Dermagraft, Apligraf, and TransCyte. These products are generally comprised of a polymer scaffold patch comprised of materials such as collagen or glycosaminoglycan polysaccharides that are seeded with human fibroblasts and cultured *in vitro* prior to application. Unfortunately, because of the technical and regulatory requirements involved in culturing and maintaining these grafts prior to use, these grafts are expensive to produce, and as they are also allografts due to incorporation of non-autologous cells, they can suffer from the same immunological drawbacks of traditional skin allografts.

3.2 *Bioprinting cells*

The cell source employed in cellularized wound healing therapies is an important component that has direct implications on cost, speed, and effectiveness of patient therapies. The widespread application of skin grafts, have lead researchers to explore delivering the cellular components of the grafts over a wider wound area, here by facilitating reduced donor site requirements through increased efficiency. Specifically, the harvesting of keratinocytes, fibroblasts, and other primary skin cell types and the distribution over the wound site through spraying or bioprinting has produced promising results in preclinical and clinical trials.[59,60] Human skin keratinocytes are one of the most obvious cell types used for wound healing, as one of the main goals of a given therapy is to regenerate the keratinocyte-based epithelial skin layer. However, autologous and allogeneic keratinocytes suffer from the same drawbacks and limitations as their autologous and allogeneic skin graft counterparts. Secondary surgical sites are required to harvest autologous cells and allogeneic cell-based treatment still has potential for rejection by the immune system, respectively.

Based on these limitations, researchers have explored the use of other cells that might be beneficial to wound healing, but may be either non-immunogenic or immunoprivileged, despite originating from allogeneic sources. With the rapid advances in stem cell and progenitor cell biology in recent years, new cell sources that have some of these beneficial characteristics have been identified.

3.2.1 *Skin-derived stem cells*

Stem cells have an essential role in the maintenance of tissue homeostasis and regeneration following injury. More than 20 different cell types have been identified residing within the skin, each performing a number of critical functions necessary for tissue function. These terminally differentiated, functional cell types are maintained by a pool of tissue resident stem cells, which are capable of proliferating and differentiating to contribute new cells to the tissue. Skin stem cells are more quiescent than other skin cells, including keratinocytes, have a high proliferative potential, and can terminally differentiate into multiple differentiation lineages.[61]

In the context of wound healing, stem cells capable of regenerating the epidermis have obvious therapeutic value. Epidermal stem cells have been identified and are characterized by prolonged self-renewal and differentiation capacity into multiple skin cell lineages.[62] Isolation and functional assessment of epidermal stem cells have been evaluated in multiple studies.[63,64] Epidermal stem cells can be selected through their rapid adherence to culture ECM-coated culture plates and are characterized by high levels of α6-integrin and low levels of CD71.[65] These skin-derived stem cells have been shown to differentiate into all hair follicle lineages, including sebaceous gland, neuron, glial cells, Schwann cells, smooth muscle cells, dermal fibroblasts, melanocytes, and other cell types.[66] Epidermal stem cells have found some application in the development of skin equivalents.[67]

An important repository of skin stem cells appears to be located at the bulge of the hair follicle, which does not degenerate during the hair cycle. The hair follicle contains multipotent stem cells that are activated at the start of the hair cycle and mobilized upon injury to stimulate regeneration of the damaged epidermis. Bulge stem cells respond rapidly to epidermal wounding during the acute wound repair phase and acquire an epidermal phenotype but interestingly, they are eliminated from the epidermis over several weeks after injury. This suggests that bulge stem cells contribute to the wound repair but not to the homoeostasis of the epidermis.[68] *In vitro,* bulge stem cells maintain their stem cell characteristics after propagation and can contribute to hair follicles, epidermis and sebaceous gland formation when combined with neonatal dermal cells in culture.[69]

Multipotent stem cells with mesodermal progeny have been isolated and characterized from human and rodent dermis. These skin-derived precursors (SKP) take approximately 2 weeks to isolate and expand,[70,71] and have been shown to differentiate into skeletogenic cell types,[72] Schwann cells,[71] neuron-like cells,[73] islet-like insulin-producing cell clusters,[74,75] and hepatocytes.[76]

Some recent studies have showed that SKP may promote angiogenesis in ischemic regions after cerebral infraction.[77] The implications of this finding may have a major impact for the treatment of chronic wounds such as diabetic ulcers.[78]

Tissue-specific stem cells hold great promise to contribute to all phases of wound healing. However, several limitations remain. Some limitations include difficulties in isolating, expanding, and differentiating these stem cells *in vitro*, as well as the potential that stem cells isolated from compromised patients, such as diabetics or the elderly may have limited effectiveness. Additionally, data on the application of skin stem cells for wound healing is not conclusive and further investigation is warranted to ensure safety and effectiveness in wound healing therapies.

3.2.2 *Bone marrow-derived mesenchymal stem cells*

Bone marrow-derived mesenchymal stem cells (MSCs) are capable of self-renewing and differentiating into at least osteocytes, chondrocytes, and adipocytes, adhering to culture plates and expression of a well defined set of surface antigens.[79,80]

MSCs have shown therapeutic potential for repair and regeneration of tissues damaged by injury or disease. A range of studies have demonstrated MSCs contribute to wound healing through multiple mechanisms, including; (i) direct differentiation into keratinocytes,[81–83] (ii) inducing angiogenesis through paracrine release vascular endothelial growth factor (VEGF), (iii) recruiting endothelial cells and endothelial progenitor cells (EPCs) to the wound, (iv) promoting regeneration of appendages by releasing cytokines such as epidermal growth factor (EGF) and keratinocyte growth factor (KGF), and (v) through anti-inflammatory mechanisms, such as releasing macrophage inflammatory protein-1 (MIP-1) and monocytes chemoattractant protein (MCP).[84,85]

In particular, MSC treatment of both acute and chronic wounds can result in accelerated wound closure, increases in re-epithelialization, formation of granulation tissue, and improved angiogenesis and neovascularization in the regenerated tissue[86] MSCs play an important role in the host defense and inflammatory processes, matrix deposition — principally collagen Types I and III, angiogenesis, and recruitment of endogenous cells for dermal and epidermal reconstruction.[87] Others showed that the major source of deposited collagen is coming from the resident cells in the skin, and bone marrow-derived stem cells are contributing partially to collagen production during fibrogenisis.[88,89] MSCs have been found to accelerate wound healing in patients with chronic and acute wounds when delivered in a fibrin hydrogel.[90] Bone marrow-derived stem cells have been shown to improve dermal rebuilding and lead to closure of non-healing chronic wounds, whether applied directly[91] or injected into the wound periphery.[92] Similar observations were seen in patients with critical limb ischemia who received MSCs injected intramuscularly or systemically.[93,94]

In addition to their role in wound healing, bone marrow- derived MSCs may play an important role in enhancing the fibrotic behavior of deep dermal fibroblasts and have a possible involvement in the pathogenesis of hypertrophic scarring.[95,96] For example, MSCs from bone marrow secrete high levels of VEGF, which is linked to excessive keloid scarring in patients who are susceptible to keloid scarring.[97] Several clinical trials have been conducted with different bone marrow-derived cells for chronic wounds and showed promising results.[98,99] MSCs have also recently been shown to be useful for improving *in vivo* skin expansion, an approach often used to establish space for surgical implants that induces tissue damage and requires treatment to heal the damaged tissue.[100] The specific mechanisms of how bone marrow-derived cells contribute to wound healing and regeneration are still under investigation. Additionally, further studies are needed to evaluate the optimal selection, expansion and delivery mechanisms for these cells to acute and chronic wounds.

3.2.3 *Perinatal tissues and stem cells*

Gestational tissue is a rich source of therapeutic biological material, with the placenta and placental membranes first used for medicinal applications for over 100 years, with the use of placental membranes for wound healing applications described in 1910 by William Thornton and Staige Davis, at Johns Hopkins Hospital.[101] Gestational tissue is often discarded following birth and is readily available without an invasive biopsy or the manipulation of embryos or somatic cells.[90–92] This means that there are minimal ethical and legal considerations associated with collection and use for clinical therapies.

The amnion is the innermost of the placental membranes lining the amniotic cavity and consists of five layers. A cell monolayer of amnion epithelial cells is in contact with the fetus during the first stages of development, and is known to have important roles in fetal development, immune regulation, and defense against infection.[102] Clinical trials and animal experiments have documented that amnion is superior to allograft and xenograft skin as a therapeutic adjunct to accelerate healing of burns, wounds, and ulcers.[103,104] Amnion has also been found to be an effective biologic dressing for burns, as it diminishes the loss of plasma, fluid, protein, and heat.[105,106] The use of amniotic membranes has also been shown to cause fewer complications than synthetic skin grafts.[107,108] The collective evidence of the clinical trials using amnion have favored its use in burns for promotion of wound healing and for increasing patient comfort and reduced requirement for dressing changes.[109] Additionally, many of these studies demonstrated the antimicrobial effects, pain relief, reduction of fluid, and reduced scar formation with amnion treatments.

Amniotic fluid-derived stem (AFS) cells are an attractive cell source for applications in regenerative medicine due to their high proliferation capacity, multipotency, immunomodulatory activity, and lack of significant immunogenicity.[110,111] Unlike embryonic stem cells, AFS cells do not form teratomas when injected into immune-deficient mice. Furthermore, AFS cells remain stable and show no signs of transformation in culture. The isolation of AFS cells is a simpler process than that for MSCs, and large numbers of AFS cells can be harvested and expanded signficantly from as little as 2 mL of amniotic fluid. These cells proliferate rapidly for many passages with doubling times of 30–36 h, and do not require supportive feeder layers.[110] The immuno-privileged and high proliferation charateristics of AFS cells have suggested that they can be used as a semi-"off the shelf" cell therapy treatment for wound healing and skin regeneration. Indeed, as described above, we demonstrated the efficacy of AFS cells for this application using a full-thickness wound model in mice, in which treatment of AFS cells delivered in a bio-ink matrix dramatically improved wound healing kinetics and regeneration of skin.[26]

However, until recently, no one had explored the use of these human AFS cells in skin bioprinting for wound healing. We recently demonstrated the use of AFS cells for this application using a full-thickness wound model in mice.[26] We employed a drop-on-demand skin printer[25] to deliver AFS cells to wounds within a fibrin–collagen gel vehicle, and compared this treatment to MSCs or no cells delivered in the same gel delivery vehicle. AFS cells and MSCs showed similar potency in induction of wound closure, significantly accelerating wound closure compared to the cell-free control. Upon histological inspection, we observed that the stem cells induced significantly greater vascularization in the new tissue, leading to formation of mature vessels, displaying robust smooth muscle actin, and a lack of extravasated red blood cells in the surrounding tissue. These results suggested that AFS cells have the potential to be an effective therapy for treating skin wounds, and skin bioprinting is an efficient and effective manner by which to deliver the cells to the wounds. Our team is currently exploring additional delivery matrices to improve the deposition procedure during bioprinting and to provide an environment that can maximize the therapeutic potential of the delivered cells.

3.3 *The role of biomaterials in bioprinting*

Biomaterials have an integral role in interfacing between hardware and biology. The materials employed for bioprinting can vary dramatically depending on the printing hardware that is used. These materials fall within the general category of biomaterials, as they often integrate and even support biological components such as cells, growth factors or cytokines, and the actual *in vivo* wound enviroments. The term biomaterials itself, comprises a massive range of materials that continues to evolve and be redefined, ranging from cell supportive maleable hydrogels, to stiff and durable metal or ceramic implants, from nanometer-sized nanoparticles for drug delivery and imaging, to electronic medical devices such as pacemakers, hearing aids, and artificial hearts. As development in materials science and biological exploration continues, so will the number of classifications and definitions of biomaterials.[112,113]

Only a subset of the entire biomaterials portfolio are suitable to act as bio-inks for bioprinting. In the overall arena of bioprinting and biofabrication, biomaterials are generally confined to 1 of 2 primary categories. The first is that of melt curable polymers that result in mechanically durable materials that act as scaffolding in tissue or cellularized constructs. These are generally employed in 3D construct fabrication, rather than bioprinting of skin. Many such materials typically require high temperatures or toxic solvents to facilitate dissolution and shaping of the base polymers, and therefore are usually not appropriate for printing together with cells or other sensitive biological agents. Instead, these scaffolds are often formed first, after which cells are seeded onto and into the scaffolds after fabrication. The second category of biomaterials is that of soft and malleable biomaterials such as hydrogels, which generally with a high aqueous content, inside of which cells are capable of residing while remaining viable. These can be comprised of synthetic or natural polymers, and more often do not possess the same levels of mechanical properties (stiffness, tensile strength, elastic modulus, etc.) as curable polymers in mechanical support roles. The inherent characteristics of these different printing materials, including mechanical properties, melting points, and available chemistries for cross-linking and functionalization are responsible for determining whether or not one can successfully use them for bioprinting. Fortunately, these different parameters can be harnessed and manipulated to yield a virtually limitless toolbox of materials and bio-ink formulations for bioprinting, as well as other biofabrication approaches.

A common hurdle that many researchers working at the nexus of biomaterials and bioprinting is that most biomaterials were not originally designed specifically for bioprinting applications. To overcome this limitation, much of the work in early and current bioprinting has focused on adapting more traditional materials through chemical and mechanical manipulations to better integrate with bioprinting hardware, while maintaining cell-friendly and cell-supportive characteristics. To be considered such, these materials must not result in toxicity in cells and optimally should provide cell-adherence motifs, such as RGD peptides, to allow for cell attachment and migration. Techniques to generate crosslinking reactions can be designed and implemented to be non-cytotoxic, allowing efficient encapsulation of various cell types within the hydrogel matrices at time of printing. This has become an important characteristic as there has been a movement in biomedical research from traditional 2D cell culture platforms to 3D systems that better mimic *in vivo* conditions.[114] All *in vivo* tissues are 3D in nature, and bioprinting should echo this characteristic, employing a 3D approach.[115]

Hydrogel bio-ink materials fall into one of two additional sub-categories: synthetic hydrogels, which are comprised of polymers that are synthesized in the laboratory, or naturally derived hydrogels, which are comprised of polymers, often polysaccharides, but can also be comprised of peptides or proteins, purified from natural sources and are often further manipulated in the laboratory. Common examples of synthetic hydrogels include polyethylene glycol (PEG)-based materials, such as PEG diacrylate (PEGDA), and polyacrylamide (PAAm)-based gels. Examples of naturally derived hydrogels that are in common use and have been explored for bioprinting, include collagen, hyaluronic acid, gelatin, alginate, and fibrin. In general, synthetic materials allow for increased control over molecular weights and distributions, as well as crosslinking chemistries, supporting precise control of mechanical properties such as elastic modulus E' and degradation kinetics. Conversely, naturally derived materials can more difficult to modify to have specific physical properties, but often have innate bioactive characteristics via biological peptide sequences and cell attachment motifs that cells interact with, improving tissue integration and biocompatibility. However, it should be noted that this list is but an overview and is not exhaustive; many other materials exist that can be used in bioprinting or modified for bioprinting, and others are currently being developed.

As described above, we initially employed a fibrin–collagen bio-ink system to be used with the skin bioprinter developed by our group,[25] which employs pneumatic pressure and a series of valves in line with different cartridges to deliver the bio-ink. Specifically, the fibrin-precursor fibrinogen was mixed with soluble collagen Type I giving the first part of a two-part system, while a solution of thrombin comprised the second part of the system. These solutions were then printed separately, only gelling upon deposition after coming into contact with one another.[26,116] Notably, the fibrinogen–thrombin gelation mechanism facilitated relatively fast crosslinking *in situ* and the collagen component cross–linked slowly within this proto-gel. While useful, we thought that this fibrin-collagen gel was not the most optimal material for this wound healing bioprinting. It required 2 solutions, thus relying on diffusion for mixing, therefore suffering from not polymerizing immediately *in situ*. In irregular 3D wound environments or convex surfaces, this resulted in some sloughing off of the material before gelation was complete. Moreover, collagen often contracts during its gelation process, and the collagen Type I isoform is also the primary extracellular matrix component in scarring, which we thought could potentially act directly against the healthy, non-scarred skin regeneration we were trying to attain. In light of this belief, we assessed a variety of other hydrogels commonly used in various regenerative medicine and tissue engineering applications gelation times, ease of use, biocompatibility, immunogenicity, and bioprinter compatibility. As expected, the different materials exhibited differing pros and cons, again suggesting that a particular material should be chosen based on the particular target application. However, we did identify additional hydrogel formulations that were suitable for integration in our skin bioprinter and likely good choices for use as hydrogel bio-inks in wound healing treatments.[117] One such hydrogel, a UV-photopolymerizable hyaluronic acid and gelatin material, capable of elastic modulus modulation, cytokine loading, and cell delivery, is currently being employed in a variety of applications in our laboratory, including skin printing.

4. Conclusions

Bioprinting is an emerging technology that has quickly established itself as a robust approach for applications in regenerative medicine, such as wound healing. Bioprinting provides the ability to deliver cells and cytokines in hydrogel bio-inks in a fast, "off the shelf" manner, facilitating rapid and complete wound coverage and closure, which is critical in cases of large full-thickness wounds or burns. As bioprinting technology, and other technologies that interface with bioprinting, continue to advance, the potential of bioprinting to influence medical practices will grow. With the exception of LIFT printing, which is still in its infancy as applied to bioprinting, printing resolution of biological materials is still not good enough for some applications. Hardware X–Y–Z axis movement generally supports very precise and accurate printing; print-head hardware and most bio-inks do not, and need to continue to be developed to integrate seamlessly with the hardware.

Bio-ink materials are already evolving rapidly. Different types of base materials are being synthesized or modified by new techniques in the laboratory, generating countless options of now printable materials. Additionally, researchers can manipulate chemistries, concentrations, cross-linkers, and bioactive components to generate materials that can be printed, and also support cellular viability and function, as we have discussed. Some of these bio-ink systems have become quite elegant, leveraging modular designs allowing them to be tailored to a variety of specific applications, such as skin regeneration. As described above, many potential materials and technologies can be employed for bioprinting of skin to treat skin wounds. Some of these approaches utilize cells with regenerative qualities. In the future, engineered cells will likely be employed to this end. Some other approaches focus on biomaterial customization to provide a regenerative niche to the patient's endogenous cells. Importantly, bioprinting hardware is essential to ensure precise deposition into wounds *in situ*, or in 3D arrangements *ex vivo/in vitro*, as well as ensuring consistency in patient-to-patient therapies. The studies presented in this chapter argue that bioprinting can be an incredibly powerful tool for treating wounds and regenerating skin. The future of bioprinting technology is very promising, and is certainly not limited only to this one application. Rather, one can expect bioprinting to impact and become ingrained in a multitude of facets of medicine.

References

1. Cherry, D. K. Hing, E. and Woodwell, D. A. Rechtsteiner, E. A. National Ambulatory Medical Care Survey: 2006 summary. *Natl Health Stat Report* 1–39 (2008).
2. Pitts, S. R. Niska, R. W. Xu, J. and Burt, C. W. National Hospital Ambulatory Medical Care Survey: 2006 emergency department summary. *Natl Health Stat Report* 1–38 (2008).
3. Miller, S. F. Bessey, P. Lentz, C. W. Jeng J. C. Schurr M. and Browning S. National burn repository 2007 report: A synopsis of the 2007 call for data. *J Burn Care Res* 29, 862–870; discussion 71 (2008).
4. Peck, M. D. Epidemiology of burns throughout the world. Part I: Distribution and risk factors. *Burns* 37, 1087–1100 (2011).
5. Kurd, S. K. Hoffstad, O. J. Bilker, W. B. and Margolis, D. J. Evaluation of the use of prognostic information for the care of individuals with venous leg ulcers

or diabetic neuropathic foot ulcers. *Wound Repair Regen* **17**, 318–325, (2009).

6. Sen, C. K. Gordillo, G. M. Roy, S. Kirsner, R. Lambert, L. Hunt, T. K. *et al.* Human skin wounds: A major and snowballing threat to public health and the economy. *Wound Repair Regen* **17**, 763–771, (2009).

7. Lesher, A. P. Curry, R. H. Evans, J. Smith, V. A. Fitzgerald, M. T. Cina, R. A. *et al.* Effectiveness of Biobrane for treatment of partial-thickness burns in children. *J Pediatr Surg* **46**, 1759–1763 (2011).

8. Rahmanian-Schwarz, A. Beiderwieden, A. Willkomm, L. M. Amr, A. Schaller, H. E. and Lotter, O. A clinical evaluation of Biobrane((R)) and Suprathel((R)) in acute burns and reconstructive surgery. *Burns* **37**, 1343–1348, (2011).

9. Latenser, B. A. Critical care of the burn patient: The first 48 hours. *Crit Care Med* **37**, 2819–2826, (2009).

10. Holavanahalli, R. K. Helm, P. A. and Kowalske, K. J. Long-term outcomes in patients surviving large burns: The skin. *J Burn Care Res* **31**, 631–639 (2010).

11. Lloyd, E. C., Rodgers, B. C., Michener, M. and Williams, M.S. Outpatient burns: Prevention and care. *Am Fam Physician* **85**, 25–32 (2012).

12. Visconti, R. P. Kasyanov, V. Gentile, C. Zhang, J. Markwald, R. R. and Mironov, V. Towards organ printing: Engineering an intra-organ branched vascular tree. *Expert Opin Biol Ther* **10**, 409–420, (2010).

13. Fedorovich, N. E. Alblas, J. de Wijn, J. R. Hennink, W. E. Verbout, A. J. and Dhert, W. J. Hydrogels as extracellular matrices for skeletal tissue engineering: state-of-the-art and novel application in organ printing. *Tissue Eng* **13**, 1905–1925, (2007).

14. Mironov, V. Boland, T. Trusk, T. Forgacs, G. and Markwald, R. R. Organ printing: Computer-aided jet-based 3D tissue engineering. *Trends Biotechnol* **21**, 157–161, (2003).

15. Mironov, V. Kasyanov, V. Drake, C. and Markwald, R. R. Organ printing: promises and challenges. *Regen Med* **3**, 93–103 (2008).

16. Boland, T. Mironov, V. Gutowska, A. Roth, E. A. and Markwald, R. R. Cell and organ printing 2: Fusion of cell aggregates in three-dimensional gels. *Anat Rec A Discov Mol Cell Evol Biol* **272**, 497–502, (2003).

17. Jakab, K. Damon, B. Neagu, A. Kachurin, A. and Forgacs, G. Three-dimensional tissue constructs built by bioprinting. *Biorheology* **43**, 509–513 (2006).

18. Marga, F. Neagu, A. Kosztin, I. and Forgacs, G. Developmental biology and tissue engineering. *Birth Defects Res C Embryo Today* **81**, 320–328 (2007).

19. Norotte, C. Marga, F. S. Niklason, L. E. and Forgacs, G. Scaffold-free vascular tissue engineering using bioprinting. *Biomaterials* **30**, 5910–5917 (2009)

20. Skardal, A. Zhang, J. and Prestwich, G. D. Bioprinting vessel-like constructs using hyaluronan hydrogels crosslinked with tetrahedral polyethylene glycol tetracrylates. *Biomaterials* **31**, 6173–6181 (2010).

21. Jakab, K. Damon, B. Marga, F. Doaga, O. Mironov, V. Kosztin, I. *et al.* Relating cell and tissue mechanics: Implications and applications. *Dev Dyn* **237**, 2438–2449 (2008).

22. Jakab, K. Neagu, A. Mironov, V. Markwald, R. R. and Forgacs, G. Engineering biological structures of prescribed shape using self-assembling multicellular systems. *Proc Natl Acad Sci USA* **101**, 2864–2869 (2004).

23. Skardal, A. Zhang, J. McCoard, L. Oottamasathien, S. and Prestwich, G. D. Dynamically crosslinked gold nanoparticle — hyaluronan hydrogels. *Adv Mater* **22**, 4736–4740 (2010).

24. Skardal, A. Zhang, J. McCoard, L. Xu, X. Oottamasathien, S. and Prestwich, G. D. Photocrosslinkable hyaluronan-gelatin hydrogels for two-step bioprinting. *Tissue Eng Part A* **16**, 2675–2685 (2010).

25. Yoo, J. J. Atala, A. Binder, K. W. Zhao, W. Dice, D. and Xu, T. Delivery System. In: Office USPaT, (ed.) United States of America. p. 22, (2011).

26. Skardal, A. Mack, D. Kapetanovic, E. Atala, A. Jackson, J. D. Yoo, J. *et al.* Bioprinted amniotic fluid-derived stem cells accelerate healing of large skin wounds. *Stem Cells Transl Med* **1**, 792–802 (2012).

27. Xu, T. Gregory, C. A. Molnar, P. Cui, X. Jalota, S. Bhaduri, S. B. *et al.* Viability and electrophysiology of neural cell structures generated by the inkjet printing method. *Biomaterials* **27**, 3580–3588 (2006).

28. Binder, K. Arthur, A. Yoo, J. Atala, and A. Drop-on-demand inkjet bioprinting: A primer. *Gene Therapy and Regulation* **6**, 33–49 (2011).

29. Boland, T. Xu, T. Damon, B. and Cui, X. Application of inkjet printing to tissue engineering. *Biotechnol J* **1**, 910–917 (2006).

30. Nakamura, M. Kobayashi, A. Takagi, F. Watanabe, A. Hiruma, Y. Ohuchi, K. *et al.* Biocompatible inkjet printing technique for designed seeding of individual living cells. *Tissue Eng* **11**, 1658–1666 (2005).

31. Xu, C. Inai, R. Kotaki, M. and Ramakrishna, S. Electrospun nanofiber fabrication as synthetic extracellular matrix and its potential for vascular tissue engineering. *Tissue Eng* **10**, 1160–1168, (2004).

32. Yoo, J. J. W.-S., N. C., US, Atala, Anthony (Winston-Salem, NC, US), Binder, Kyle W. (Winston-Salem, NC, US), Zhao, Weixin (Winston-Salem, NC, US), Dice, Dennis (Yadkinville, NC, US), Xu, Tao (El Paso, TX, US). Delivery System. United States 2011.

33. Binder, K. *In situ* bioprinting of the skin [Dissertation]: Wake Forest University; 2011.

34. Roth, E. A. Xu, T. Das, M. Gregory, C. Hickman, J. J. and Boland, T. Inkjet printing for high-throughput cell patterning. *Biomaterials* **25**, 3707–3715, (2004).

35. Parzel, C. A. Pepper, M. E., Burg, T. Groff, R. E. and Burg, K. J. EDTA enhances high-throughput two-dimensional bioprinting by inhibiting salt scaling and cell aggregation at the nozzle surface. *J Tissue Eng Regen Med* **3**, 260–268 (2009).

36. Cui X. Dean, D. Ruggeri, Z. M. and Boland, T. Cell damage evaluation of thermal inkjet printed Chinese hamster ovary cells. *Biotechnol Bioeng* **106**, 963–969 (2010).

37. Chang, R. Nam, J. and Sun, W. Effects of dispensing pressure and nozzle diameter on cell survival from solid freeform fabrication-based direct cell writing. *Tissue Eng Part A* **14**, 41–48 (2008).

38. Nair, K. Gandhi, M. Khalil, S. Yan, K. C. Marcolongo, M., Barbee, K. *et al*. Characterization of cell viability during bioprinting processes. *Biotechnol J* **4**, 1168–1177 (2009).

39. Saunders, R. E. Gough, J. E. and Derby, B. Delivery of human fibroblast cells by piezoelectric drop-on-demand inkjet printing. *Biomaterials* **29**, 193–203, (2008).

40. Demirci, U. and Montesano, G. Single cell epitaxy by acoustic picolitre droplets. *Lab Chip* **7**, 1139–1145, (2007).

41. Tasoglu, S. and Demirci, U. Bioprinting for stem cell research. *Trends Biotechnol* **31**, 10–19, (2013).

42. Kim, J. D. Choi, J. S. Kim, B. S. Choi, Y. C. and Cho, Y. W. Piezoelectric inkjet printing of polymers: Stem cell patterning on polymer substrates. **51**, 2147–2154 (2010).

43. Murphy, S. V. and Atala, A. 3D bioprinting of tissues and organs. *Nat Biotechnol* **32**, 773–785 (2014).

44. Hennink, W. E. and van Nostrum, C. F. Novel crosslinking methods to design hydrogels. *Adv Drug Deliv Rev* **54**, 13–36 (2002).

45. Bohandy, J. and Kim, B. Adrian, F. Metal deposition from a supported metal film using an excimer laser. *Journal of Applied Physics* **60**, 1538 (1986).

46. Barron, J. A. Ringeisen, B. R. Kim, H. Spargo, B. J. and Chrisey, D. B. Application of laser printing to mammalian cells. *Thin Solid Films* **453**, 383–387, (2004).

47. Chrisey, D. B. MATERIALS PROCESSING: The Power of Direct Writing. *Science* **289**, 879–881 (2000).

48. Colina, M. Serra, P. Fernandez-Pradas, J. M. Sevilla, L. and Morenza, J. L. DNA deposition through laser induced forward transfer. *Biosens Bioelectron* **20**, 1638–1642 (2005).

49. Dinca, V. Kasotakis, E. Catherine, J. Mourka, A. Ranella, A., Ovsianikov, A, *et al*. Directed three-dimensional patterning of self-assembled peptide fibrils. *Nano Letters* **8**, 538–543 (2008).

50. Hopp, B. Smausz, T. Kresz, N. Barna, N. Bor, Z. Kolozsvari, L. *et al*. Survival and proliferative ability of various living cell types after laser-induced forward transfer. *Tissue Eng* **11**, 1817–1823 (2005).

51. Gruene, M. Deiwick, A. Koch, L. Schlie, S. Unger, C. Hofmann, N. *et al*. Laser Printing of Stem Cells for Biofabrication of Scaffold-Free Autologous Grafts. *Tissue Eng Part C Methods* 2010.

52. Koch, L. Kuhn, S. Sorg, H. Gruene, M. Schlie, S. Gaebel, R. *et al.* Laser printing of skin cells and human stem cells. *Tissue Eng Part C Methods* **16**, 847–854 (2010).

53. Guillotin, B. Souquet, A. Catros, S. Duocastella, M. Pippenger, B. Bellance, S. *et al.* Laser assisted bioprinting of engineered tissue with high cell density and microscale organization. *Biomaterials* **31**, 7250–7256 (2010).

54. Guillemot, F. Souquet, A. Catros, S. and Guillotin, B. Laser-assisted cell printing: Principle, physical parameters versus cell fate and perspectives in tissue engineering. *Nanomedicine* (Lond) **5**, 507–515 (2010).

55. Pescosolido, L. Schuurman, W. Malda, J. Matricardi, P. Alhaique, F. and Coviello, T., *et al.* Hyaluronic acid and dextran-based semi-IPN hydrogels as biomaterials for bioprinting. *Biomacromolecules* **12** 1831–1838, (2011).

56. Jakab, K. Norotte, C. Damon, B. Marga, F. Neagu, A. Besch-Williford, C. L. *et al.* Tissue engineering by self-assembly of cells printed into topologically defined structures. *Tissue Eng Part A* **14**, 413–421 (2008).

57. Marga, F. Jakab, K. Khatiwala, C. Shepherd, B. Dorfman, S. Hubbard, B. *et al.* Toward engineering functional organ modules by additive manufacturing. *Biofabrication* **4**, 022001, (2012).

58. Bertassoni, L. E. Cardoso, J. C. Manoharan, V. Cristino, A. L. Bhise, N. S. Araujo, W. A. *et al.* Direct-write bioprinting of cell-laden methacrylated gelatin hydrogels. *Biofabrication* **6**, 024105 (2014).

59. Gerlach, J. C. Johnen, C. Ottomann, C. Brautigam, K. Plettig, J. Belfekroun, C. *et al.* Method for autologous single skin cell isolation for regenerative cell spray transplantation with non-cultured cells. *Int J Artif Organs* **34** 271–279, (2011).

60. Sood, R. Roggy, D. E. Zieger, M. J. Nazim, M. Hartman, B. C. and Gibbs, J. T. A comparative study of spray keratinocytes and autologous meshed split-thickness skin graft in the treatment of acute burn injuries. *Wounds* **27** 31–40, (2015) .

61. Tumbar, T. Guasch, G. Greco, V. Blanpain, C. Lowry, W. E. Rendl, M. *et al.* Defining the epithelial stem cell niche in skin. *Science* **303** 359–363, (2004).

62. Blanpain, C. and Fuchs, E. Epidermal Stem Cells of the Skin. *Ann Rev Cell Dev Biol* **22**, 339–373 (2006).

63. Doucet, Y. S. and Owens, D. M. Isolation and functional assessment of cutaneous stem cells. *Methods Mol Biol* 1785–1783_**13**, (2015).

64. Reiisi, S. Esmaeili, F. and Shirazi, A. Isolation, culture and identification of epidermal stem cells from newborn mouse skin. *In Vitro Cell Dev Biol Anim* **46**, 54–59 (2010).

65. Fujimori, Y. Izumi, K. Feinberg, S. E. and Marcelo, C. L. Isolation of small-sized human epidermal progenitor/stem cells by Gravity Assisted Cell Sorting (GACS). *J Dermatol Sci* **56**, 181–187 (2009).

66. Lau, K. Paus, R. Tiede, S. Day, P. and Bayat, A. Exploring the role of stem cells in cutaneous wound healing. *Exp Dermatol* 18, 921–933 (2009).

67. Kim, D. S. Cho, H. J. Choi, H. R. Kwon, S. B. and Park, K. C. Isolation of human epidermal stem cells by adherence and the reconstruction of skin equivalents. *Cell Mol Life Sci* 61, 2774–2781 (2004).

68. Ito, M. Liu, Y. Yang, Z. Nguyen, J. Liang, F. Morris, R. J. *et al.* Stem cells in the hair follicle bulge contribute to wound repair but not to homeostasis of the epidermis. *Nat Med* 11, 1351–1354 (2005).

69. Cotsarelis, G. Epithelial stem cells: a folliculocentric view. *J Invest Dermatol* 126, 1459–1468 (2006).

70. Toma, J. G. McKenzie, I. A. Bagli, D. Miller, F. D. Isolation and characterization of multipotent skin-derived precursors from human skin. *Stem Cells* 23, 727–737 (2005).

71. Biernaskie, J. A. McKenzie, I. A. Toma, J. G. and Miller, F. D. Isolation of skin-derived precursors (SKPs) and differentiation and enrichment of their Schwann cell progeny. *Nat Protoc* 1, 2803–2812 (2006).

72. Lavoie, J. F. Biernaskie, J. A. Chen, Y. Bagli, D. Alman, B. Kaplan, D. R. *et al.* Skin-derived precursors differentiate into skeletogenic cell types and contribute to bone repair. *Stem Cells Dev* 18, 893–906, (2009).

73. Fernandes, K. J. Kobayashi, N. R, Gallagher, C. J. Barnabe-Heider, F. Aumont, A. Kaplan, D. R. *et al.* Analysis of the neurogenic potential of multipotent skin-derived precursors. *Exp Neurol* 201, 32–48 (2006).

74. Mehrabi, M. Mansouri, K. Hosseinkhani, S. Yarani, R. Yari, K. Bakhtiari, M. *et al.* Differentiation of human skin-derived precursor cells into functional islet-like insulin-producing cell clusters. *In Vitro Cell Dev Biol Anim* 51, 595–603 (2015).

75. Guo, W. Miao, C. Liu, S. Qiu, Z. Li, J. and Duan, E. Efficient differentiation of insulin-producing cells from skin-derived stem cells. *Cell Prolif* 42, 49–62 (2009).

76. Rodrigues, R. M. De Kock, J. Branson, S. Vinken, M. Meganathan, K. and Chaudhari, U., *et al.* Human skin-derived stem cells as a novel cell source for *in vitro* hepatotoxicity screening of pharmaceuticals. *Stem Cells Dev* 23, 44–55 (2014).

77. Mao, D. Yao, X. Feng, G. Yang, X. Mao, L. Wang, X. *et al.* Skin-derived precursor cells promote angiogenesis and stimulate proliferation of endogenous neural stem cells after cerebral infarction. *Biomed Res Int* 945846, 26 (2015).

78. Vasudeva, V. S. Abd-El-Barr, M. M. and Chi, J. H. Implantation of neonatal skin-derived precursor schwann cells improves outcomes after incomplete cervical spinal cord injury in rats: *Neurosurgery*. 77(2), N15–N17 (2015).

79. Minguell, J. J. Conget, P. and Erices, A. Biology and clinical utilization of mesenchymal progenitor cells. *Braz J Med Biol Res* 33, 881–887 (2000).

80. Boxall, S. A. and Jones, E. Markers for characterization of bone marrow multipotential stromal cells. *Stem Cells Int* 975871, **14**, (2012).

81. Wu, Y. Chen, L. Scott, P. G. and Tredget, E. E. Mesenchymal stem cells enhance wound healing through differentiation and angiogenesis. *Stem Cells* **25**, 2648–2659 (2007).

82. Wu, Y. Wang, J. Scott, P. G. and Tredget, E. E. Bone marrow-derived stem cells in wound healing: A review. *Wound Repair Regen* **15**, (2007).

83. Sasaki, M. Abe, R. Fujita, Y. Ando, S. Inokuma, D. and Shimizu, H. Mesenchymal stem cells are recruited into wounded skin and contribute to wound repair by transdifferentiation into multiple skin cell type. *J Immunol* **180**, 2581–2587 (2008).

84. Fathke, C. Wilson, L. Hutter, J. Kapoor, V., Smith, A. Hocking, A. *et al.* Contribution of bone marrow-derived cells to skin: Collagen deposition and wound repair. *Stem Cells* **22**, 812–822 (2004).

85. Javazon, E. H. Keswani, S. G. Badillo, A. T. Crombleholme, T. M. Zoltick, P. W. Radu, A. P. *et al.* Enhanced epithelial gap closure and increased angiogenesis in wounds of diabetic mice treated with adult murine bone marrow stromal progenitor cells. *Wound Repair Regen* **15**, 350–359 (2007).

86. Maxson, S. Lopez, E. A. Yoo, D. Danilkovitch-Miagkova, A. and LeRoux, M. A. Concise Review: Role of Mesenchymal Stem Cells in Wound Repair. *Stem Cells Transl Med* **1**, 142–149 (2012).

87. Rea, S. Giles, N. L. Webb, S. Adcroft, K. F. Evill, L. M. Strickland, D. H. *et al.* Bone marrow-derived cells in the healing burn wound — more than just inflammation. *Burns* **35**, 356–364 (2009).

88. Higashiyama, R. Nakao, S. Shibusawa, Y. Ishikawa, O. Moro, T. Mikami, K. *et al.* Differential contribution of dermal resident and bone marrow-derived cells to collagen production during wound healing and fibrogenesis in mice. *J Invest Dermatol* **131**, 529–536 (2011).

89. Higashiyama, R. Moro, T. Nakao, S. Mikami, K. Fukumitsu, H. Ueda, Y. *et al.* Negligible contribution of bone marrow-derived cells to collagen production during hepatic fibrogenesis in mice. *Gastroenterology* **137**, 1459–1466 (2009).

90. Falanga, V. Iwamoto, S. Chartier, M. Yufit, T. Butmarc, J. Kouttab, N. *et al.* Autologous bone marrow-derived cultured mesenchymal stem cells delivered in a fibrin spray accelerate healing in murine and human cutaneous wounds. *Tissue Eng* **13**, 1299–1312 (2007).

91. Badiavas, E. V. and Falanga, V. Treatment of chronic wounds with bone marrow-derived cells. *Arch Dermatol* **139**, 510–516 (2003).

92. Rogers, L. C. Bevilacqua, N. J. and Armstrong, D. G. The use of marrow-derived stem cells to accelerate healing in chronic wounds. *Int Wound J* **5**, 20–25 (2008).

93. Klepanec, A. Mistrik, M. Altaner, C. Valachovicova, M. Olejarova, I. Slysko, R. *et al.* No difference in intra-arterial and intramuscular delivery of autologous bone marrow cells in patients with advanced critical limb ischemia. *Cell Transplant* **21**, 1909–1918 (2012).

94. El Sadik, A. O. El Ghamrawy, T. A. and Abd El-Galil, T. I. The Effect of Mesenchymal Stem Cells and Chitosan Gel on Full Thickness Skin Wound Healing in Albino Rats: Histological, Immunohistochemical and Fluorescent Study. *PloS One* **10**, (2015).

95. Ding, J. Ma, Z. Shankowsky, H. A. Medina, A. and Tredget, E. E. Deep dermal fibroblast profibrotic characteristics are enhanced by bone marrow-derived mesenchymal stem cells. *Wound Repair Regen* **21**, 448–455 (2013).

96. Wang, J. Dodd, C. Shankowsky, H. A. Scott, P. G. and Tredget, E. E. Deep dermal fibroblasts contribute to hypertrophic scarring. *Lab Invest* **88**, 1278–1290, (2008).

97. Wilgus, T. A. Ferreira, A. M. Oberyszyn, T. M. Bergdall, V. K. and Dipietro, L. A. Regulation of scar formation by vascular endothelial growth factor. *Lab Invest* **88**, 579–590 (2008).

98. Yang, M. Sheng, L. Zhang, T. R. and Li, Q. Stem cell therapy for lower extremity diabetic ulcers: Where do we stand? *Biomed Res Int* 462179, **18**, (2013).

99. Kirby, G. T. Mills, S. J. Cowin, A. J. and Smith, L. E. Stem Cells for Cutaneous Wound Healing. *Biomed Res Int* 285869, **2** (2015).

100. Wang, X. Li, C. Zheng, Y. Xia, W. Yu, Y. and Ma, X. Bone marrow mesenchymal stem cells increase skin regeneration efficiency in skin and soft tissue expansion. *Expert Opin Biol Ther* (2012).

101. Davis, J. S. Skin transplantation. *Johns Hopkins Hospital Reports* **15**, 307–396 (1910).

102. Bowen, J. M., Chamley, L., Keelan, J. A. and Mitchell, M. D. Cytokines of the placenta and extra-placental membranes: Roles and regulation during human pregnancy and parturition. *Placenta* **23**, 257–273, (2002).

103. Bose, B. Burn wound dressing with human amniotic membrane. *Ann R Coll Surg Engl* **61**, 444 (1979).

104. Bennett, J. Matthews, R. and Faulk, W. P. Treatment of chronic ulceration of the legs with human amnion. *The Lancet* **315**, 1153–1156 (1980).

105. Atanasov, W. Mazalova, J. Todorod. R, S. K. and Trencheva, W. Use of amniotic membrane as biological dressings in contemporary treatment of burns. *Ann Medi Burns Club* **7**, (1994).

106. Sawhney, C. P. Amniotic membrane as a biological dressing in the management of burns. *Burns* **15**, 339–342 (1989).

107. Fisher, J. C. Amniotic membranes as a temporary wound dressing. *Plast Reconstr Surg* **52**, 601 (1973).

108. Ward, D. and Bennett, J. The long-term results of the use of human amnion in the treatment of leg ulcers. *Br J Plast Surg* **37**, 191–193 (1984).

109. Kesting, M. R. Wolff, K.-D. Hohlweg-Majert, B. and Steinstraesser, L. The role of allogenic amniotic membrane in burn treatment. *J Burn Care Res* **29**, 907–916 (2008).

110. De Coppi, P. Bartsch, G. Jr. Siddiqui, M. M. Xu, T. Santos, C. C. Perin, L. *et al.* Isolation of amniotic stem cell lines with potential for therapy. *Nat Biotechnol* **25**, 100–106 (2007).

111. Moorefield, E. C. McKee, E. E. Solchaga, L. Orlando, G. Yoo, J. J. Walker, S. *et al.* Cloned, CD117 selected human amniotic fluid stem cells are capable of modulating the immune response. *PLoS One* **6**, e26535 (2011).

112. Williams, D. The continuing evolution of biomaterials. *Biomaterials* **32**, 1–2 (2011).

113. Williams, D. F. On the nature of biomaterials. *Biomaterials* **30**, 5897–5909 (2009).

114. Prestwich, G. D. Evaluating drug efficacy and toxicology in three dimensions: Using synthetic extracellular matrices in drug discovery. *Acc Chem Res* **41**, 139–148 (2008).

115. Mironov, V. Reis, N. and Derby, B. Review: Bioprinting: A beginning. *Tissue Eng* **12**, 631–634 (2006).

116. Xu, T. Binder, K. W. Albanna, M. Z. Dice, D. Zhao, W. Yoo, J. J. *et al.* Hybrid printing of mechanically and biologically improved constructs for cartilage tissue engineering applications. *Biofabrication* **5**, 015001, (2013).

117. Murphy, S. V. Skardal, A. and Atala, A. Evaluation of hydrogels for bioprinting applications. *J Biomed Mater Res A* **101**, 272–284 (2013).

13. Electroporation Applications in Wound Healing

Laure Gibot*, Tadej Kotnik[†] and Alexander Golberg[‡]

*Institut de Pharmacologie et de Biologie Structurale,
Université de Toulouse, CNRS, UPS, France
[†]Faculty of Electrical Engineering, University of Ljubljana, Slovenia
[‡]Porter School of Environmental Studies, Tel Aviv University, Israel

Abstract

Controlling the process of wound healing is necessary to prevent abnormal healing processes such as chronic wounds and scarring. In this chapter, we discuss the use of electroporation, and it's various applications treating wounds. Electroporation is an increase of cell membrane permeability in a temporary or permanent way by exposing the cells to a high voltage pulsed electric field. Temporary increase of cell membrane permeability by external pulsed electric fields is known as reversible electroporation and permanent increase in the permeability, which leads to cell death, is known as irreversible electroporation. Both modes of electroporation have been used for the wound healing applications. We start the chapter with the basic explanation of electroporation phenomena. Next, we discuss the use of reversible electroporation in the wound healing for DNA delivery to wounds. We continue with examples of the use of irreversible electroporation for burn wounds disinfection. Next, we show the use of both reversible and irreversible electroporation in scar treatment. We conclude with future direction that could lead this non-thermal, chemical free intervention by pulsed electric fields described for practical clinical applications in wound management.

1. Brief Introduction to Electroporation Phenomena

The first reports of pulsed electric fields applied for medical purposes date back to antiquity, with Greeks and Romans using electric rays and torpedo fish to numb pain and treat headaches.[1] Still, the physical nature of these effects then remained unexplained for almost two millennia, until the middle of the 18th century, when Galvani demonstrated that electric pulses act on nerves and muscles, while Nollet performed the first systematic investigation of the disruptive effects of electric discharges on the skin,[2] most notably the occurrence of "red spots" that we now identify primarily as damage to the capillaries due to electroporation of the plasma membranes of their endothelial cells. The electrical conductivity changes in nerves damaged by electric fields, reported by Ritter in 1802,[3] Frankenhaeuser and Widén in 1954,[4] and Stampfli and Willi in 1957,[5] are now also known to be the consequence of electroporation — in this case of the plasma membrane of the neurons.

In 1898, Fuller investigated sterilization of water samples by high voltage discharges,[6] reporting what was likely the first evidence on non-thermal electric-pulse inactivation of microbes, nowadays understood as irreversible electroporation of their envelopes (membranes and walls). The first systematic studies of microbe inactivation by electroporation was published by Sale and Hamilton in a series of three papers in 1967–1968.[7–9] The modern medical applications of electroporation technology probably started in 1982 with the seminal work of Neumann and colleagues in which they used pulsed electric fields to temporarily permeabilize a cell membrane to foreign DNA molecules which can thus be delivered into cells[10]; a month later, the same group named this method "electroporation".[11]

Since the discovery of electroporation, several theoretical descriptions of this phenomenon have been proposed, assuming either a certain type of deformation of the membrane lipids,[12–14] their phase transition,[15] breakdown of interfaces between the domains with different lipid compositions,[16] or denaturation of membrane proteins.[17] However, all these descriptions suffer from serious flaws.[18] Today's consensus describes electroporation as formation of aqueous pores in the lipid bilayer,[18–20] hence the nowadays prevalent use of the term "electroporation" as opposed to the broader term "electropermeabilization", which is also applicable to all the alternative explanations of the electrically induced or enhanced membrane permeability.

The theory of aqueous pore formation, largely based on thermodynamical considerations, describes the formation of aqueous pores as started by penetration of water molecules into the lipid bilayer of the membrane, which leads to reorientation of the adjacent lipids with their polar headgroups

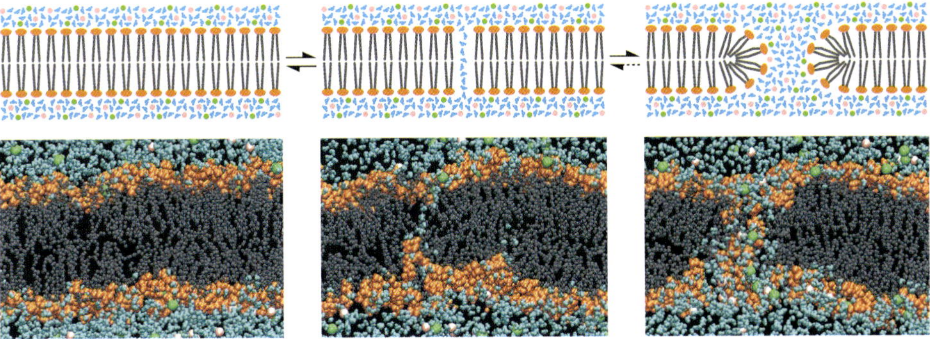

Figure 1. An idealized molecular-level scheme (top) and an atomic-level molecular dynamics simulation (bottom) of electroporation, with the electric field perpendicular to the bilayer plane. In the simulation, a POPC bilayer surrounded by a saline solution is exposed to a field of 4 MV/cm and snapshots are taken 0, 0.15, and 0.50 ns after the field is turned on. Left: the intact bilayer. Middle: water molecules start penetrating the bilayer, forming a water wire. Right: the lipids adjacent to the water wire start reorienting towards the water molecules with their polar headgroups, stabilizing the pore and allowing further water, as well as other polar molecules and ions to enter. The atoms of the lipid headgroups and tails are shown in orange and gray, respectively, water molecules in cyan, sodium ions in green, and chloride ions in pink. Top panel reprinted with permission from Ref. 26; © 2013 Elsevier. Bottom panel reprinted with permission from Ref. 20, © 2012 IEEE.

towards these water molecules (Figure 1). Unstable pores with nanosecond lifetimes can form even in the absence of an external electric field, but presence of such a field induces a voltage across the lipid bilayer (i.e. a difference between the electric potentials on the inner and the outer side of the membrane), thus reducing the energy required for penetration of water into the bilayer. This increases both the probability of pore formation and the pores' average lifetime, resulting in a larger number of pores formed in the bilayer per unit of area and per unit of time, with the pores more stable than in the absence of the electric field. For transmembrane voltages in hundreds of millivolts, the number of pores becomes large enough, and their average lifetimes long enough (from milliseconds up to minutes) for detectable increase of membrane permeability and transport of molecules across it.

Aqueous pores in the bilayer have radii of up to several nanometers, which is too small for observation under optical microscopes, while sample preparation required for electron microscopy of soft matter (vacuumization, fixation, and/or metallic coating) is too aggressive for reliable preservation of semi-stable bilayer structures, and electrically induced pores observed in this manner cannot be clearly distinguished from artifacts.[19] Somewhat in contrast, the pores in the cell wall, particularly in the peptidoglycan wall of

Gram-positive bacteria, are much more stable and visualizable under electron microscopes.[21,22] Still, during the last decade, the theory of aqueous pore formation was corroborated rather convincingly by molecular dynamics simulations (Figure 1), which largely confirm the theoretical view of the sequence of molecular-scale events, and also show a clear increase in pore formation and stability with the increase of the external electric field, and hence of transmembrane voltage induced by this field.[23–25]

In electroporation, the formation of pores is governed by statistical thermodynamics,[18] so it is strictly speaking not a threshold event, in the sense that the pores could only form at electric field strengths exceeding a certain fixed value. Nonetheless, electroporation-mediated transport across the membrane is strongly correlated with the transmembrane voltage induced by the electric field delivered to the cells, which is in turn proportional to the strength of this field.[27] The correlation between the induced transmembrane voltage and the electroporation-mediated transport is demonstrated particularly clearly by combining potentiometric measurements and monitoring of transmembrane transport on the same cell.[27]

There are four general contiguous ranges of electric pulse amplitude and duration, each characterized by different properties of the electropores formed and transport through them. For weak and short pulses, the pores, even if formed, are too small and short-lived for measurable transport, hence there is no detectable electroporation. For stronger and/or longer pulses, the pores formed provide a temporary pathway for transport, but after the pulses they reseal, the transport ceases, and most cells retain their viability. As the amplitude and/or duration of the pulses is increased further, an increasing fraction of pores reseals too slowly to preserve cell viability, or not at all, leading to cell death, but without thermal damage — this is non-thermal irreversible electroporation.[28] Finally, for even stronger and/or longer pulses, irreversible electroporation becomes accompanied by thermal damage. These four ranges partly overlap, both because pore formation is stochastic and because the exposed cells generally vary in size and shape. The range boundaries also depend on the cell type and the properties of the cells' surrounding medium. As the exposure duration increases, the transitions between adjacent regions occur at lower pulse amplitudes, but the range of detectable poration has a lower bound, and below a certain pulse amplitude, no transmembrane transport is detected no matter how long the applied pulses.

Similarly to pore formation, pore resealing is a stochastic process, but it proceeds on a much longer time scale: formation of electropores takes nano to microseconds, while their resealing is often only completed seconds or even minutes after the end of the exposure.[29–31] More detailed measurements reveal that the resealing proceeds in several stages with time constants ranging from micro- and/or milliseconds up to tens of seconds.[30,31] Unfortunately, neither the existing theory nor the experiments can currently provide a reliable elucidation of the specific events characterizing each of these distinctive stages, while reliable molecular dynamics simulations, even in their most simplified coarse-grained versions, cannot yet cover time scales that long.

Although the effects of pulsed electric fields on biological matter have been investigated scientifically for more than 250 years, only in the last three decades the practical applications of electroporation system broke into food processing, pharmaceutical industry, and medicine. Electroporation systems of different scales and for various applications have been commercialized; ranging from $1\,\mu l$ multipurpose bench systems for 10 *t/h* processing facilities for sugar extraction (Figure 2).

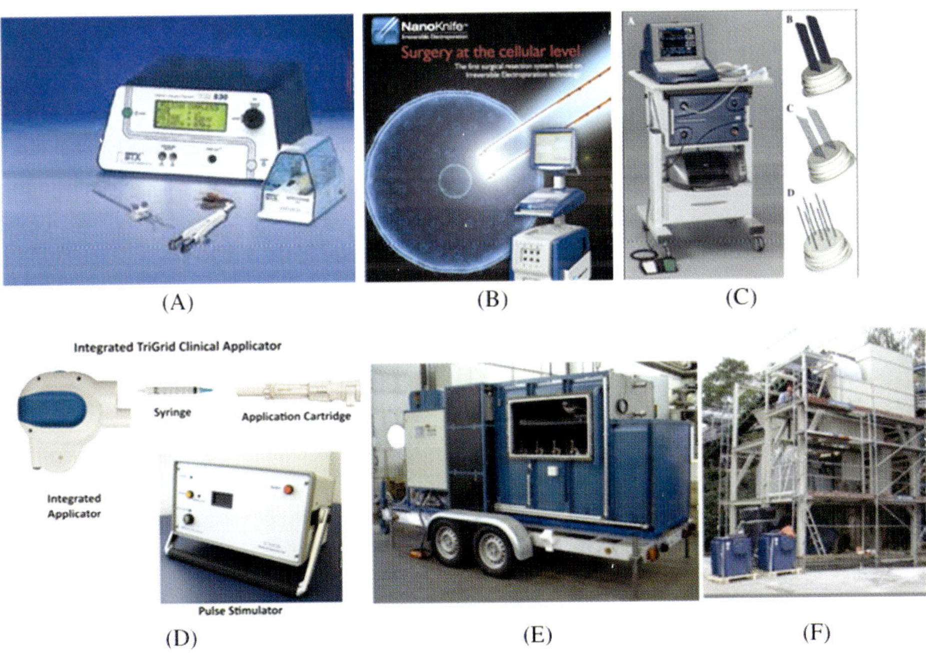

Figure 2. Examples of commercialized electroporation systems. (A) Bench scale, multipurpose electroporator, BTX®, Harvard Apparatus; (B) NanoKnifeTM tissue ablation system, Angiodynamics; (C) Clinoporator for gene and small molecules delivery; (D) DNA vaccination gun, Ichor Medical Systems; (E) transportable electroporation device for grapes processing, The Karlsruhe Institute of Technology; (F) 10 *t/h* electroporation system for sugar extraction, The Karlsruhe Institute of Technology.

In medicine, reversible electroporation technologies and methods have been developed into novel clinical applications: electrochemotherapy (ECT), first demonstrated by Okino and Mohri in 1987[32] and gene electrotherapy that followed.[33] Lee and colleagues suggested that the tissue electric trauma has a non-thermal, irreversible electroporation nature of damage, in addition to thermal effects.[34] Davalos, Mir, and Rubinsky in 2005 proposed that pulsed electric fields can non-thermally ablate clinically relevant volumes of tissue, for example solid tumors.[35] Miklavcic et al. showed that electric field distribution correlates with effectiveness of ECT[36] — this serves as the basis for careful treatment planning of electroporation technologies in tissues. Treatment planning is a critical part of the successful electroporation, treatment protocol.[37] The applications of electroporation, based technologies and treatments have been reviewed extensively. The goal of this chapter is to discuss recent advances of the electroporation technologies and treatments in the wound healing.

2. Gene Electrotransfer in Wound Healing

Gene delivery by electroporation, also named gene electrotransfer (GET), is a promising physical tool to safely and locally deliver genetic information into a targeted tissue. The first published clinical trial originated in 2004, and by now more than 50 trials have used electroporation for plasmid delivery.[38] This method of vectorization, based on electric field application, is highly promising in both cancer[39] and gene therapy[40] as well as for infectious diseases vaccination.[41] It has also been adapted to improve cutaneous wound healing.[42] Indeed, in some pathological situations such as in diabetes, infection, or venous diseases, non-healing chronic wounds are observed, and requires the development of innovative therapeutic strategies.[43] Instead of topically applying recombinant growth factors, which is expensive and can jeopardize wound care,[44] the administration of a gene encoding these therapeutic growth factors appears as a promising approach.[45] Indeed, it allows the continuous local production of therapeutic factors during transgene expression. Since the transfection is not stable, plasmid expression stops classically in a few days or at the latest weeks after GET, which is sufficient enough for helping wound closure while being totally safe.

Even if pig skin anatomy and wounding processes are the closest ones to those of humans, the cost per animal as well as for housing facilities led to the widely used rodent model to study GET in skin.[46] Interestingly, murine genetic mutants such as leptin receptor-deficient diabetic mice $Lepr^{db}/Lepr^{db}$ allow study of GET in a model of delayed wound healing (Table 1). Different

Table 1. Studies on therapeutic GET promoting cutaneous wound healing.

Models	Gene	Electrode	Electrical Parameters	References
Rat	VEGF	Needles	3 pulses lasting 50 ms, 160 V/cm, 1 Hz	51
Rat	VEGF	Needles	8 pulses lasting 20 ms, 200 V/cm, 1 Hz	50
Rat	VEGF	Posts	72 pulses lasting 150 ms, 500 V/cm,6.6 Hz	49
Rat	bFGF	Needles	3 pulses lasting 50 ms, 50 V/cm, 1 Hz, followed by 3 pulses of the opposite polarity	65
Diabetic mouse	KGF-1	Pins	6 pulses lasting 100 μs, 1800 V/cm,8 Hz	66
Septic rat	KGF-1	Pins	6 pulses lasting 100 μs, 1800 V/cm, 8 Hz	67
Diabetic mouse	TGF-β1	Syringe	6 pulses lasting 20 ms,100 V/cm, 1 Hz	58
Diabetic mouse	HIF-1α	Pins	10 pulses lasting 20 ms, 800 V/cm, 8 Hz	68
Normal and diabetic mouse	hCAP-18/LL-37	Plates	1 pulse lasting 100 μs, 700 V/cm, immediately followed by 1 pulse lasting 400 ms at 200 V/cm	69
Rat hindlimb ischemia	bFGF	Posts	Pulses lasting 150 ms, 250 V/cm	70

Source: Reproduced with permission from Ref. [42].

types of wounds were considered in *in vivo* experiments both in normal or pathological context. In addition to the incisional wound model, in which a sharp instrument incises the skin,[47] excisional wounds removed a piece of full-thickness skin creating a large defect.[48] Furthermore, wounds created during reconstructive surgery such as skin flap[49,50] or the transverse rectus abdominis myocutaneous (TRAM) flap [51] were also investigated, principally in order to prevent skin necrosis after surgical intervention.

Reporter genes such as luciferase, green fluorescent protein (GFP), DsRed and Beta-galactosidase (LacZ) have been widely used to decipher GET mechanisms in cutaneous context.[52] They allowed identifying efficiently transfected cell types as well as level and duration of transgene expression in skin, depending on electrode design and electrical parameters applied,[47,53–55] (Table 1). Since GET is dependent on DNA electrophoresis,[56,57] trains of pulses lasting several milliseconds are generally applied (Table 1).

To improve wound healing, therapeutic relevant plasmids have been electrotransferred to injured skin. Plasmids encoding molecules recognized to induce cell proliferation, to promote cell migration or differentiation, and to stimulate angiogenesis were used (Table 1). Vascular endothelial growth factor (VEGF), keratinocyte growth factor (KGF), and transforming growth factor (TGF) are the most represented in *in vivo* experiments.

Some authors demonstrated in GET context that electric field applied alone significantly improved wound closure compared to non-treated wounds.[58] Indeed, even if underlying mechanisms are not fully elucidated,[59] electrical stimulation is known for decades to improve wound healin.[60] These interesting information pave the way for a better comprehension and exploitation of electrostimulation potential of electric field in wound healing.

Even though murine skin is recognized to be biologically and anatomically distinct from human skin,[61,62] it is encouraging to obtain animal data suggesting that wound healing benefits from GET with therapeutic plasmids. Future developments of GET will be to study gene electrotherapy mechanisms on human skin substitutes reconstructed by tissue engineering. Indeed, some innovative tissue-engineered approaches such as self-assembly[63] allow to produce *in vitro* human skin substitutes adapted to GET studies.[64] It appears to be a promising approach that could provide novel findings on GET mechanisms in human context of wound healing.

Wound care is becoming a major societal issue with aging of the population. Indeed, aging skin undergoes gradual structural and physiological changes which lead to global skin fragility and susceptibility to injuries such as pressure ulcers and wounds.[71] Thus, it is of utmost importance to develop innovative therapeutic strategies both to promote tissue regeneration in aged skin and to improve wound healing processes. The combination of GET with electric therapy[72,73] seems to be an interesting and promising field of research to improve wound healing, especially in infected chronic wounds.

3. Wounds Disinfection with Irreversible Electroporation

Wound infection management still remains challenging, and choosing an appropriate treatment is a difficult task.[74] Although >6000 types of wound dressings exist,[75] the problem of wound infections has yet to be solved. In the case of burn wounds, approximately 500,000 people seek medical treatment for every year in the US; infection remains a major cause of morbidity and mortality in these patients.[76] In addition to the extent and nature of the thermal injury affecting the susceptibility to infection, the type and amount of the microbial burden colonizing the wound appear to influence the risk of morbidity and mortality. Pathogens that infect burn wounds are primarily *Acinetobacter baumannii*, methicillin-resistant *Staphylococcus aureus, Pseudomonas, and Klebsiella* — pathogens that are increasingly resistant to various antimicrobial agents.[77]

Irreversible electroporation is under scrutiny for the last 50 years. During this time multiple parameters were reported for a wide range of bacteria. Most of the reports, however, focused on the bacteria which are common food pathogens; parameters were optimized for food and liquid processing facilities. According to the FDA the typical electric field used for food disinfection applications is 200–800 Vmm^{-1} with pulse duration of several microseconds and pulse number of up to 30.[75] Recently IRE was also applied to treat the bacterial wound infections (Figure 3). IRE efficiently disinfected contaminated burned murine skin. Using 80 pulses of 5000 Vcm^{-1}, stable disinfection with 4.91 ± 0.71 Log10 reduction of *Acinetobacter baumannii*, 3 h after treatment was achieved (Figures 3–5).[78,79]

The magnitude of disinfection was correlated with both the electric field strength and the number of delivered pulses. The increase in pulse number led to a larger reduction in bacterial load and bioluminescent signal immediately after treatment, as compared to the increase in the field strength. Increasing the pulse number from 40 to 80, led to a 255% increase in the reduction of bacterial load in the wound, from 1.49 ± 0.07 Log10 to 5.30 ± 0.85 Log10. Increasing the field, however, from 2500 Vcm^{-1} to 5000 Vcm^{-1}, while keeping the number of pulses at 40, led to only a ~37% increase in the log reduction of bacterial load in the wound, from 1.49 ± 0.07 Log10 to 2.04 ± 0.29 Log10.

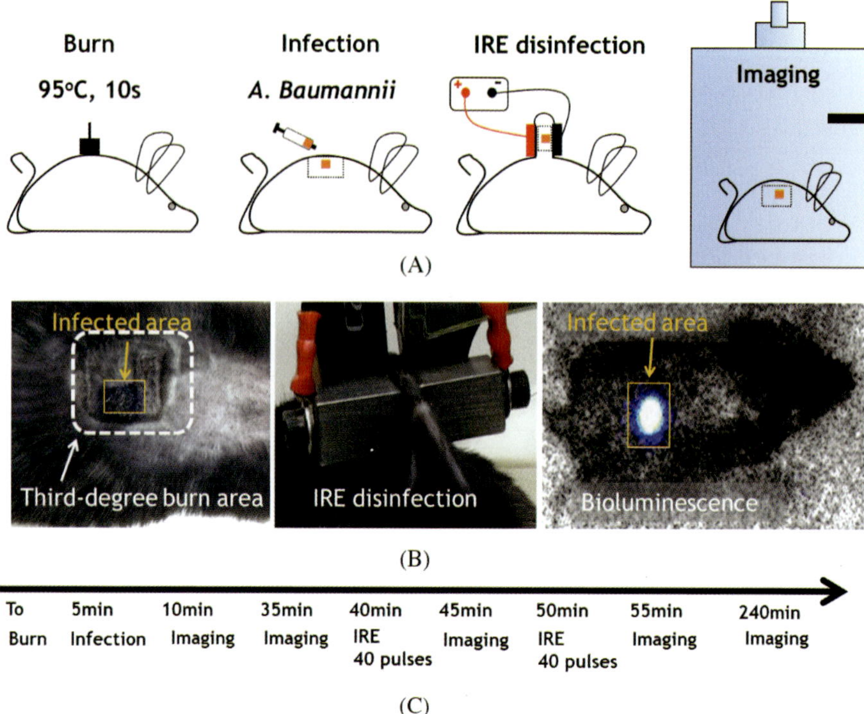

Figure 3.   Pulsed electric field disinfection *in vivo*: C57BL/6 black mice model. (A) Schematic illustration for procedures we used in this study. The 1 cm^2 burn injury was followed by dispersion of *A baumannii* on part of the wound. Next pulsed electric field was applied using two plate electrodes. *A baumannii* infection load was quantified using bioluminescent imaging. (B) Left panel shows digital photography of the burned (white frame) and infected (orange frame) areas of the skin. Central panel shows digital photography of the applied electrodes. Right panel shows the images of the mice as observed inside the dark imaging box. Orange frame shows the infected area as detected by a strong bioluminescent signal emitted from bacteria. (C) The performed experiment timeline. Reprinted with permission from Ref. 79.

Surprisingly, the increase in IRE-disinfection capability did not correlate with the increase of energy delivered. For delivery of 80 pulses with 2500 Vcm^{-1} at 1 Hz, ~9 J are needed. For delivery of 40 pulses with 5000 Vcm^{-1} at 1 Hz, ~18 J are needed. These energy consumption findings are interesting: they show that increasing the number of pulses from 40 to 80, leads to a significantly larger bacterial load reduction than increasing the electric field strength from 2500 Vcm^{-1} to 5000 Vcm^{-1}, therefore bacterial killing does not necessarily depend on delivered energy. These findings are strikingly different from heat/radiation-based disinfection where the bacterial load reduction directly correlates with consumed energy. Nevertheless, these findings are consistent with the current electroporation theory. According to the current theory,

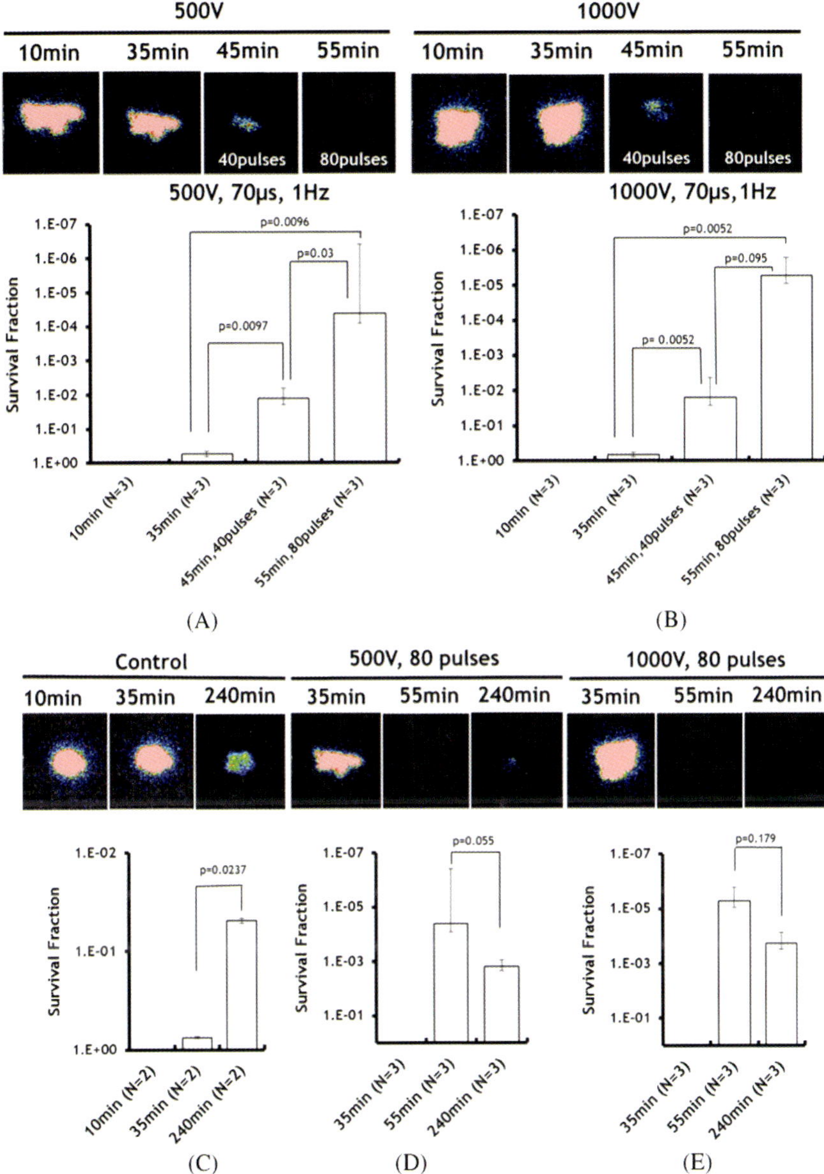

Figure 4. The effect of pulse number and electric field strength on *A. baumannii* infection load reduction. (A) Applied voltage 500 V, 2 mm gap between electrodes. (B) Applied voltage 1000 V, 2 mm gap between electrodes. Top panel shows the post-burn time when the images were taken. The bottom panel shows the survival fraction of microorganisms as detected by the top panel images (logarithmic, inversed scale). N shows the number of animals per group. (C) Control: not treated, burned and infected skin. (D) Applied voltage 500 V, 2 mm gap between electrodes. (E) Applied voltage 1000 V, 2 mm gap between electrodes. Top panel shows the post-burn time when the images were taken. The bottom panel shows the survival fraction of microorganisms as detected by the top panel images (logarithmic, inversed scale). N shows the number of animals per group. Error bar ± standard deviation of the mean. Reprinted with permission from Ref. 79.

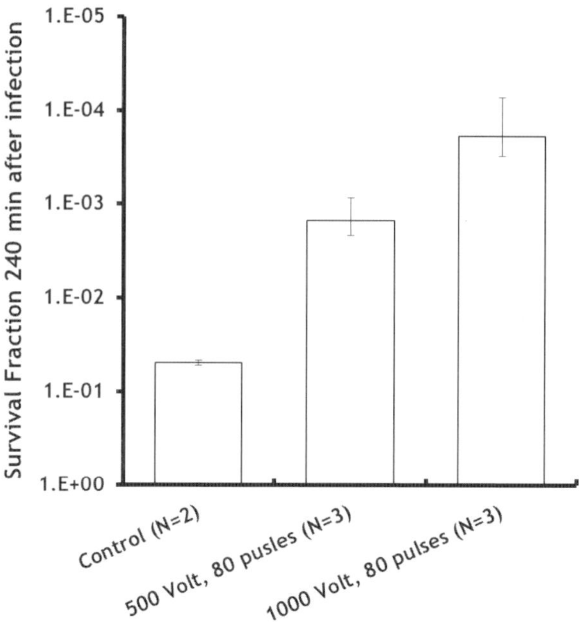

Figure 5. Summary of *A. baumannii* survival three hours after IRE, the effect of electric field intensity. *N* shows the number of animals per group. Error bar ± standard deviation of the mean. Gap between electrodes 2 mm. Reprinted with permission from Ref. 79.

increasing the field strength increases the total electroporated surface of the cell membrane. Increasing number of pulses, after the electroporation threshold potential is reached, increases the number and size of the aqueous pores of the membranes at the electroporated site.[80–82] Additional studies showed that normal IRE ablated skin regenerates without scars.[72,83]

4. Burn Scars Treatment with Partial Irreversible Electroporation

Hypertrophic scarring (HTS) after trauma and burn injury remains a major clinical challenge that leads to physical, aesthetic, functional, psychological, and social stresses in thousands of patients.[84] There is a critical need for new approaches for HTS research and therapy. Complex interaction between cells, extracellular matrix, signaling factors and environmental factors, such as oxygenation, are most often ignored in many proposed HTS therapies, which usually focus on a single target. This focus on single targets can lead to failure in clinical trials as recently shown in TGF-β3 based therapy in 2011.[85] Recently, fractional laser ablation has been proposed to treat scars.[86] However, light penetration in skin is low due to the high scattering of light in skin.[87] In contrast, for electroporation-based therapies, the

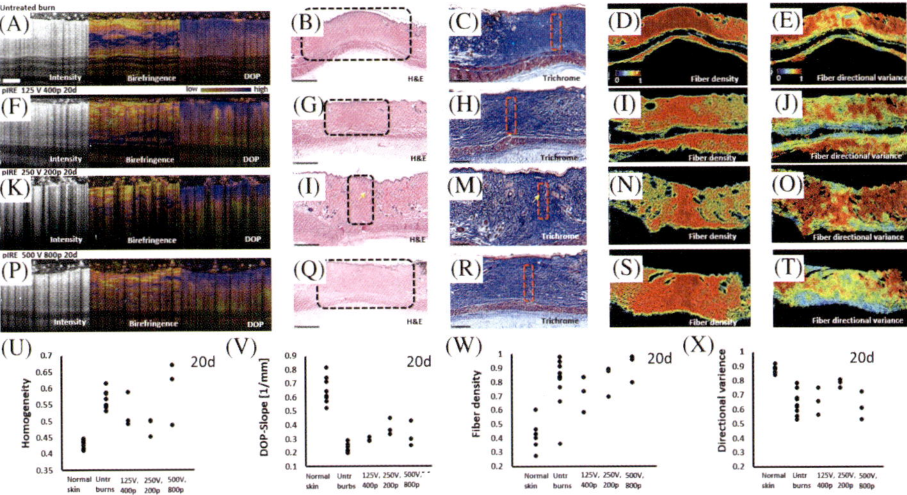

Figure 6. Skin regeneration following partial irreversible electroporation of third degree burns six months after the injury. Untreated burns ($n = 9$): (A) PS-OCT; (B) histology H&E. (C) Histology Masson's trichrome; (D) fiber density; (E) fiber directional variance. Burns treated with 125 V, 400 pulses every 20 days ($n = 3$): (F) PS-OCT; (G) histology H&E. (H) Histology Masson's trichrome; (I) fiber density; (J) fiber directional variance. Burns treated with 250 V, 200 pulses every 20 days ($n = 3$): (K) PS-OCT; (L) histology H&E. (M) Histology Masson's trichrome; (N) fiber density; (O) fiber directional variance. Burns treated with 500 V, 800 pulses every 20 days ($n = 3$): (P) PS-OCT; (Q) histology H&E. (R) Histology Masson's trichrome; (S) fiber density; (T) fiber directional variance. In all pIRE protocols pulse duration was 70 µs and pulse delivery frequency was 3 Hz. (U) Quantification of PS-OCT homogeneity ($n = 3$ for normal skin and pIRE, $n = 9$ for untreated burns). (V) Quantification of PS-OCT DOP-slope ($n = 3$ for normal skin and pIRE, $n = 9$ for untreated burns). (W) Quantification of fiber density ($n = 3$ for pIRE and $n = 9$ for untreated controls). All individual measurements are reported. Scale bar: (A), (C), (F), (H), (K), (M), (P), (R) = 500 µm, (B), (G), (L), (Q) = 1 mm. Reprinted from Ref. 90 with permission.

distribution of the electric fields depends on the electrical properties of tissues such as resistance and capacitance.[88] In 2013, a therapy combining delivery of bleomycin with electroporation has been evaluated for the treatment of HTS and keloids in humans (Figure 6).[89]

Recently, a new therapeutic procedure based on intermittently delivering partial irreversible electroporation (pIRE) to burn wounds in order to prevent scarring was introduced.[90] The treatments were delivered at various specified time intervals for three months and the wounds were left to heal for an additional three months after the delivery of the last treatment. Nine combinations of the possible pIRE parameters were tested. The wound healing was monitored by visual inspection and with polarization-sensitive optical coherence tomography (PS-OCT)[91] for the entire six months after the burn injury.

To evaluate the impact of pIRE on the wound healing, the scar area, features from the analysis of PS-OCT and histology were used. From histological sections the collagen fiber density and the fiber directional variance at the center of the original wound were measured. Fully developed scars are expected to have a higher density of fibers, paired with a lower directional variance, i.e. the fibers are more organized and less intercalated than in normal skin. For PS-OCT, a texture analysis of the imaged birefringence to express its homogeneity, and extraction of the slope of the degree of polarization (DOP) along the depth of the scar were performed. Fully developed scars have homogenous birefringence with a slowly decreasing DOP, whereas normal skin has a very heterogeneous birefringence with a rapidly decreasing DOP. The impact of pIRE on these features is summarized in Table 2. PS-OCT results and histology of a representative subset of the evaluated parameters appear in Figure 6. The smallest scars were observed in the group of animals treated every twenty days (Figure 6 shows the results for *in vivo* PS-OCT imaging and histological analysis for this group).

The largest reduction in the scar size (19.6 ± 5.1 mm^2 vs. 46.6 ± 9.1 mm^2 in the untreated burns) was achieved with 250 volts, 2 mm gap between electrodes and 200 pulses, applied every 20 days for 3 months after the burn injury.[90] We mention here applied voltage and a gap between electrodes, as normal skin is a heterogeneous organ and, thus, field distribution varies at different layers. A total of five treatments were delivered for this setting. This 57.9% reduction in scar size of pIRE treated wounds in comparison with untreated scars was accompanied by PS-OCT (Figure 6K, U, V) and histological (Figure 6L, M, N, O, W, X) features approaching those of normal skin, indicating the regenerative effect of pIRE. Wounds treated with this pIRE setting showed a homogeneity of 0.486 ± 0.027 (normal skin 0.427 ± 0.013, untreated burns 0.570 ± 0.027); DOP-slope of 0.383 ± 0.060 (normal skin 0.653 ± 0.097, untreated burns 0.226 ± 0.034), fiber density 0.821 ± 0.111 (normal skin 0.421 ± 0.011, untreated burns 0.796 ± 0.178); and fiber directional variance 0.781 ± 0.028 (normal skin 0.879 ± 0.027, untreated burns 0.637 ± 0.085), indicating enhanced wound regeneration and resulting smaller scars (Figure 6).

Six months after the third degree burn injury, histopathological analysis of untreated scars revealed large scars with a rectangular shape, as seen in histological cross-sections in the plane orthogonal to epidermis (Figure 6B, C). The uniformly dense dermis (with a fiber density of 0.796 ± 0.178, almost double the density of normal skin, 0.421 ± 0.011, Figure 6D) overlaid with a plaque-like epidermis featured collagen fibers that ran predominantly parallel to the surface without evidence of any normal pattern of

Table 2. The impact of the pIRE of the scar formation and properties six months after the third degree burn. Reprinted from Ref. 90 with permission.

Voltage	Number of pulses	Treatment frequency	Scar area (mm^2)	Fiber density	Fiber directional variance	Homogeneity	DOP-slope [1/mm]
Normal skin			0	0.42 ± 0.05	0.88 ± 0.01	0.43 ± 0.01	0.65 ± 0.030
Untreated burn			46.6 ± 9.1	0.85 ± 0.032	0.64 ± 0.030	0.57 ± 0.009	0.22 ± 0.020
125	200	10	39.0 ± 13.4	0.87 ± 0.032	0.64 ± 0.030	0.55 ± 0.009	0.33 ± 0.011
125	400	20	38.67 ± 5.2	0.72 ± 0.021	0.66 ± 0.029	0.53 ± 0.007	0.30 ± 0.010
125	800	30	37.9 ± 4.9	0.81 ± 0.072	0.61 ± 0.054	0.55 ± 0.0031	0.29 ± 0.008
250	200	20	19.6 ± 2.5	0.82 ± 0.064	0.78 ± 0.017	0.49 ± 0.016	0.38 ± 0.035
250	400	30	35.2 ± 25.5	0.92 ± 0.027	0.70 ± 0.050	0.55 ± 0.020	0.32 ± 0.066
250	800	10	23.5 ± 11.2	0.90 ± 0.015	0.73 ± 0.024	0.55 ± 0.014	0.33 ± 0.035
500	200	30	34.4 ± 2.7	0.96 ± 0.016	0.74 ± 0.059	0.53 ± 0.042	0.36 ± 0.047
500	400	10	32.9 ± 14.9	0.85 ± 0.028	0.63 ± 0.025	0.61 ± 0.020	0.30 ± 0.032
500	800	20	31.9 ± 2.2	0.91 ± 0.058	0.62 ± 0.056	0.60 ± 0.055	0.33 ± 0.054

intercalated collagen (fiber directional variance 0.637 ± 0.085, in comparison to 0.879 ± 0.027 in normal skin, Figure 6E). There were no hair follicles or other skin appendages. The capillaries and venules were oriented perpendicular to surface. In contrast, pIRE treated wounds presented smaller scars with small hair follicles in the scar area (Figure 6L, M, yellow arrows). Different from untreated, rectangular burn scars, pIRE treated scars featured a cross-sectional shape of a trapezoid, with the shorter base oriented towards the epidermis. Nevertheless, in the center of the treated scars the fiber density was still larger than in normal skin (0.821 ± 0.111). However, striking intercalation of collagen fibers in contrast to the linear arrangement in the untreated scars was observed (fiber directional variance in the pIRE treated wound was 0.781 ± 0.028, in comparison to 0.637 ± 0.085 in the untreated burn, and 0.879 ± 0.027 in normal skin). The fiber architecture was close to normal, except that the fibers were thicker than in normal skin (Figure 6N, O, W, X). The capillaries and venules in the dermis showed a close to normal, predominantly horizontal orientation.[90]

pIRE is a non-invasive HTS treatment therapy affecting directly the cell membranes, preserving the wound, and without using chemicals or light. The best identified pIRE protocol includes three essential components: (1) an electric field strength corresponding to an applied voltage of 250 V across electrodes separated by 2 mm, (2) 200 pulses, and (3) four treatments delivered every 20 days starting at the time of injury. Using Taguchi Robust experiment design, it was shown that this combination of parameters lead to the most significant improvement of the wound regeneration after third degree burns in terms of scar size, fiber orientation, and partial regeneration of skin appendages. pIRE effects on tissue are very different from laser-induced tissue damage, because IRE allows for special and temporal control of viable cell density without any significant effect on the surrounding ECM, tissue oxygenation, and mechanical properties.[72]

5. Future Directions on the use of Electroporation in Wound Healing

An ultimate goal in the use of electroporation technologies in medicine is the improvement of human health. A major obstacle in preclinical research of wound healing therapies is the lack of validated animal models that correspond closely to human adults. To date, animal models of tissue regeneration and repair have focused on amphibians, rabbits, pigs, and small rodents.[92,93] Most of the proof of the concept animal studies in electroporation discussed here were performed in small rodents, a model which obviously has its limitations in comparison to human wounds. While human wound healing is based on cell migration, rodents have *panniculus carnosus* in dorsal skin, a muscle which contributes to wound healing by contraction. Additional work in large models will have to be performed in order to reduce the impact of self-contraction on the wound. The best would be to take advantage of tissue engineering technological progress to develop relevant human skin substitutes for wound healing studies. The next major step for IRE applied to regenerative medicine should be a first clinical trial to demonstrate the safety and efficacy of this new method in humans. An additional important issue to be addressed in the following work is pain. A dense grid of electrodes could provide precise delivery of the electric fields to the upper layers of the papillary dermis in humans, avoiding the exposure of large nerves in the skin. The next step should be the development of flexible electrode grids to allow precise delivery of various electric fields to the various locations in the wound. Further optimization of the IRE protocol could improve additional metrics of skin regeneration, particularly the reduction of collagen fiber density in the center of the wound and complete elimination of the infecting organisms.

References

1. Bullock, T. H. Hopkins, C. D. and Popper, A. N. *Electroreception* New York: Springer, (2005).
2. Nollet, J. Recherches sur les causes particulieres des phénoménes électriques, *Paris Chez H. L. Guerin L. F. Delatour* (1754).
3. Noad, H. *Lectures on Electricity:Comprosing Galvnism, Magnetism, Electromagentism, Maneto- and Thermo- Electricity, and Electro-Physiology.* London: George Knoght and Sons (1849).
4. Frankenhaeuser, B. and Widén, L. Anode break excitation in desheathed frog nerve. *J Physio* **131**, 243–247 (1956).
5. Stampfli, R. and Willi, M. Membrane potential of a Ranvier node measured after electrical destruction of its membrane. *Experientia* **13**(7), 297–298 (1957).
6. Fuller, G. *Report on the investigations into the purification of the Ohio river water at Lousville Kentuki* New York: Company DVN (1898).
7. Hamilton, W. and Sale, J. H. Effects of high electric fields on microorganismsII. Mechanism of action of the lethal effect. *Biochim Biophys* **148**(3), 789–800 (1967).
8. Sale, J. and Hamilton, W. Effects of high electric fields on micro-organisms. 3. Lysis of erythrocytes and protoplasts. *Biochim Biophys Acta* **163**(1), 37–43 (1968).
9. Sale, J. H. and Hamilton, W. Effects of high electric fields on microorganisms: I. Killing of bacteria and yeasts. *Biochim Biophys Acta* **148**(3), 781–788 (1967).

10. Wong, T.-K. and Neumann, E. Electric field mediated gene transfer. *Biochem Biophys Res Commun* **107**(2), 584–587 (1982).

11. Neumann, E. Schaefer-Ridder, M. Wang, Y. and Hofschneider, P. H. Gene transfer into mouse lyoma cells by electroporation in high electric fields. *Eur Mol Biol Organ J* **1**(7), 841–845 (1982).

12. Michael, D. and O'Neill, M. Electrohydrodynamic instability in plane layers of fluid. *J Fluid Mech* **41**, 571–580 (1970).

13. Crowley, J. M. Electrical breakdown of bimolecular lipid membranes as an electromechanical instability. *Biophys J* **13**(7), 711–724 (1973).

14. Steinchen, A. Gallez, D. and Sanfeld, A. A viscoelastic approach to the hydrodynamic stability of membranes. *J Colloid Interface Sci* **85**, 5–15 (1982).

15. Sugár, I. A theory of the electric field-induced phase transition of phospholipid bilayers. *Biochim Biophys Acta* **556**, 72–85 (1979).

16. Cruzeiro-Hansson, L. and Mouritsen, O. G. Passive ion permeability of lipid membranes modelled via lipid-domain interfacial area. *BBA — Biomembranes* **944**(1), 63–72 (1988).

17. Tsong, T. Y. Electroporation of cell membranes. *Biophys J* **60**(2), 297–306 (1991).

18. Weaver, J. C. and Chizmadzhev, Y. A. Theory of electroporation: A review. *Bioelectrochemistry Bioenerg* **41**(2), 135–160 (1996).

19. Spugnini, E. P. Arancia, G. Porrello, A. Colone, M. Formisano, G. Stringaro, A. Citro, G. and Molinari, A. Ultrastructural modifications of cell membranes induced by 'electroporation' on melanoma xenografts. *Microsc Res Tech* **70**(12), 1041–1050 (2007).

20. Kotnik, T. Kramar, P. Pucihar, G. Miklavčič, D. and Tarek, M. Cell membrane electroporation — Part 1: The phenomenon. *IEEE Electr Insul M* **28**(5), 14–23 (2012).

21. Yeo, S. K. and Liong, M. T. Effect of electroporation on viability and bioconversion of isoflavones in mannitol-soymilk fermented by lactobacilli and bifidobacteria. *J Sci Food Agric* **93**(2), 396–409 (2013).

22. Pillet, F. Formosa-Dague, C. Baaziz, H. Dague, E. and Rols, M.-P. Cell wall as a target for bacteria inactivation by pulsed electric fields. *Sci Rep* **6**, 19778 (2016).

23. Delemotte, L. and Tarek, M. Molecular dynamics simulations of lipid membrane electroporation. *J Membr Biol* **245**(9), 531–543 (2012).

24. Tarek, M. Membrane electroporation: A molecular dynamics simulation. *Biophys J* **88**, 4045–4053 (2005).

25. Böckmann, R. A. de Groot, B. L. Kakorin, S. Neumann, E. and Grubmüller, H. Kinetics, statistics, and energetics of lipid membrane electroporation studied by molecular dynamics simulations. *Biophys J* **95**(4), 1837–1850 (2008).

26. Kotnik, T. Lightning-triggered electroporation and electrofusion as possible contributors to natural horizontal gene transfer. *Phys Life Rev* **10**(3), 351–370 (2013).

27. Kotnik, T. Pucihar, G. and Miklavčič, D. Induced transmembrane voltage and its correlation with electroporation- mediated molecular transport. *J Membr Biol* **236**(1), 3–13 (2010).

28. Golberg, A. and Yarmush, M. L. Nonthermal irreversible electroporation: Fundamentals, applications, and challenges. *IEEE Trans Biomed Eng* 60(3) 707–714 (2013).

29. Saulis, G. Venslauskas, M. S. and Naktinis, J. Kinetics of pore resealing in cell membranes after electroporation. *J Electroanal Chem* 321(1), 1–13 (1991).

30. Hibino, M. Itoh, H. and Kinosita, K. Time courses of cell electroporation as revealed by submicrosecond imaging of transmembrane potential. *Biophys J* 64(6), 1789–1800 (1993).

31. Pucihar, G. Kotnik, T. Miklavcic, D. and Teissie, J. Kinetics of transmembrane transport of small molecules into electropermeabilized cells. *Biophys J* 95(6), 2837–2848 (2008).

32. Okino, M. and Mohri, H. Effects of a high-voltage electrical impulse and an anticancer drug on *in vivo* growing tumors. *JPn J Cancer Res* 72(12), 1319–1321 (1987).

33. Nomura, M. Nakata, Y. Inoue, T. Uzawa, A. Itamura, S. Nerome, K. Akashi, M. and Suzuki, G. *In vivo* induction of cytotoxic T lymphocytes specific for a single epitope introduced into an unrelated molecule. *J Immunol Methods* 193(1), 41–49 (1996).

34. Lee, R. C. and Kolodney, M. S. Electrical injury mechanisms: electrical breakdown of cell membranes. *Plast Reconstr Surg* 80(5), 672–679 (1987).

35. Davalos, R. V. Mir, L. M. and Rubinsky, B. Tissue ablation with irreversible electroporation. *Ann Biomed Eng* 33(2), 223–231 (2005).

36. Miklavcic, D. Beravs, K. Semrov, D. Cemazar, M. Demsar, F. and Sersa, G. The importance of electric field distribution for effective *in vivo* electroporation of tissues. *Biophys J* 74(5), 2152–2158 (1998).

37. Kos, B. Zupanic, A. Kotnik, T. Snoj, M. Sersa, G. and Miklavcic, D. Robustness of treatment planning for electrochemotherapy of deep-seated tumors. *J Membr Biol* 236(1), 147–153 (2010).

38. Heller, R. and Heller, L. C. Gene Electrotransfer Clinical Trials. *Adv Genet* 89, 235–262 (2015).

39. Sersa, G. Teissie, J. Cemazar, M. Signori, E. Kamensek, U. Marshall, G. and Miklavcic, D. Electrochemotherapy of tumors as *in situ* vaccination boosted by immunogene electrotransfer. *Cancer Immunol Immunother* 64(10), 1315–1327 (2015).

40. Bettan, M. Emmanuel, F. Darteil, R. Caillaud, J. M. Soubrier, F. Delaere, P. Branelec, D. Mahfoudi, A. Duverger, N. and Scherman, D. High-level protein secretion into blood circulation after electric pulse-mediated gene transfer into skeletal muscle. *Mol Ther* 2(3), 204–210 (2000).

41. Vandermeulen, G. Staes, E. Vanderhaeghen, M. L. Bureau, M. F. Scherman, D. and Préat, V. Optimisation of intradermal DNA electrotransfer for immunisation. *J Control Release* 124(1–2), 81–87 (2007).

42. Gibot, L. and Rols, M. P. Gene transfer by pulsed electric field is highly promising in cutaneous wound healing. *Expert Opin Biol Ther* 16(1), 67–77 (2016).

43. Mudge, E. J. Recent accomplishments in wound healing. *Int Wound J* **12**(1), 4–9 (2015).

44. Smiell, J. M. Wieman, T. J. Steed, D. L. Perry, B. H. Sampson, A. R. and Schwab, B. H. Efficacy and safety of becaplermin (recombinant human platelet-derived growth factor-BB) in patients with nonhealing, lower extremity diabetic ulcers: A combined analysis of four randomized studies. *Wound Repair Regen* **7**(5), 335–346 (1999).

45. Yao, F. and Eriksson, E. Gene therapy in wound repair and regeneration. *Wound Repair Regen* **8**(6), 443–451 (2000).

46. Sullivan, T. P. Eaglstein, W. H. Davis, S. C. and Mertz, P. The pig as a model for human wound healing. *Wound Repair Regen* **9**(2), 66–76 (2001).

47. Gao, Z. Wu, X. Song, N. Cao, Y. and Liu, W. Electroporation-mediated plasmid gene transfer in rat incisional wound. *J Dermatol Sci* **47**(2), 161–164 (2007).

48. Byrnes, C. K. Malone, R. W. Akhter, N. Nass, P. H. Wetterwald, A. Cecchini, M. G. Duncan, M. D. and Harmon, J. W. Electroporation enhances transfection efficiency in murine cutaneous wounds. *Wound Repair Regen* **12**(4), 397–403 (2004).

49. Basu, G. Downey, H. Guo, S. Israel, A. Asmar, A. Hargrave, B. and Heller, R. Prevention of distal flap necrosis in a rat random skin flap model by gene electro transfer delivering VEGF(165) plasmid. *J Gene Med* **16**(3–4), 55–65 (2014).

50. Ferraro, B. Cruz, Y. L. Coppola, D. and Heller, R. Intradermal delivery of plasmid VEGF(165) by electroporation promotes wound healing. *Mol Ther* **17**(4), 651–657 (2009).

51. Rezende, F. C. Gomes, H. C. Lisboa, B. Lucca, A. F. Han, S. W. and Ferreira, L. M. Electroporation of vascular endothelial growth factor gene in a unipedicle transverse rectus abdominis myocutaneous flap reduces necrosis. *Ann Plast Surg* **64**(2), 242–246 (2010).

52. Gothelf, A. and Gehl, J. Gene electrotransfer to skin; review of existing literature and clinical perspectives. *Curr Gene Ther* **10**(4), 287–299 (2010).

53. Gothelf, A. Eriksen, J. Hojman, P. and Gehl, J. Duration and level of transgene expression after gene electrotransfer to skin in mice. *Gene Ther* **17**(7), 839–845 (2010).

54. Heller, L. C. Jaroszeski, M. J. Coppola, D. McCray, A. N. Hickey, J. and Heller, R. Optimization of cutaneous electrically mediated plasmid DNA delivery using novel electrode. *Gene Ther* **14**(3), 275–280 (2007).

55. Roos, A. K. Eriksson, F. Timmons, J. A. Gerhardt, J. Nyman, U. Gudmundsdotter, L. Bråve, A. Wahren, B. and Pisa, P. Skin electroporation: Effects on transgene expression, DNA persistence and local tissue environment. *PLoS One* **4**(9), (2009).

56. Sukharev, S. I. Klenchin, V. A. Serov, S. M. Chernomordik, L. V. and Chizmadzhev Yu, A. Electroporation and electrophoretic DNA transfer into

cells. The effect of DNA interaction with electropores. *Biophys J* **63**(5), 1320–1327 (1992).

57. Satkauskas, S. Bureau, M. F. Puc, M. Mahfoudi, A. Scherman, D. Miklavcic, D. and Mir, L. M. Mechanisms of *in vivo* DNA electrotransfer: respective contributions of cell electropermeabilization and DNA electrophoresis. *Mol Ther* **5**(2), 133–140 (2002).

58. Lee, P. Y. Chesnoy, S. and Huang, L. Electroporatic delivery of TGF-β1 gene works synergistically with electric therapy to enhance diabetic wound healing in db/db mice. *J Invest Dermatol* **123**(4), 791–798 (2004).

59. Reid, B. and Zhao, M. The Electrical Response to Injury: Molecular Mechanisms and Wound Healing. *Adv Wound Care* **3**(2), 184–201 (2014).

60. Kloth, L. C. Electrical Stimulation Technologies for Wound Healing. *Adv Wound Care* **3**(2), 81–90 (2014).

61. Gerber, P. A. Buhren, B. A. Schrumpf, H. Homey, B. Zlotnik, A. and Hevezi, P. The top skin-associated genes: A comparative analysis of human and mouse skin transcriptomes. *Biol Chem* **395**(6), 577–591 (2014).

62. Pasparakis, M. Haase, I. and Nestle, F. O. Mechanisms regulating skin immunity and inflammation. *Nat Rev Immunol* **14**(5), 289–301 (2014).

63. Athanasiou, K. A. Eswaramoorthy, R. Hadidi, P. and Hu, J. C. Self-Organization and the Self-Assembling Process in Tissue Engineering. *Annu Rev Biomed Eng* **15**(1), 115–136 (2013).

64. Madi, M. Rols, M.-P. and Gibot, L. Efficient In Vitro Electropermeabilization of Reconstructed Human Dermal Tissue. *J Membr Biol* **248**(5), 903–908 (2015).

65. Fujihara, Y. Koyama, H. Nishiyama, N. Eguchi, T. and Takato, T. Gene transfer of bFGF to recipient bed improves survival of ischemic skin flap. *Br J Plast Surg* **58**(4), 511–517 (2005).

66. Marti, G. Ferguson, M. Wang, J. Byrnes, C. Dieb, R. Qaiser, R. Bonde, P. Duncan, M. D. and Harmon, J. W. Electroporative transfection with KGF-1 DNA improves wound healing in a diabetic mouse model. *Gene Ther* **11**(24), 1780–1785 (2004).

67. Lin, M. P. Marti, G. P. Dieb, R. Wang, J. Ferguson, M. Qaiser, R. Bonde, P. Duncan, M. D. and Harmon, J. W. Delivery of plasmid DNA expression vector for keratinocyte growth factor-1 using electroporation to improve cutaneous wound healing in a septic rat model. *Wound Repair Regen* **14**(5), 618–624 (2006).

68. Liu, L. Marti, G. P. Wei, X. Zhang, X. Zhang, H. Liu, Y. V. Nastai, M. Semenza, G. L. and Harmon, J. W. Age-dependent impairment of HIF-1α expression in diabetic mice: Correction with electroporation-facilitated gene therapy increases wound healing, angiogenesis, and circulating angiogenic cells. *J Cell Physiol* **217**(2), 319–327 (2008).

69. Steinstraesser, L. Lam, M. C. Jacobsen, F. Porporato, P. E. Chereddy, K. K. Becerikli, M. Stricker, I. Hancock, R. E. Lehnhardt, M. Sonveaux, P. Préat, V. and Vandermeulen, G. Skin Electroporation of a Plasmid Encoding hCAP-18/LL-37

Host Defense Peptide Promotes Wound Healing. *Mol Ther* **22**(4), 734–742 (2014).

70. Ferraro, B. Cruz, Y. L. Baldwin, M. Coppola, D. and Heller, R. Increased perfusion and angiogenesis in a hindlimb ischemia model with plasmid FGF-2 delivered by noninvasive electroporation. *Gene Ther* **17**(6), 763–769 (2010).

71. Farage, M. A. Miller, K. W. Berardesca, E. and Maibach, H. I. Clinical implications of aging skin: Cutaneous disorders in the elderly. *Am J Clin Dermatol* **10**(2) 73–86 (2009).

72. Golberg, A. Khan, S. Belov, V. Quinn, K. P. Albadawi, H. Felix Broelsch, G. Watkins, M. T. Georgakoudi, I. Papisov, M. Mihm, M. C. Austen, W. G. and Yarmush, M. L. Skin rejuvenation with non-invasive pulsed electric fields. *Sci Rep* **5**, 10187 (2015).

73. Kloth, L. C. Electrical stimulation for wound healing: A review of evidence from *in vitro* studies, animal experiments, and clinical trials. *Int J Low Extrem Wounds* **4**(1), 23–44 (2005).

74. Gottrup, F. and Apelqvist, J. EWMA Document: Antimicrobials and Non-healing Wounds-Evidence, controversies and suggestions. *J Wound Care* **22**(5), 1–89 (2013).

75. Metzger, S. Clinical and financial advantages of moist wound management. *Home Healthc Nurse* **22**, 586–590 (2004).

76. Church, D. Elsayed, S. Reid, O. Winston, B. and Lindsay, R. Burn wound infections. *Clin Microbiol Rev* **19**(0893–8512), 403–434 (2006).

77. Keen, E. F. Robinson, B. J. Hospenthal, D. R. Aldous, W. K. Wolf, S. E. Chung, K. K. and Murray, C. K. Prevalence of multidrug-resistant organisms recovered at a military burn center. *Burns* **36**(6) 819–825 (2010).

78. Golberg, A. Broelsch, G. F. Vecchio, D. Khan, S. Hamblin, M. R. Austen, W. G. Sheridan, R. L. and Yarmush, M. L. Pulsed Electric Fields for Burn Wound Disinfection in a Murine Model. *J Burn Care Res*, (2014).

79. Golberg, A. Broelsch, G. F. Vecchio, D. Khan, S. Hamblin, M. R. Austen, W. G. Sheridan, R. L. and Yarmush, M. L. Eradication of multidrug-resistant A. *baumannii* in burn wounds by antiseptic pulsed electric field. *Technology* **2**(2), 153–160 (2014).

80. Escoffre, J.-M. Portet, T. Wasungu, L. Teissié, J. Dean, D. and Rols, M.-P. What is (still not) known of the mechanism by which electroporation mediates gene transfer and expression in cells and tissues. *Mol Biotechnol* **41**(3), 286–295 (2009).

81. Gabriel, B. and Teissié, J. Direct observation in the millisecond time range of fluorescent molecule asymmetrical interaction with the electropermeabilized cell membrane. *Biophys J* **73**(5), 2630–2637 (1997).

82. Gabriel, B. and Teissié, J. Time courses of mammalian cell electropermeabilization observed by millisecond imaging of membrane property changes during the pulse. *Biophys J* **76**(4), 2158–2165 (1999).

83. Golberg, A. Broelsch, G. F. Bohr, S. Mihm, M. C. Austen, W. G. Albadawi, H. Watkins, M. T. and Yarmush, M. L. Non-thermal, pulsed electric field cell ablation: A novel tool for regenerative medicine and scarless skin regeneration. *Technology* **1**(1), 1–8 (2013).

84. Sen, C. K. Gordillo, G. M. Roy, S. Kirsner, R. Lambert, L. Hunt, T. K. Gottrup, F. Gurtner, G. C. and Longaker, M. T. Human Skin Wounds: A Major and Snowballing Threat to Publich Health and the Economy. *Wound Repair Regen* **17**(6), 763–771 (2009).

85. Gauglitz, G. G. Management of keloids and hypertrophic scars: Current and emerging options. *Clin Cosmet Investig Dermatol* **6**, 103–114 (2013).

86. Choi, J. E. Oh, G. N. Kim, J. Y. Seo, S. H. Ahn, H. H. and Kye, Y. C. Ablative fractional laser treatment for hypertrophic scars: comparison between Er:YAG and CO2 fractional lasers. *J Dermatolog Treat* **25**(4), 299–303 (2014).

87. Preissig, J. Hamilton, K. and Markus, R. Current laser resurfacing technologies: A review that delves beneath the surface. *Semin Plast Surg* **26**(3), 109–116 (2012).

88. Golberg, A. Bruinsma, B. G. Uygun, B. E. and Yarmush, M. L. Tissue heterogeneity in structure and conductivity contribute to cell survival during irreversible electroporation ablation by 'electric field sinks'. *Sci Rep* **5**, 8485 (2015).

89. Manca, G. Pandolfi, P. Gregorelli, C. Cadossi, M. and de Terlizzi, F. Treatment of keloids and hypertrophic scars with bleomycin and electroporation. *Plast Reconstr Surg* **132**, 621e–630e (2013).

90. Golberg, A. Villiger, M. Khan, S. Quinn, K. P. Lo, W. Bouma, B. E. Mihm, W. G. J. Martin C. Jr. Austen, and Yarmush, M. L. Preventing Scars after Injury with Partial Irreversible Electroporation. *J Invest Dermatol* (2016).

91. Lo, W. C. Y. Villiger, M. Golberg, A. Broelsch, G. F. Khan, S. Lian, C. G. Austen, W. G. Yarmush, M. and Bouma, B. E. Longitudinal, 3D *in vivo* imaging of collagen remodeling in murine hypertrophic scars using polarization-sensitive optical frequency domain imaging. *J Invest Dermatol* (2015).

92. Ramos, M. L. C. Gragnani, A. and Ferreira, L. M. Is there an ideal animal model to study hypertrophic scarring? *J Burn Care Res* **29**(2), 363–368, (2008).

93. Song, F. Li, B. and Stocum, D. L. Amphibians as research models for regenerative medicine. *Organogenesis* **6**, 141–150 (2011).

Index